AF334431

Biological and Medical Physics, Biomedical Engineering

BIOLOGICAL AND MEDICAL PHYSICS, BIOMEDICAL ENGINEERING

This series is intended to be comprehensive, covering a broad range of topics important to the study of the physical, chemical and biological sciences. Its goal is to provide scientists and engineers with textbooks, monographs, and reference works to address the growing need for information. The fields of biological and medical physics and biomedical engineering are broad, multidisciplinary and dynamic. They lie at the crossroads of frontier research in physics, biology, chemistry, and medicine.

Books in the series emphasize established and emergent areas of science including molecular, membrane, and mathematical biophysics; photosynthetic energy harvesting and conversion; information processing; physical principles of genetics; sensory communications; automata networks, neural networks, and cellular automata. Equally important is coverage of applied aspects of biological and medical physics and biomedical engineering such as molecular electronic components and devices, biosensors, medicine, imaging, physical principles of renewable energy production, advanced prostheses, and environmental control and engineering.

More information about this series at http://www.springer.com/series/3740

Michael O. Dada • Bamidele O. Awojoyogbe

Computational Molecular Magnetic Resonance Imaging for Neuro-oncology

 Springer

Michael O. Dada (iD)
Department of Physics
Federal University of Technology
Minna, Nigeria

Bamidele O. Awojoyogbe (iD)
Department of Physics
Federal University of Technology
Minna, Nigeria

ISSN 1618-7210   ISSN 2197-5647   (electronic)
Biological and Medical Physics, Biomedical Engineering
ISBN 978-3-030-76727-3   ISBN 978-3-030-76728-0   (eBook)
https://doi.org/10.1007/978-3-030-76728-0

This Springer imprint is published by the registered company Springer Nature Switzerland AG
The registered company address is: Gewerbestrasse 11, 6330 Cham, Switzerland

Preface

Molecular imaging is promising for two main reasons. First, it enables human diseases to be detected early, that is, before noticeable symptoms manifest. Second, it makes precision drug delivery possible. This technique has the potential of demonstrating changes in tissue physiology, biochemistry, and biology on image scanners. The implication of this is that we can now perform patient-specific treatment, precise follow-up after treatment, and patient monitoring using computational techniques.

Noninvasive medical interventions that effectively increase physician performance in arresting or curing disease, that reduce risk, pain, complications, and reoccurrence for the patient, and that decrease healthcare costs are now within reach. What is yet required is focused reduction of recent and continuing advances in visualization technology to the level of practice, so that they can provide new tools and procedures for smarter healthcare. While magnetic resonance imaging (MRI) is one of the most developed of all the techniques of molecular imaging, it is also unfortunately one of the most expensive diagnostic tools anywhere. It is therefore necessary to develop mathematical concepts based on the fundamental Bloch nuclear magnetic resonance (NMR) flow equation for simple, cost-effective computational MRI to be used in the diagnosis and therapy of brain-related diseases at the molecular level.

The rapid development of innovations including Internet of Things (IoT), big data analysis techniques, and miniature wearable biosensors is generating new opportunities for healthcare systems. Many challenges in the emerging technology can be addressed by the development of consistent, suitable, safe, flexible, and real-time healthcare systems based on the Bloch NMR flow equation.

This book presents mathematical and computational concepts (generally applicable to the analysis of biological and non-biological systems) specifically applied for the analysis of brain tissues and neuro-oncology.

Brain tissues can be likened to complex systems and are often dominated by large numbers of processes. When deviations occur in these processes, human disease conditions are produced. Understanding these processes is important not just in unraveling the causes of diseases, but also the manner of disease propagation and the best plan for treatment. The inadequate understanding of molecular dynamics of diseases is one reason why many diseases remain incurable and become life-threatening. Molecular MRI now provides new ways of visualizing molecular dynamics and their roles in human diseases.

However, current molecular imaging techniques suffer from low sensitivity as well as problems in temporal and spatial resolutions. The available imaging equipment no longer matches the increasing number of patients requiring diagnosis. The few available imaging machines are costly to maintain while financial difficulties are making acquisition of new ones nigh impossible. Obviously, experimental methods alone are no longer enough for efficient diagnosis, and these challenges may now require development of appropriate mathematical models and sophisticated computer simulations based on the Bloch NMR flow equations to complement laboratory and clinical observations. Such mathematical models have the potential to provide insight into the imaging of the molecular interactions through the analysis of the behavior of relaxation processes as observed on magnetic resonance (MR) scans.

The goal is to explore how patients or hospitals can benefit from using novel situation-aware technologies in real-world settings through the exploitation of big data technologies in the context of smart healthcare: wearable sensors, body area sensors, Internet of Things (IoT), machine learning, and artificial intelligence. Based on this theoretical innovation, it may be possible to develop training software to simulate MRI experiments and provide visual training tools to help understand MRI technology for future generations.

The book is organized as follows. Chapter 1 presents a general introduction to the molecular or cellular processes associated with disease conditions. The development of model solutions to the Bloch NMR flow differential equations which can be applied to describe specific clinical problems is presented in Chap. 2. Chapters 3 and 4 present mathematical analyses developed for radio frequency identification (RFID) systems for computational MRI, based on Bloch NMR flow equations and Hermite functions for detailed studies of processes taking place at molecular level in living tissues (particularly for MRI neuro-oncology).

Chapters 5–10 document analytical expressions obtained for the Bloch NMR flow equations and the solutions developed into the various computer programs used. The nuclei of general interest, hydrogen (proton), and fluorine are very abundant, and they have particularly strong NMR/MRI signals. Hence, Chap. 6 is devoted to quantum and classical mechanical analyses of Bloch NMR flow equations so that the wave function of hydrogen atom and hydrogen-like ions can be represented in terms of NMR/MRI parameters. Chapters 7, 8, and 9 present specific analytical methods and computer programs that may be needed to perform MRI tissue diagnosis and therapy at the molecular level, based on relaxometric data. Chapter 10 focuses on computational analysis of MRI contrast agents and their physico-chemical variables. This is considered very important because of the need to improve the sensitivity and specificity of MR imaging using new MR contrast agents designed to probe specific molecules or cells.

The final chapter, Chap. 11, presents general conclusions.

This book is intended for students, scientists, engineers, medical personnel, and researchers who are interested in developing new concepts for deeper appreciation of computational MRI for medical diagnosis, prognosis, therapy, and management of tissue diseases. Researchers who are interested in developing new concepts for the application of artificial intelligence, machine learning, deep learning, and radiomics to solve real-life biological and medical problems at the molecular level utilizing fundamental mathematical concepts will also find it useful. We hope that this book will help a new generation of researchers who want to be involved in this new field of science which is expected to become of great practical importance. We also expect that this book will provide a new and deeper appreciation of computational MRI.

Minna, Nigeria   Michael O. Dada
Minna, Nigeria   Bamidele O. Awojoyogbe

Acknowledgments

Due to the ability of magnetic resonance imaging (MRI) to probe right to the fundamental level, scientists may be able to image human cellular functions and such imaging modalities would definitely help in the understanding of human disease conditions and thus permit the physician to make more specific diagnosis, prognosis, and the appropriate therapy. The basic challenge in this direction is finding the right mathematical frameworks and the physics which appropriately describe the processes involved. This book was written to meet this challenge. Particularly essential to the development of the initial concepts of the book were the research meetings and seminars as well as the symposia, workshops, conferences, schools, and colleges held at the International Center for Theoretical Physics (ICTP), Trieste, Italy, that Awojoyogbe attended as an ICTP regular associate between the year 2000 and 2008. We would like to take this opportunity to thank the following senior academics: Professor D. K. De of Sustainable Green Power Technologies, USA, Professor P. Boesiger of the Institute of Biomedical Engineering and Medical Informatics, University and ETH, Zurich, Switzerland, Professor Paola Fantazzini of the University of Bologna, Italy, Professor Silvio Aime of University of Torino, Italy, Professor Bill Price of Western Sydney University, Australia, and Professor Julian Chela-Flores of ICTP, Trieste, Italy, for their academic guidance and encouragement. We thank Dr. Udunna Anazodo of Western University, Canada, for her determination to promote MRI training and research among younger scientists. We would like to thank our immediate past and present Vice Chancellors of our University: Professor M. A. Akanji and Professor Abdullahi Bala for providing the needed academic leadership and encouragement. We appreciate Professor Funmi O. Olubode-Sawe of Federal University of Technology, Akure, for her support on English language editing of this book. We thank the unanimous reviewers of this book and the editorial team of the distinguished Biological and Medical Physics/Biomedical Engineering series of Springer Nature for supporting the publication of this book. Finally, we thank the students, scientists, engineers, medical personnel, researchers, and research administrators from all over the world who are interested in reading this book for developing new concepts for deeper appreciation of computational MRI for medical diagnosis, prognosis, therapy, and management of tissue diseases.

Kindly send your feedback to us through abamidele@futminna.edu.ng or awojoyogbe@yahoo.com.

Contents

List of Figures

List of Tables

Book Review

This book presents mathematical and computational concepts that can be applied to the analysis of brain tissues, to complement noninvasive imaging for efficient and early diagnosis of disease. The basis of these concepts is the Bloch NMR flow equations.

Chapter 1 establishes the biological bases of disease and the need for early diagnosis for prompt intervention. It highlights the drawbacks of invasive medical procedures which include tissue damage and false diagnosis. In Chap. 2, the authors describe how magnetic resonance works and how knowledge of the expected behavior of normal tissue helps to determine the presence of disease when abnormal behavior is noticed. Specifically, it describes the principles and physics of magnetic resonance imaging (MRI), the design of MRI equipment, and how images are acquired and enhanced for analysis. Then it goes on to show how Bloch's flow equations can mathematically model MRIs of different disease conditions. Chapter 3 presents a new mathematical formulation for diffusion MRI developed by solving the Bloch NMR flow equation. This would probably be the most opaque chapter for people without a strong grounding in physics and mathematics, but will be interesting to diffusion MRI specialists.

In Chap. 4, the Bloch NMR flow equation is developed into two second-order differential equations which can predict fluid flow at suction points analytically and precisely. The flow properties of the time-independent Bloch NMR flow equations describe the dynamics of blood flow under the influence of a radio frequency identification (RFID) system subject to the resonance condition at Larmor frequency. These equations depend on existing knowledge of fluid dynamics applied to predict analytically and precisely the blood flow behavior at suction points in the circulatory system.

Chapter 5 presents practical ways of using computational MRI to provide innovative data-driven algorithms to predict, simplify, and characterize brain tumors and delineate their mechanisms. The chapter first establishes the inadequacies of current imaging techniques to differentiate between primary and metastatic malignant brain tumors for better treatment planning. To address the problem, the chapter presents a computational MR model developed from Bloch's NMR flow equation to classify brain tumors using MRF-derived relaxometric data. The performance of eight machine learning models as diagnostic tools was assessed. Developments in computational MRI could lead to the invention of interactive brain tumor classification apps and the design/development of wearable devices in the future.

Chapter 6 focuses on quantum and classical mechanical analyses of the Yukawa potential which is based on a differential equation developed from Bloch's NMR flow equations, with the wave function of hydrogen atoms and hydrogen-like ions being represented in terms of NMR parameters. As a method of analyzing hydrogen atom and hydrogen-like ions, three advantages put this model above others: its easier visualization, being based on the analytical solution of the fundamental Bloch NMR flow equations, the formulation of a weighting parameter as an interesting physical mechanism that provides a switch between classical and quantum mechanism, and the combination of the quantum and the classical models which can complement each other to provide an excellent innovation for the NMR study of atomic structure of hydrogen-like particles in general for patient-specific and time-unique solutions, an important aspect of personalized medicine. This would pave the way for the next generation of molecular medicine.

While previous chapters focused on solid tissue, Chap. 7 focuses on the flow of blood and presents methods of applying known quantum mechanical formulations and models to the Bloch NMR flow equations, providing a theoretical foundation that may enhance accurate understanding of the transport of nanodevices in microscopic blood vessels used in nanomedicine. Nanomachines can be developed to deliver specially designed quantum drugs carried by blood to the site of infections. The question is: "how much is known of how blood flows to these sites and how to exercise control over the nanomachines carrying the drugs?" The chapter answers these questions by presenting a mathematical algorithm to describe in detail the dynamical state of the hydrogen atom starting from the NMR flow equation, using basic quantum mechanical tools. The mathematical approach may be applied to any situation of restricted fluid flow in small blood vessels.

The premise of Chap. 8 is that experimental methods must be complemented by developing appropriate mathematical models and innovative computer simulation for efficient diagnosis. One method applied in the chapter is the use of R machine learning, which can be used for intensity visualization, intensity separation, and neighborhood information. The works of different researchers in the areas of noise removal and image enhancement were reviewed. Then, an algorithm to classify human brain tumors hierarchically is presented. Using R machine language with various classifiers, distinctions are made between glioblastoma multiforme, metastasis, and lower grade glioma. Chapter 9 demonstrates the use of Avizo, an imaging processing software for distinguishing between radiation necrosis and tumor progression. By the use of several computer images generated by the software, the differences between necrosis and tumor images are established.

Chapter 10 provides a detailed discussion of MR contrast agents GADOVIST, MULTIHANCE, and Gadomer, and their physicochemical basis, which had been previously mentioned in Chap. 2. For the various contrast agents, a Wolfram Mathematica computer program successfully distinguished between the performances of the different MR contrast agents. Solutions to the time-independent NMR Bloch flow equation feature in the writing of the computer program, and the magnetization power of various contrast agents within various laboratory solvents or human tissues was compared quite easily. This can prepare the reader for future innovations in computational MRI.

Chapter 11 briefly highlights the overall goal of the book, especially to facilitate innovation in advanced MRI technologies that could be applied to fundamental basic and preclinical research problems in neuro-oncology and to develop training software that can simulate MRI experiments and provide visual training tools to help understand MRI technology. These might provide support for the development, maintenance, and operation of appropriate MRI devices.

The book will make an interesting read for its audience. The language is accessible, and the equations provide detailed and step-by-step procedures of the mathematical developments for the readers.

Federal University of Technology,   Funmi O. Olubode-Sawe
Akure, Nigeria

Chapter 1
General Introduction

Abstract This Chapter establishes the biological bases of disease and the need for early diagnosis for prompt intervention. It highlights the drawbacks of invasive medical procedures which include tissue damage and false diagnosis. The chapter reveals that Molecular imaging is important not only in disease diagnosis and treatment but is also crucial for drug development by facilitating an improved knowledge of *in vivo* pharmacokinetics, which employs radiolabelled and optical compounds as pharmaceutical agents. Moreover, with molecular imaging, therapeutic effects can be monitored quantitatively at the molecular levels with precision.

Keywords Molecular imaging · Disease diagnosis and treatment · Pharmaceutical agents · Therapeutic effects

1.1 Molecular or Cellular Processes Associated with Disease Conditions

Biological systems are made from the efficient organization of a collection of specialized and unique molecules. Life is built from these molecules which are the materials with which DNAs are built. When DNA is built into cells, and cells into organs, a complete biological system is efficiently built. These building blocks are associated with processes which make up living systems (Gawad, 2005; Modo & Bulte, 2007).

To gain insights into the molecular or cellular processes associated with disease conditions in physiological systems, extensive knowledge of the processes through which cells and molecules develop into organs is required. The most helpful medical interventions are those in which molecular anomalies are detected at the very early stages of diseases and interventions executed. This kind of medical intervention is referred to as molecular medicine, the essence of which is to treat symptoms of diseases at the molecular level. Molecular medicine is more successful when clinical scientists detect signs of human diseases and treat them before they become life-threatening to the patients (Dietel & Sers, 2006; Steel, 2005).

M. O. Dada, B. O. Awojoyogbe, *Computational Molecular Magnetic Resonance Imaging for Neuro-oncology*, Biological and Medical Physics, Biomedical Engineering, https://doi.org/10.1007/978-3-030-76728-0_1

Before recent research developments, molecular and cellular studies have mainly been centred on invasive (medical treatment involving cutting into the body), irreversible histological and molecular biological techniques. Histological investigations rely on the utilization of a group of antibodies which can detect highly specific molecules (Modo & Bulte, 2007). Such an investigation could, for example, demonstrate differences between various cellular phenotypes. Molecular biological investigations could be used to provide detailed descriptions of biological molecules and their components, thus allowing for the definition of the molecular processes associated with specific diseases (Silva et al., 2004; Bailey & Ulrich, 2004).

Despite the successes of these techniques, the inability to access intact physiological systems has been a drawback. One downside of histological investigations is light scattering, a process where incident light of energy ($\hbar\omega_i$) is absorbed by a system, causing light of energy ($\hbar\omega_s$) to be emitted. This problem makes histological investigations impractical for the assessment of deep tissues. Molecular biological techniques typically entail tissue disintegration.

These techniques require surgical procedures, usually leading to tissue damage. Therefore, these surgical procedures are usually avoided because of complications that could result in the course of treatment. For example, molecular changes associated with early signs of Alzheimer's disease or epilepsy has been found around the hippocampus. An examination of this tissue through surgical procedures normally requires a deep penetration of the brain with surgical instruments. This is problematic for a number of reasons. First, the instruments may cause injury to surrounding healthy tissues. Second, the precision of such surgical procedures is dependent on detailed knowledge of the brain region under examination. Third, if the disease is located in large organs, surgical procedures may help to examine a tiny healthy part of the organ but present a false diagnosis for a much larger chunk of the organ (Modo & Bulte, 2007).

An increased presence of a specific protein (acting as peripheral biomarkers) in the blood may seem to justify surgical procedures. However, the damage caused to the organ may outweigh the benefits of such a process. These limitations may preclude the use of surgical procedures for early disease diagnosis. In other words, invasive surgical methods are not efficient for tissue monitoring to determine whether or not the molecules are working properly. Improper behaviour is usually an early sign of the emergence of molecular diseases.

In contradistinction to histopathology techniques, conventional molecular imaging methods are techniques for obtaining knowledge-based molecular information in molecular medicine for patient care. It is noteworthy that bio-molecular biology and histopathology are invaluable in the recognition of specific imaging targets for cellular and molecular imaging. Histopathology and molecular biology are uniquely dependent on each other. Molecular biology not only requires histopathology to delineate tissue regions for advanced bio-molecular analysis, it also requires histopathology to develop molecular probes for the assessment of specific tissue sections. This implies that every analytical technique in molecular medicine requires input from other imaging techniques. The inter-dependence must be appreciated and explored to advance molecular medicine. To develop new applications of molecular imaging, various analytical techniques should be employed such that complementary features are provided to ensure significant advancements in medical technology.

1.2   Molecular Imaging

As already pointed out above, the use of molecular biology and histology in human medicine has a shortcoming: the inability to provide a satisfactory non-invasive imaging of deep tissues. Consequently, these techniques have not provided impressive results in the assessment of molecular and cellular dynamics involved in the onset of diseases in living organisms. Molecular imaging techniques are now being developed to address these problems and provide a window for the timely detection of molecular processes and associated interactions within physiological systems. Molecular imaging refers to techniques for exploring the physics and chemistry of molecular dynamics in physiological systems.

Non-invasive visualization of cellular and molecular dynamics in an intact animal may prove to be very important in providing better understanding of the development of life and in unraveling processes that cause human pathological conditions. Molecular imaging provides a visual representation of the dynamics of certain molecules in intact animals while cell-based (cellular)imaging is the imaging of particular cells in living animals (Modo & Bulte, 2007).

It is worthy of note that from the perspective of biology and chemistry, molecular imaging is different from cellular imaging. However, from the perspective of the physics of imaging techniques, the situation is different. Physics deals mostly with interaction of energy and matter (Callaghan, 1994). When energy is applied to a physiological system, its constituent molecules absorb energy in a unique manner. After some time, the energy is given out and is captured for analysis because it has the footprints of the molecules. However, cells develop and operate based on molecular interactions, without which these cells will never exist. Therefore, cellular imaging may be too broad if we are to study the detailed events that lead to disease development. That is, in the determination of dynamic events in physiological systems, the nature of molecules is not the only factor considered but, more importantly, the interactions of the molecules. Therefore, from the perspective of physics, molecular imaging is essentially cellular imaging in which we study the behaviour of cells by observing the dynamics of the constituent biomolecules (Lodish, 2004). In this study henceforth, both techniques will be considered as magnetic resonance (MR)molecular imaging.

The visualization of molecular and cellular dynamics in living animals without the need for invasive biopsies is the difference between molecular imaging, molecular biology and histology. Traditional imaging methods like magnetic resonance imaging (MRI) and functional magnetic resonance imaging (fMRI) are different from molecular imaging because of its sensitivity to specific molecules or cells. We may therefore say that the difference between molecular imaging and conventional imaging is the imaging target and not the methods employed.

1.3 Importance of Molecular Imaging in Human Medicine

Molecular biology has proved to be very important in providing the targets for molecular interventions; but its diagnostic capability has not been impressive (Caravan et al., 1999; Gries, 2002). It has been unable to provide deep tissue information for the determination of the location of a developing pathology (Martell & Motekaitis, 1992; Rudin & Weissleder, 2003). In recent times, researchers and practitioners alike have been emphasizing the need for more specialized imaging methods which can enhance *in-vivo* assessment of biomolecules, which can contribute to the advancement of personalized molecular medicine (Heckl et al., 2004; Rudin & Weissleder, 2003; Weissleder, 1999).

The main goal of molecular medicine is to address anomalies in molecular dynamic functions at a very early stage; hence, molecular imaging is an important diagnostic tool for clinical scientists to image deviant biomolecules at early stages within physiological systems (Calvo & Semelka, 1999). Therefore, for any technique to be regarded as a remarkable diagnostic tool for molecular imaging, it must be able to connect molecular medicine to molecular biological processes (Schwaiger & Weber, 2004). It must also deal with spatial resolution problems which may arise when the sizes of image voxels are reduced in order to image smaller objects. The reduction leads to the image signals becoming too weak due to the influence of background noise from the surroundings. A good molecular imaging technique is expected to mitigate spatial resolution challenges (Hemmer, 2013).

In addition to the potential application of molecular imaging in human medicine, it could lead to the development of integrative technology for *in vivo* studies of biological processes and their influence on biomolecular behaviour in living systems (Vander Elst et al., 1994; Cacheris et al., 1990). These biomolecular behaviours usually alter biochemical processes, which largely anticipate the anatomical changes that are the basis of current diagnostic modalities (Aime et al., 2005).

The integration of cellular behaviour, molecular biology, histopathology and clinical pharmacology, through medical imaging of physiological systems, will lead to development of novel tissue imaging approaches for improved health care (Kirik et al., 2005). Developments in molecular imaging have been found to be crucial to the advancement of drug development as molecular targets become increasingly specific and clinical methods are fine-tuned (Rudin & Weissleder, 2003; Herschman, 2003). In fact, molecular imaging techniques are invaluable in treatment planning and in monitoring drugs administered within human tissues. With molecular imaging technologies, clinical and physical monitoring of a whole animal is facilitated in the course of *in vivo* studies of the effects of molecular drugs on various organs (Piwnica-Worms et al., 2004; Gheysens & Gambhir, 2005; Massoud & Gambhir, 2003).

Current clinical pharmacogenetic matching processes are not adequate for medical care. These processes need to be so refined that pharmacological agents are perfected and tuned to be capable of determining specific gene expression profiles in various parts of the brain (Staddon et al., 2002). As the production of chemical compounds which are specifically developed to monitor localized neuro-chemical changes often observed in brain diseases is becoming a reality (Staddon et al., 2002; Herschman, 2003), these pharmacological agents need to be monitored, considering that they occasionally require occasional guidance to target points. Molecular imaging is the best way to conduct such monitoring.

With developments in molecular imaging, the investigation of elemental constituents of tissues may be performed to explore tissue processes such as gene expression and functional cell connections (Modo & Bulte, 2007). A molecular imaging technique that tactically provides links between brain functional connectivity and human behaviour while providing information on the associated anatomical/molecular changes has the potential to give incredible integration of various organizational levels with which the effects of gene expression on human behaviour may be demonstrated (Ryan et al., 2001; Jasanoff, 2005).

Since signs of diseases are modifications of normal body functions, new research efforts indicate the possibility of clinical diagnosis moving beyond mere emphasis on disease symptoms towards detailed analyses of the specific molecular dynamics and associated deviations leading to disease emergence (Ryan et al., 2001). Currently, many medical disciplines rely on molecular pathology for disease diagnosis. However, their applications and use have been successful only for easily accessible tissues; their uses have been quite difficult for brain and heart diseases (Modo & Bulte, 2007). This shows that medicine of internal organs will have to significantly depend on molecular imaging and as new approaches to molecular imaging are developed, clinical practice is definitely made better. Once clinical scientists are able to understand and visualize the details of disease dynamics, they would be able to make better diagnosis and offer better treatment plans (Ryan et al., 2001).

It is quite interesting to note that the advantages of molecular medicine are not limited to earlier disease diagnosis, better drug manufacture and treatment of diseases. Molecular medicine will also lead to a complete change in current definitions of and perspectives on tissue diseases (Pomper, 2005). However, such impressive promises and possibilities will depend on the technological advances in tissue imaging (Jager et al., 2005; Pomper, 2005; Levin, 2005; Frost, 2003).

1.4 Molecular Imaging Techniques

Tamaki and Kuge (2010) and Aime et al. (2005) provide a succinct summary of the various molecular imaging techniques currently in use as follows:

1. bioluminescent and fluorescent dyes for optical imaging
2. target-dependent paramagnetic and superparamagnetic ligands for magnetic resonance imaging and magnetic resonance spectroscopy
3. specialized probes using positron emission tomography (PET) or single photon emission computed tomography (SPECT) for radionuclide imaging.
4. microbubbles for ultrasound imaging.

An outline of the strengths and drawbacks of each molecular imaging technique are presented in Table 1.1.

Optical imaging has impressive strengths in temporal and spatial resolution with high target-to-background activity. This technique has wide applications in cellular and molecular imaging for clinical diagnosis. However, it has large optical signal attenuation through the body making it relatively poor in human tissue imaging.

MRI has high spatial resolution, with very low signal attenuation through body tissues. Despite these advantages over other techniques, its low sensitivity makes human application with paramagnetic or super-paramagnetic ligands a bit challenging. MR sensitivity challenge is made worse by the difficulties in measuring physiological properties such as the concentration of specific ions, temperature, redox potential, enzymatic activity and pH. This is because once imaging probes have been ingested into physiological systems, monitoring their local concentrations become extremely difficult (Aime et al., 2005). New research efforts are now addressing this challenge. First, new responsive ligands, which are safe for administration in physiological systems, have been developed. Second, novel types of MRI paramagnetic contrast media known as chemical paramagnetic exchange saturation transfer (PARACEST) agents have also been developed as concentration-independent responsive agents (Zhang et al., 2003). Third, new imaging equipment being developed may prove to be efficient in dealing with this limitation; this, however, requires theoretical formulations to serve as the foundations for these new instruments.

Imaging with radionuclide (employing either PET or SPECT) has an edge over other molecular imaging techniques because of its high affinity for detecting radiolabeled ligands (Aime et al., 2005; Zhang et al., 2003). Some of the downsides of radionuclide imaging are low temporal and temporal resolutions, increased background activity and ionizing radiations. Despite these challenges, this imaging technique has been widely used in clinical applications for studies of various disease processes with minimal safety concerns (Aime et al., 2005).

Ultrasound as an imaging technique relies on the use of micro-bubbles for molecular imaging. Ultrasound has been demonstrated to possess a potential for optimizing molecular drug delivery to specific regions of a damaged tissue (Tamaki & Kuge, 2010). The imaging technique has impressive spatial resolution (50–500 μm) without the use of ionizing radiation (Beer & Schwaiger, 2011). However, it suffers from its relatively poor depth penetration and limited sensitivity. Another disadvantage of the ultrasound imaging technique is the shortage of skilled operators. In addition, not all regions of the body can be assessed. Regions such as the lungs, bones and adult brain are difficult to assess with ultrasound. Although with ultrasound, documentation for comparison of clinical results of examinations carried out at different times is problematic, its complete non-invasiveness makes it ideal for repeat examinations for assessment of therapeutic response (Beer & Schwaiger, 2011).

In summary, molecular imaging is an impressive and speedily growing biomedical imaging method for clinical molecular applications. The purpose of molecular imaging is to monitor and image fundamental physiological processes at molecular,

Table 1.1 Medical imaging techniques and their characteristics (Tamaki & Kuge, 2010)

	Radiation exposure	Spatial resolution	Temporal resolution	Target-to-background	Attenuation	Human imaging	Treatment application
Optical imaging	No	1	1	1	1	Possible	No
MRI	No	1	0	1/2	No	Possible	No
SPECT-PET	Yes	0	0	0	0	Yes	Yes
Ultrasound	No	1	1	1/2	1	Yes	Yes

$0 \rightarrow$ low, $1/2 \rightarrow$ medium, $1 \rightarrow$ high

sub-cellular and cellular levels. As pointed out above, molecular imaging is important not only because it provides increased knowledge of known physiological processes but also because it enables the investigation of unknown cellular and molecular processes. These events are often at the centre of the initiation and evolution of diseases (Rausch et al., 2007). In comparison to conventional *invitro* cell or tissue culture, *in vivo* molecular imaging in animal studies gives room for non-invasive and repetitive cellular imaging of targeted physiological processes in biological tissues. With this method, clinical scientists are able to focus specifically on targeted dysfunctional biological processes for medical diagnosis and prognosis (Tamaki & Kuge, 2010).

1.5 Significance of MRI Techniques

As discussed above, different techniques are used for visualizing molecules and their associated *in vivo* molecular dynamics. The imaging techniques have similar physical properties such as radionuclides, detection of resonant frequencies, emitted light and sound signals. However, their features contrast in several other ways (Rudin & Weissleder, 2003).

The determining factor for the choice of a molecular imaging technique is not only its capacity to detect physiological molecules or cells but also, and more importantly, its applicability in clinical diagnosis. In other words, pre-clinical success is not enough. Any molecular imaging technique must pass the test of clinical application. Another factor considered in choosing a particular technique is its capacity for repetitive clinical assessments of patients. Of all the techniques of molecular imaging discussed previously in this chapter, positron emission tomography and magnetic resonance imaging are the most developed. They are readily available clinical diagnosis and patient monitoring after treatments (Modo & Bulte, 2007).

Unlike optical imaging technique, PET and MRI are capable of penetrating deeply into the tissues and consequently, they can image a live animal. Radioligands employed in PET for detecting specific molecules are very impressive for molecular imaging because the ligands can easily cross the blood–brain barrier being quite small (Hoh et al., 1997). These radioisotopes can easily be mounted with target molecules for molecular imaging (Conti, 1995; Gibson et al., 2000; Halldin et al., 2001; Kegeles & Mann, 1997). Consequently, PET possesses a strong ability to visualize specific molecules in the physiological systems (Phelps, 2000; Czernin & Phelps, 2002).

However, the reliance of PET on radioisotopes for image production has a significant shortcoming: the safety issue. It is unsafe to expose a single subject to several sessions of PET activity. In fact, radioisotope treatment may lead to new and unexpected health challenges because they deal with ionizing radiations (Tamaki & Kuge, 2010).

MRI depends on the detection of nuclear magnetic resonance signals from hydrogen atoms through their magnetic moment just after the introduction of radiofrequency electromagnetic pulse (Chatham & Blackband, 2001). The non-invasiveness of MRI and non-existence of ionizing radiation make it very useful for continuous studies of the same subject; there is no limit to the number of MR sessions a patient may be subjected to. With this feature comes the opportunity to conduct series of studies on the same patient for the investigation of disease progression, monitoring of drug effects and patient response to therapy. Imaging scientists may learn from their mistakes via feedback methods and be able to perfect old imaging procedures and apply new ones. This shows that MR imaging is a versatile technique because nuclear resonance-dependent methods could be combined to make a more robust molecular and cellular MR imaging method. Therefore, MRI is flexible, easy and safe for molecular imaging of tissues and associated disease dynamics (Modo & Bulte, 2007).

Magnetic resonance spectroscopy (MRS), interventional magnetic resonance imaging (iMRI) and functional/pharmacological magnetic resonance imaging (f/phMRI) are examples of imaging methods that can be implemented on the same hardware as complements to molecular MRI (Callaghan, 1994). The spatial resolution of MRI is dependent on the magnetic field strength (B_0), a function of the signal-to-noise ratio (SNR), the ratio of the strength of electrical or other signal carrying information to that of unwanted interference. This parameter can show the extent to which a laboratory scan is able to demonstrate contrast between various tissues (Haacke et al., 1999).

Tissue contrasts depend on the distribution of hydrogen atoms within the tissues and their chemical microenvironment. Whether these atoms are free or bound in bio-molecules, MRI is capable of visualizing their dynamics (Callaghan, 1994). Clinical scanners in common use currently have magnetic field strengths of 1.5 tesla (or 1.5 T, $\omega_0 = 64$ MHz) and 3.0 tesla (or 3.0 T, $\omega_0 = 128$ MHz). Meanwhile, experimental MR studies using living animals are conducted at field strengths of 4.7 T ($\omega_0 = 170$ MHz) and above; the smaller the target volume, the better the observed signal (Haacke et al., 1999). Different parts of a physiological system can easily be scanned, but the magnetic field strength and the imaging sequences must properly be selected for the physiological system (Cowan, 1997; Sasaki et al., 2003).

The high spatial resolution of MRI (ten times higher than that of PET), impressive tissue contrast, non-invasiveness for continuous investigation and great flexibility combine to make magnetic resonance imaging a more appealing diagnostic tool for imaging molecular processes; making it much better than other imaging methods (Modo & Bulte, 2007). Moreover, apart from large molecular complexes, which MR spectroscopy is able to easily detect (Kwock, 1998), MR signal detection demonstrate general behaviour for several molecules or cells.

Molecular imaging is important not only in disease diagnosis and treatment but is also crucial for drug development by facilitating an improved knowledge of *in vivo* pharmacokinetics, which employs radiolabelled and optical compounds as pharmaceutical agents. Moreover, with molecular imaging, therapeutic effects can be monitored quantitatively at the molecular levels with precision (Isin et al., 2012). In clinical settings, molecular-targeted drugs have been developed and administered for treatment because they can be used as monitoring agents and to characterize biological tissues (Kiessling et al., 2004). Therefore, this imaging method has the potential of making the manufacture of new drugs much easier and optimizing disease treatment for patient-friendly medical care.

The focus of this book is computational molecular magnetic resonance imaging (CMMRI).

Chapter 2
Fundamental Physics of Nuclear Magnetic Resonance

Abstract This Chapter presents fundamental physics of nuclear magnetic resonance imaging (NMR), how it works and how knowledge of the expected behaviour of normal tissue helps to determine the presence of disease which suggests abnormal behaviour of NMR transverse magnetizations and signals. Specifically, it describes the principles and physics of magnetic resonance imaging (MRI), the design of magnetic resonance imaging equipment and how images are acquired and enhanced for analysis. Then it shows how Bloch's flow equations can mathematically model MRIs of different disease conditions.

Keywords Bloch NMR flow equation · Physics of nuclear magnetic resonance (NMR) · Physics of magnetic resonance imaging (MRI) · Contrast agents and their physicochemical basis

2.1 Overview of Magnetic Resonance Imaging

Magnetic resonance imaging is a technique for non-invasive visualization of internal regions of structures. The components of a typical magnetic resonance imaging device are a radio frequency (RF) transmitter and receiver, a polarization magnet, magnetic gradient coils, and a computer for signal acquisition and computation of the magnetic resonance images (Dawson & Lauterbur, 2008). In order to avoid misconceptions related to techniques involving ionizing radiations, nuclear magnetic resonance (NMR) Imaging is usually shortened to magnetic resonance imaging (MRI).

Nuclear magnetic resonance is a phenomenon through which atomic nuclei under the influence of a static magnetic field resonate after the application of an oscillating electromagnetic field at the Larmor frequency or resonant frequency. Because the frequency and phase of the resonance signals contain spatial information, images are easily reconstructed from them (Cowan, 1997). Magnetic resonance imaging is a special imaging technique because it is safe, is non-invasive and produces detailed information that other techniques are unable to give. Most of the widely used MRI

M. O. Dada, B. O. Awojoyogbe, *Computational Molecular Magnetic Resonance Imaging for Neuro-oncology*, Biological and Medical Physics, Biomedical Engineering, https://doi.org/10.1007/978-3-030-76728-0_2

applications have been in diagnostic medicine but it is becoming an invaluable technique in fields as diverse as materials science applications, oil exploration, process control and process optimization in oil refineries, non-destructive testing, and analytical chemistry (Webb, 2007).

2.1.1 Principles of Magnetic Resonance Imaging

Atoms of different types of nuclei have magnetic fields and behave like microscopic magnets. This behaviour is very noticeable in the hydrogen atom. Microscopic magnetism is a consequence of isotopes with an odd number of neutrons and/or protons possessing intrinsic magnetic moment and angular momentum (nonzero spin), whereas nuclides with even numbers of neutrons and protons have zero total spin (Slichter, 2013; Sprawls, 1987). Within a geomagnetic field, these bio-magnets line up to some extent in a particular direction (sometimes called polarization). Nuclear spins behave like microscopic compass needles always precessing (such as gyroscopes spinning) in a precise direction (Haacke et al., 1999). In order to reconstruct MR images, the propensity of the nuclear spins to be polarized in the direction of the static magnetic field is perturbed by another oscillatory magnetic field and the excitation energy measured.

It is important to note that nuclear spins from various parts of living systems can be made to precess at unique resonant frequencies such that energy due to the electromagnetic field excitations at these frequencies produce spatial signals. So, the reconstructed maps are images that are easily calculated, improved (sometimes by signal amplification) and displayed on computer screens for analysis. MRI normally requires a small amount of resonant energy to be absorbed and later emitted. According to the rules of nuclear resonance, the energy involved in this process is directly proportional to the energy extracted from the applied radio waves. As a result of this small energy absorption, MRI is considered safe for observation of human subjects (Dawson & Lauterbur, 2008). In addition to this positive feature, MRI neither affects any chemical processes nor alters bio-molecules.

It is very interesting that the atomic nuclei within precessing molecules have a memory of their unique nature and their environment. This is the information that they give out as energy and which is reconstructed as images. Once placed within a static magnetic field (B_o), depending on their orientation, atomic nuclei undergo excitations with energy difference (ΔE). The change in energy, ΔE is a factor on which the strength of the resonance signals depends. The relationship between this energy and the resonance frequency (f_0) is expressed by the following equation (Slichter, 2013):

$$\Delta E = hf_0 \qquad\qquad (2.1)$$

where h represents Planck's constant. The static magnetic field (B_o) and the constant gyromagnetic ratio (γ) determine the magnitude of ΔE and f_0. Gyromagnetic ratio is a constant, which is unique for different kinds of atomic nucleus.

Equation (2.2) is known as the Larmor relationship and it relates this unique constant to the magnetic field and resonant frequency. This equation is the basis on which the principles of magnetic resonance are based (Cowan, 1997):

$$f_0 = \left(\frac{\gamma}{2\pi}\right) B_0 \tag{2.2}$$

A combination of Eqs. (2.1) and (2.2) leads to:

$$\Delta E = h f_0 = h\left(\frac{\gamma}{2\pi}\right) B_0 \tag{2.3}$$

The phenomenon of magnetic nuclear resonance was first observed and shown by I. I. Rabi (Rabi, 1937). Edward Purcell and Felix Bloch, working independently, demonstrated how the detection of nuclear magnetic resonance signals can be achieved with the use of radio waves. The 1952 Nobel Prize for Physics was awarded to these scientists in recognition of this achievement (Castelli et al., 2011). The discovery of NMR led to development of nuclear magnetic spectroscopy or magnetic resonance spectroscopy (MRS), without which chemistry would not be modernized (Dawson & Lauterbur, 2008). Within the static magnetic field, B_0 the atomic nuclei are able to respond to a time-dependent magnetic field B_1 according to their unique Larmor resonance frequency (f_0) with which their corresponding nuclear energy is changed. When the applied oscillatory field is removed (turned off), the atomic nuclei give off radio waves as the nuclear spins return to their equilibrium states (Haacke et al., 1999). These emitted signals lead to the NMR spectra used for imaging. As mentioned above, various atomic nuclei can be studied with NMR since their unique resonant frequencies lead to the production of distinct spectra for diagnostic imaging.

The hydrogen atom has the largest value of gyromagnetic ratio. This means that according to Eq. (2.3), the hydrogen nuclear spins produce the largest energy and hence, the largest resonant signal at any specific magnetic field strength (H_0). This is due to the abundance of hydrogen atoms in nature. In addition, the human body is made up of about two-thirds of water and the incredible convergence of signal strength and natural abundance is responsible for the significantly high resolution associated with the hydrogen nucleus (Callaghan, 1994). In fact, most clinical applications of MRI are reliant on proton resonance. Meanwhile, in the studies and imaging of tissue metabolism, other nuclei such as ^{31}P and ^{23}Na have proved to be very important and useful (Cowan, 1997).

Spatial magnetic field gradients (**G**) are usually required to alter the polarizing magnetic field such that the fields seen by each nucleus within an object are location-dependent (the vector G represents gradient field that is dependent on its three components G_x, G_y and G_z). For instance, if G_x has linear behaviour along the x-axis, an extra magnetic field (G_xx) is produced at a given point x. In a three-dimensional space, gradient fields G_x, G_y and G_z at unique point (x, y, z) influence Eq. (2.3) as follows (Callaghan, 1994; Cowan, 1997; Haacke et al., 1999):

$$f_0 = \frac{\gamma}{2\pi}(B_0 + G) \tag{2.4}$$

$$E = h\frac{\gamma}{2\pi}(B_0 + G) \tag{2.5}$$

where for two-dimensional imaging, $\mathbf{G}$ is usually G_x and G_y, with G_z added for three-dimensional imaging. Since the use of spatial gradients results in atomic nuclei in various spatial locations experiencing distinct values of the term (B_0 + G), the problem of reconstructing an image becomes a problem of interpreting the spectral frequencies (Dawson & Lauterbur, 2008). New MRI sequences and methods developed over time with each possessing diverse strengths and weaknesses according to the clinical image required or desired. Although current practices of MRI rely on two-dimensional image acquisitions, the methodology employed can be used for three or more dimensions. However, the only hindrance to achieving this is the number of resonance signals, which can be obtained and how well the imaging hardware and image sequence software are able to cope with the requirements (Abragam, 1961; Haacke et al., 1999).

2.1.2 Physics of Magnetic Resonance Imaging

The phenomenon and mode of operation of magnetic resonance imaging are different from those of all other clinical imaging techniques (Sprawls, 1987; Caravan, 2007). For an image to be formed the object is required to interact with an oscillatory electromagnetic field whose wavelength is not more than the smallest feature of interest to be contrasted. This marks out the region of interest and may be used in the definition of image voxels with the computation of the image to be generated.

Electron microscopy, infrared imaging, light microscopy and ultraviolet imaging have phenomena analogous to that observed in MRI. Magnetic resonance signals usually have the values of their wavelengths lying in the kilometre range so that the resonant frequencies can be measured in megahertz. Paul Lauterbur (Lauterbur, 1979) initially called the technique *Magnetic Resonance Zeugmatography* to underscore the novelty of this method of producing MR images. MR imaging is possible with the long wavelength and low energy magnetic resonance signals because of the coupling of the two magnetic fields (Dawson & Lauterbur, 2008). The principle of zeugmatographic techniques is consistent with most types of energy couplings. This principle is currently being applied to other related fields such as electron paramagnetic resonance imaging.

2.1.3   Magnetic Resonance Imaging Equipment

The MRI machine is the equipment for excitations of objects and acquisition of the corresponding image. The constituents of MRI equipment are the following: the magnet, the gradients and the RF system.

2.1.3.1   The Magnet

Most magnets used for MRI are configured. Configurations such as open MRI give room for MRI-guided surgical procedures and may drastically reduce claustrophobia. Stand up MRI are also being tried for addressing this issue (Dawson & Lauterbur, 2008).

The resolution of an MR image is a function of the strength of the magnetic field such that the higher the field strength, the larger the signal and the better the imaging resolution (Sprawls, 1987; Haacke et al., 1999). The first MRI systems used permanent magnets that can achieve field strengths up to about 0.5 T. While MRI can be done with permanent iron-core magnets or with ordinary electromagnets or even, under very special circumstances (Devkota & Imberger, 2009), the earth's magnetic field, the images obtained are not useful for medical diagnostic purposes because the signal strength is too small for mapping. Clinical MRI systems use superconducting electromagnets that achieve field strengths of up to 10 Tesla, typically 0.2 to 4 Tesla. The use of superconducting magnets may have been the single most significant contribution to making medical MRI possible (Sprawls, 1987).

2.1.3.2   Gradients

Three electromagnets (back-to-front, side-to-side, and head to toe of the patient) generate known variations of the constant magnetic field (G_x, G_y and G_z) for spatial encoding according to Eq. (2.5). The gradients can be mixed freely, so MRI allows completely flexible orientation of images. These gradient coils are boosted by amplifiers for rapid and precise adjustments to field strengths and direction. The power and precision of these adjustments are the determinants of image quality. Applying stronger gradient fields would lead to higher resolution or faster imaging; gradient systems capable of faster switching can also allow for faster imaging. However, the extent to which higher gradient fields can be used and the speed of use are determined by safety. This is the reason why strong gradient MRI applications are used mostly for non-physiological systems.

2.1.3.3   RF System

The varying electromagnetic field (B_1) is introduced into magnetic system using an RF transmission system which is made up of an RF synthesizer, a power amplifier, and a transmitting coil (Dawson & Lauterbur, 2008; Haacke et al., 1999). Although the transmitting coil can be used to receive MR signals, better images are obtained by using a separate receiving coil that fits closely to the part of the body to be imaged. These coils are built for specific imaging purposes. The more sophisticated receiving coils are made of multi-element phased arrays which are able to acquire multiple channels of data in parallel, a phenomenon known as parallel imaging. Parallel imaging utilizes unique signal acquisition schemes to replace some of the spatial coding provided by the magnetic gradients with the spatial sensitivity of the different coil elements, and this leads to a significant improvement in image resolution and a faster image acquisition time.

2.1.4   Image Acquisition and Computation

For the original MR images, Lauterbur attached a resistor to a wire and another wire to a capacitor, with a vacuum tube in between (Dawson & Lauterbur, 2008). Numbers were read and pencilled onto a grid and this was used to create the image. The first computer used to generate MR images had 14 kilobytes of memory. Nowadays, computers have large memories and it has not been surprising that progress in MRI has been going hand in hand with advances in computer technology (Haacke et al., 1999). Image computation algorithms and signal acquisition methods have been numerous and have advanced at an incredible pace. Specific image processing methods such as *k-space* are now popular in contemporary MRI. K-space is covariant with real space, so such space and physical spaces are interchangeable (Clare, 1997). K-space has offered a simplified means of describing the observed signals, a task that is quite difficult in real space.

2.1.5   Image Contrast

One of the most crucial features of MRI is the high quality of image contrast. This is the reason why it is different from all other human imaging techniques, especially when used for imaging soft tissues (Modo & Bulte, 2007). MR image contrasts are typically given by the concentration or density of the nucleus being investigated. They are also dependent on the relaxation times (T_1 and T_2) of the signals on application of a transient B_1 pulse. Different tissues usually have different measures of water proton relaxation times; when a tissue is affected by disease, these relaxation times change in a unique manner. This is the reason why they have lent themselves to tissue imaging and diagnosis. Raymond Damadian was the first scientist who noted this when he discovered that the relaxation times are able to differentiate normal tissues from cancers (Damadian, 1971).

T_1 (Longitudinal or spin-lattice relaxation, approximately equal to 1 s for biological water protons) is the time-constant for the resonant system to return to thermal equilibrium. This takes place as the relaxing atomic nuclei lose the energy acquired during the B_1 pulse to their environment (Modo & Bulte, 2007). However, the signals are usually lost before T_1 relaxation is completed due to spin dephasing or spins becoming incoherent; leading to the NMR signal disappearing. This process is known as spin-spin or T_2 relaxation (T_2 time-constant is approximately equal to 100 ms for biological water protons). In order to perform disease diagnosis with NMR, image acquisition parameters are chosen such that the relative weights of T_1 and T_2 relaxation are enhanced. The use of external contrast agents to enhance the normal contrast among soft tissues has become common place in laboratory studies and clinical investigations. It was first introduced experimentally in 1978 and has been advancing rapidly ever since. These contrast agents may consist of a paramagnetic substance such as gadolinium, conjugated to other moieties to make it safer in the body (Aime et al., 2005). Paramagnetic contrast agents have a strong magnetic influence on the nearby molecules and greatly enhance the image of the regions within which it is confined.

Currently, there are numerous types of MRI contrast agents which are designed for specialized purposes (Berry, 2009; Castelli et al., 2011; Terreno et al., 2013). The invention and introduction of these agents are not just aimed at providing the strength of contrast but also to show contrast for specific disease conditions. An example of this is the contrast agents that link to receptors on the surface of cancer cells for better visualization of the cancer proliferation and possible hyperthermia treatment and monitoring (Frank et al., 2007). Despite this important advancement in the usefulness of contrast agents, the effects of their use in short term or long term must be constantly evaluated.

2.1.6   Applications of MR Imaging

Magnetic resonance imaging has been successfully applied to different fields and problems. Apart from medical imaging, MRI has found use in chemical analysis (structure determination), purity determination, non-destructive testing, acquisition of dynamic information, petrophysics, flow probes for NMR spectroscopy, process control, quantum computing (http://en.wikipedia.org/wiki/Nuclear_magnetic_reso nance) and so on. However, as regards MR molecular imaging, there is a need to outline few applications which are important to this study. These applications are given as follows.

2.1.6.1 Physiological MR Imaging

Magnetic resonance signals are results of the dynamics of atomic nuclei and the surrounding magnetic field it experiences (Gupta et al., 2014). Other determinants of magnetic response of atoms include blood flow, chemistry, chemical exchange, diffusion and other physiological phenomena. These are molecular signatures which are very important for molecular imaging. An image containing information about these molecular parameters would definitely prove to be very important in providing detailed understanding of how tissues and organs function, both normally and in disease conditions. Molecular MRI has been at the centre of different MRI techniques which have been developed and are rapidly advancing with the aim of elucidating changes in these phenomena, with emphasis on different physiological states or differential diagnoses of disease (Dawson & Lauterbur, 2008).

2.1.6.2 Magnetic Resonance Angiography (MRA)

MRA is used to generate images of arteries for the purpose of evaluation for possible ruptures or abnormal narrowing due to the presence of plaques (Hartung et al., 2011). This technique has excited the interest of many physicians working on cardiovascular disease because of its ability to noninvasively visualize vascular diseases (Yucel et al., 1999). This is strongly related to cardiovascular magnetic resonance (CMR), the computational aspect of which we hope to emphasize in the course of this study. It is noteworthy that molecular imaging in this context requires the imaging of the molecular flow and dynamics involving the components of blood in vascular systems.

2.1.6.3 Functional Imaging

This is conventionally known as functional magnetic resonance imaging (fMRI). fMRI is based on the fact that the brain uses more oxygen when it is active and the local blood flow increases to supply even more oxygen than is required (Belliveau et al., 1991; Cohen & Bookheimer, 1994). This over-supply leaves hemoglobin (oxygen carrier) more oxygenated and the magnetic distortion by deoxyhemoglobin decreases. This is the basis of the blood oxygen level dependent (BOLD) effect of fMRI (Ogawa et al., 1998), a non-invasive technique for assessing brain function. It has been demonstrated that the BOLD effect correlates directly with electrical communication among nerve cells, the synaptic activity (Le Bihan et al., 2006; Ogawa et al., 1998).

2.1.6.4 Spectroscopic Imaging

Spectroscopy technically combines the effect of molecular structure on the magnetic field experienced by an atomic nucleus and the chemical shift with the effects of magnetic field gradients used in MRI (Kwock, 1998; Staudacher et al., 2013). Chemical shifts show spectral behaviour of different chemical compounds and form the basis of NMR in analytical chemistry (Kwock, 1998; Staudacher et al., 2013). Chemical shifts are combined with MRI to produce physical maps of molecules that are important to cellular function and thus lend their support to molecular imaging. Therefore, Spectroscopic MRI or chemical shift MRI has great potential because of the possibility of direct observation of the chemical basis of tissue diseases.

2.1.6.5 Diffusion Imaging

Diffusion imaging measures the rate and direction of water or metabolite diffusion within body tissues (Basser & Jones, 2002). Diffusing imaging has proved to be very useful in observation of strokes (in which the water of edema has been found diffuse very freely). A modified form of this technique is known as diffusion tensor imaging or diffusion tractology and it provides spectacular images of tracts of muscle or nerve fiber bundles, because water diffusion is much faster along the length of the fibers than across them (Basser & Jones, 2002). These images are clinically useful in demonstrating interruption of normal fiber anatomy normally found to be caused by tumors or trauma.

2.2 MR Contrast Agents and Their Physicochemical Basis

Increasing the sensitivity and specificity of magnetic resonance imaging has been a challenge for molecular MR imaging. These challenges are centered on the detection of molecular and cellular events which are usually required to demonstrate specificity for the particular biological event (Modo & Bulte, 2007). Hence, specificity consists of being able to reliably contrast a particular molecule from noise and other molecules due to their unique MR relaxation properties. For example, in order to achieve high specificity in an antibody system targeting particular antigen, the system will selectively bind to the molecule being investigated. Meanwhile, the molecular properties of the antibody system are highly modulated (Guccione et al., 2003; Artemov, 2003).

When the antibody is combined with a magnetic contrast agent, the molecular processes that it undergoes alter the relaxation properties of the contrast agent such that the molecules of interest can be selectively detected with MR imaging (Lorusso et al., 2005). The provision of high specificity MR contrast agents is a very flexible process which can be easily adapted to specialized systems (Laurent et al., 2008; Aime et al., 2005, 2007). The chemical features of the MR contrast agents will determine their binding characteristics to the molecule of interest, their tissue penetration and circulation, and potentially their cellular uptake (Lorusso et al., 2005).

It is possible to modify MR contrast agents into multifunctional assemblies for improving their attractiveness to molecular cellular imaging (Aime et al., 2007; De Vries et al., 2014). For example, contrast agents designed to cross the blood–brain barrier and selectively detect amyloid plaques have been designed for molecular cardiovascular imaging (Podulso et al., 2002). In recent times, gadolinium, Gd (III) units have been functionally designed in order to include them in the formation of large molecular aggregates like liposomes or micelles (Aime et al., 2005). The principles of MR contrast agents will now be discussed.

2.2.1 *Methods of Generating Contrast in MR Agents*

The basic principle of MR imaging is the generation of contrast between different tissues from the distribution and movement of hydrogen protons. These distributions are usually reflected in the spin-lattice relaxation time (T_1) and the spin-spin relaxation time (T_2) (Cowan, 1997).

Endogenous contrast methods are able to differentiate between tissues within a physiological system. The contrast agents are to distinguish white from grey matter (Modo & Bulte, 2007). In some situations, exogenously administered contrast agents are quite useful in highlighting particular tissue aspects. For example, many MR contrast agents are unable to cross the intact blood-brain barrier (BBB) and hence, leakage of contrast agents into the brain can therefore be used to assess damage to the BBB. Contrast agents may be specially designed with peptides such that they are able to cause active transport across the BBB and are therefore useful for visualizing targets that normally cannot be accessed by these agents.

Contrast agents can either be eliminated quickly by the body's reticuloendothelial system or they can evade all detection methods and generate MR contrast for times longer than planned (Caravan, 2007). Contrast agents can also be designed such that their particles bind to specific molecular targets. Whenever they are unspecific, they may be used as blood pool agents; and are then termed intravascular contrast agents. Blood pool agents can be used for delayed steady state and can be combined with pass arterial imaging (Geraldes & Laurent, 2009). Their extended circulation time makes them useful in both arterial and venous mapping. They can also be used for the detection of gastrointestinal bleeding, visualization of the vasculature associated with certain tumors, measuring tissue blood volume and perfusion and detection of endovascular leaks (Geraldes & Laurent, 2009).

2.2.2 *Atomic Basis of Magnetic Resonance*

In Sect. 2.1.1, we introduced the basic principles of magnetic resonance imaging. However, it is very important to outline the atomic background of these principles as this is necessary for the understanding of the molecular perspective of magnetic resonance. Atomic nuclei possess magnetic moments which are proportional to their nuclear spin (I); for a hydrogen atom, $I = {}^1/_2$ (Caravan, 2007). On application of an external magnetic field, the nuclear moments align themselves with only certain allowed orientations (Callaghan, 1994; Cowan, 1997; Haacke et al., 1999). For example, $I = {}^1/_2$ signifies a state where there are only two possible spin orientations and is expressed as the magnetic quantum number $m_I = +{}^1/_2, -{}^1/_2$. The spins of hydrogen atom align either parallel or antiparallel to the applied magnetic field. When an appropriate resonant energy (ΔE) is applied to this system, spins undergo transition between parallel and antiparallel states. This energy is given by Eq. (2.3).

If there is a difference in the population of spins between the $+{}^1/_2$ and $-{}^1/_2$ states, there will be a net absorption of energy when RF f_0 is applied. The ratio of the population ($N_{-1/2}/N_{+1/2}$) between these two states is given by the Boltzmann equation (Caravan, 2007):

$$\frac{N_{-1/2}}{N_{+1/2}} = \exp\left(-\frac{hf_0}{k_B T}\right) \tag{2.6}$$

where k_B is the Boltzmann constant, T is the absolute temperature of the population and h is the Planck constant. Since the applied energy is very small compared to the thermal energy due to collisions within the population, we write:

$$hf_0 << k_B T \tag{2.7}$$

Equation (2.6) now becomes:

$$\frac{N_{-1/2}}{N_{+1/2}} = 1 - \frac{hf_0}{k_B T} \tag{2.8}$$

The implication of Eq. (2.8) is that in NMR frequency range, very few spins are excited and hence, only about 0.001% of the hydrogen is detected. This has implication for sensitivity in MRI. Eqs. (2.3), (2.7) and (2.8) imply that in order to increase sensitivity, nuclei with high γ must be employed and considerably high B_1 fields. The achievable frequency value is directly proportional to sensitivity (Callaghan, 1994). This is the reason why the hydrogen atom is preferable and why NMR research is rapidly tending towards higher-field MR machines. For example, water is typically imaged because it is the most concentrated of all hydrogen-containing molecules.

Molecular sensitivity can be increased if the ratio in Eq. (2.8) is reduced to the barest minimum. For example, in nuclei such as ${}^{13}C$, ${}^{129}Xe$ and ${}^3He$, the material may be polarized at low temperatures and then the temperature is increased to room temperature without changing the hyperpolarization (Haacke et al., 1999; Callaghan, 1994). Thus, we see that whenever nuclei are placed in a magnetic field, it takes a certain time for them to align to be parallel or antiparallel to the field. The time constant at which this rate alignment is achieved is T_1 (Griffiths, 1986; Liang & Lauterbur, 2000; Gibby, 2005).

When the radiofrequency (RF) field is applied, spins absorb energy in order to undergo transitions between the $+^1/_2$ and $-^1/_2$ states; and on removing the RF pulse, the spins emit energy and return to their initial equilibrium state. This emitted energy is what determines the signals detected in the imaging experiments (Liang & Lauterbur, 2000). The rate at which the nuclei in the population return to their initial state is termed relaxation: for the component of magnetization parallel to the external field, the time constant is the longitudinal or spin-lattice (T_1) relaxation time while for the component of magnetization perpendicular to the external field, the time constant is the transverse or spin-spin (T_2) relaxation time (Cowan, 1997).

Transverse relaxation takes place because of local magnetic field inhomogeneities that are caused by an interchange of the chemistry and physics of molecules. These processes are given as follows (Caravan, 2007):

1. The chemistry of NMR relaxation is demonstrated by the microscopic effects caused by magnetic interactions between neighbouring molecules, while
2. The physics of relaxation is centred on the macroscopic effects related to the spatial variation of the external field.

The aggregate effect is termed T_2^*, while relaxation as a result of molecular effects is known as T_2. It very interesting to note that even within a compound, the relaxation due to protons is dependent on their binding state; for example, water in different tissues has different relaxation times (Hazlewood et al., 1974; Sprawls, 1987). Therefore, if image acquisition sequences are made sensitive to differences in T_1, T_2, and T_2^*, excellent tissue contrast can be obtained in normal and disease conditions.

2.2.3 *Relaxivity and T_1, T_2, T_2^* Contrast Agents*

All contrast agents shorten T_1, T_2, and T_2^*. However, it is quite important to classify MRI contrast agents into two broad groups based on whether the substance increases the transverse relaxation rate ($R_2 = 1/T_2$) by roughly the same amount that it increases the longitudinal relaxation rate ($R_1 = 1/T_1$) or whether $1/T_2$ is altered to a much greater extent (Caravan, 2007).

(a) The first category is called T_1 agents. The reason for their behaviour is that these agents alter $1/T_1$ (on percentage basis) of tissue more than $1/T_2$ as a result of the fast endogenous transverse relaxation in tissue. With most pulse sequences, this dominant T_1 lowering effect leads to increases in signal intensity and hence, they are positive contrast agents (Sprawls, 1987).

(b) The second category is T_2 agents which largely increase the relaxation rate $1/T_2$ of tissue selectively and lead to a reduction in signal intensity; these are negative contrast agents. Paramagnetic gadolinium- and manganese-based contrast agents are examples of T_1 agents (Caravan, 2007; Aime et al., 2005, 2007), while ferromagnetic and superparamagnetic iron oxide particles are examples of T_2 agents.

In T_1-weighted images, the repetition time (TR) is set short relative to T_1. Within these circumstances, protons with long T_1 are not given enough time to relax (by giving off the absolved RF energy) before the next pulse of radiofrequency energy (Sprawls, 1987), thus leading to the detection of low signals from these protons. If T_1 is short, NMR relaxation is fast and most of the signals can be detected. This is the reason why short T_1 results in positive image contrast.

For a T_2-weighted image, the echo repetition time (TE) is long relative to T_2 (Sprawls, 1987; Caravan, 2007); a process in which fast transverse relaxation leads to signal attenuation and eventual signal loss. A tissue with long T_2 leads to positive contrast while regions with short T_2 leads to negative contrast and will appear dark. Similarly, in T_2*-weighted images (typically gradient echo images), tissue with short T_2* will appear dark (Sprawls, 1987).

There are many mechanisms through which T_1 and T_2 are shortened by contrast agents. However, in most cases, the effect of these mechanisms can be reduced to a single constant, known as relaxivity. The simple way to quantify this effect is to consider the rate of relaxation, $1/T_1$. In medical imaging, the contrast agent increases the relaxation rate proportional to the amount of contrast agent according to the following expression (Aime et al., 2007; Furlani, 2010):

$$\frac{1}{T_i} = \frac{1}{T_{io}} + r_i[CA]; \quad i = 1, 2, \tag{2.9}$$

where T_i ($i = 1$) is the observed spin-lattice relaxation time, T_i ($i = 2$) is the observed spin-spin relaxation time with contrast agent in the tissue. T_{io} is the T_i relaxation time prior to addition of the contrast agent (Aime et al., 2005; Furlani, 2010), [CA] is the concentration of contrast agent, and r_i is the longitudinal relaxivity (when $i = 1$) or transverse relaxivity (when $i = 2$). The MRI parameter r_i (per millimolar per second or $mM^{-1}\ sec^{-1}$) is usually called relaxivity. Hence, the meaning of Eq. (2.9) is that if the relaxation rate $1/T_i$ is a function of contrast agent concentration, r_i is the slope of graph.

Relaxivity has proved to be a very useful parameter for an *in-vitro* ranking of different contrast agents. Increased relaxivity usually translates into greater contrast at an equivalent dose or equivalent contrast at a lower dose (Aime et al., 2005; Furlani, 2010). However, signal behaviour in physiological systems (*in-vivo*) is more complex than the simple linear relationship demonstrated in Eq. (2.9). Physical and chemical reasons are responsible for this observation. Several attempts have been made to address this problem (such as the use of PARACEST agents), all of which are centred on the development of responsive contrast agents. One of the tasks of this study is to address this problem by reformulation of transverse magnetization functions with memory of *in-vivo* complexities encountered by contrast agents. Meanwhile, we shall discuss the chemistry of some of these agents and their functionalization.

2.2.4   *Chemistry of T_1 Agents*

MRI contrast agents must be biocompatible pharmaceuticals in addition to nuclear relaxation probes. Normally, relatively high doses are used; it is required that they should have good water solubility, be nontoxic at the appropriate dose (according to expected imaging effects) and $1/T_1$ changes as small as 10% to 20% can be detected by MRI (Caravan, 2007). As mentioned above, T_1 agents are typically gadolinium (III) complexes, manganese (II) complexes or as in several animal studies, just the Mn^{2+} cation (Silva et al., 2004; Aoki et al., 2004).

In terms of chemical properties, gadolinium compounds exhibit an eight-coordinate ligand binding to Gd and a single water molecule coordinated to Gd. The multidentate ligand is required for safety (Caravan et al., 1999; Aime et al., 2005). The ligand encapsulates the gadolinium, resulting in a high thermodynamic stability and kinetic inertness with respect to metal loss (Caravan, 2007). This allows the contrast agent to be excreted intact; this is very crucial since these contrast agents tend to be much less toxic than their individual components. For example, the diethylenetriaminepentaacetate (DTPA) ligand and gadolinium chloride both have a median Lethal Dose (LD_{50}) (a dose that causes death in 50% of the animals) of 0.5 mmol/kg in rats, while the Gd-DTPA complex has nearly a factor of 20 higher safety margin, with an LD_{50} of 8 mmol/kg for the Gd-DTPA complex (Gries, 2002).

Metal complex stability can be assessed by determining the metal-ligand stability constant or formation constant (Martell & Motekaitis, 1992). Stability constants are typically very high (Caravan et al., 1999) for Gd complexes used as contrast agents ($K > 10^{17}$ M^{-1}) and their determination requires knowledge of the protonation constants or pKa values, of the ligand. A less rigorous but still practical approach is to determine the relative stability compared to that of a known agent.

Chemical kinetics is also important. This is related to how fast the complex gets into the gadolinium. It should be noted that metal ion release is catalyzed by acids, as well as competing metal ions such as zinc, or other coordinating anions like phosphate (Caravan, 2007). Dissociation rates can be measured absolutely using, for instance, radiochemical (Caravan et al., 2001) or optical methods (Sarka et al., 2002) for detection. Alternately, some relative rates can be measured and compared to a benchmark compound like Gd-DTPA or Gadolinium Tetraazacyclododecanetetraacetate (Gd-DOTA).

Vander Elst and colleagues have monitored the change in P-31 relaxation rate as a function of time when a Gd complex is subjected to a cocktail of phosphate groups

(Vander Elst et al., 1994, 1997). They have also ranked the relative inertness of various contrast agents. As a general rule, macrocyclic ligands like DOTA tend to give more kinetically inert complexes than acyclic ligands like DTPA. Thus, this study develops a method for measuring this feature within the MRI domain without the need to resort to other techniques because processing time is usually high. We shall attempt to solve the problem computationally in this study.

2.2.5 T_2 Agents

Paramagnetism involves the magnetism of small isolated ions that only behave as local magnets in the presence of an external magnetic field. For paramagnetic materials that contain multiple ions, the total magnetic susceptibility is directly proportional to the number of ions in the material. Therefore, molar susceptibility (magnetic susceptibility divided by the number of ions) is constant.

Many other materials exhibit ferromagnetism and superparamagnetism (Caravan et al., 1999; Caravan, 2007). For certain materials such as ferrite (iron oxide), the individual spins of each iron cooperatively, through quantum mechanical interactions, build up to give the crystal a very large total spin, resulting in a very large molecular magnetic susceptibility that is a function of the number of spins (Laurent et al., 2008). Such materials are called ferromagnetics, and their magnetism persists outside the external magnetic field. A weaker form of this is superparamagnetism, involving small particles of iron oxide with aligned spins in a magnetic field. Since the particles are small (submicron), the magnetic susceptibility effect is smaller than for large crystals of ferrites. Superparamagnets are not magnetic outside of the external field. These iron oxide particles represent an important class of contrast agents (Bjornerud & Johansson, 2004).

The iron oxide particles consist of a core of one or more magnetic crystals of Fe_3O_4 embedded in a coating. Consideration of size distribution is very important in the design of these are materials. Ultra-small particles of iron oxide (USPIOs) have a single crystal core and a submicron diameter (Jung & Jacobs, 1995). Small particles of iron oxide (SPIOs) have cores containing more than one crystal of Fe_3O_4 and are larger than USPIOs but still submicron (for example, ferumoxide has a crystal diameter of 4.3 to 4.8 nm and a global particle diameter of ca. 200 nm) (Jung & Jacobs, 1995). USPIOs and SPIOs are small enough to form a stable suspension and can be administered intravenously. The size differences result in differences in pharmacokinetic behaviour (Caravan, 2007). There are also large particles that are used for oral applications (for example, Abdoscan, 50 nm crystals making up a 3 µm particle). USPIOs are also referred to as microcrystalline iron oxide nanoparticles (MIONs) (Muller et al., 2001; Laurent et al., 2008).

Inner-sphere water molecules are non-existent in iron particles and relaxation of water arises from the water molecules diffusing near the particle. However, the mechanism of outer-sphere relaxation is different from what has been described above (Aime et al., 2005). One feature is that the crystals have a net magnetization, and as the external field is increased, the magnetization is increased. (This is also the case for Gd, but the effect is much smaller.) The modulation of the net magnetization can cause proton relaxation (the Curie spin relaxation). The theories describing the field dependence of iron oxide relaxivity have been reported (Muller et al., 2001). Solvent relaxation induced by iron oxide systems is complex. Particles such as USPIO and MION-46L have been fully characterized (Bulte et al., 1993, 1999) through the use of a combination of variable field T_2 measurements (T_2 relaxometry), electron paramagnetic resonance (EPR) and magnetization measurements. In order to fully explain experimental findings in the characterization, the existence of three different magnetic phases for USPIOs were proposed (Bulte et al., 1993): a superparamagnetic core, an antiferromagnetic ferritin-like phase of incompletely converted iron oxyhydroxide, and a paramagnetic surface effect of ferric ions.

General relaxivity behaviour has been noted in these particles. In USPIOs, longitudinal relaxivity (r_1) can be quite high and they can function as effective T_1 agents. The r_2/r_1 ratio for USPIOs is significantly larger than for gadolinium complexes and r_2/r_1 increases with increasing magnetic field (Hardy & Henkelman, 1991; Weisskoff et al., 1994). When there is aggregation of crystals, which is the case in SPIOs, longitudinal relaxivity tends to decrease (r_1 drops) and transverse relaxivity increases (r_2 increases). Thus, for both the particles and aggregates of particles the ratio of r_2/r_1 typically increases as the size of the particles or aggregates increases, though the T_2 relaxivity as a function of particle size can be complicated. The effect of aggregation of crystals is that the aggregate itself can be considered a large magnetized sphere whose magnetic moment increases with increasing field strength. This gives rise to susceptibility effects and the SPIOs can act as T_2^* relaxation agents. This has important consequences when considering the effects of contrast agent compartmentalization on imaging (Hardy & Henkelman, 1991; Weisskoff et al., 1994).

There is also a speciation effect on relaxivity for iron particles. When these nanoparticles cluster together, transverse relaxivity, r_2 increases. This phenomenon has been exploited such that sensors based on assembling and disassembling these nanoparticle clusters have been created (Perez et al., 2002a, b).

2.2.6 *Functionalization of MR Contrast Agents*

Functionalization is simply the engineering of MR contrasts agents according to specific design purposes. This has led to significant improvements in medical diagnosis in terms of higher specificity, better tissue characterization, reduction of image artifacts and functional information (Aime et al., 2005). Advanced engineering of MR contrast agents will result in ever more sophisticated particles that can cross the intact BBB and be absorbed in one particular type of cell.

Based on the functionalization of MR contrast agents, significant developments have been made in both cellular and molecular MR imaging (Modo & Bulte, 2007).

MR contrast agents can be transported into different types of cells *in-vitro* or *in-vivo* to track these cells by MR imaging (Modo et al., 2005), or they can be engineered to be loaded unto particular molecules on tissues, such as blood vessels (Delikatny & Poptani, 2005).

The increasingly sophisticated design of these particles will result in better and more refined methods of detecting specific molecules. For example, targeted MR contrast agents will produce a signal change even if they are not bound to the molecule of interest. By designing contrast agents to produce a signal change only when bound to a molecule of interest (Louie et al., 2000; Li et al., 2002), it is possible to scan the subject quickly because there is no need to wait for unbound contrast agent to be washed out. These agents are usually called "smart" or "responsive" MR contrast agents and are a group of interesting contrast agents that change their properties according to the features the environment (Aime et al., 2005; Lowe, 2004). Environment sensing agents can be used for MR measurements of various parameters such as pH, temperature, enzymatic activity, redox potential, or molecular interactions (Aime et al., 1996, 2002; Perez et al., 2002a, b).

In genetic studies, environmentally influenced agents can be biologically regulated to reflect an up-regulation of a gene. For example, the gene responsible for ferritin transport into cells can be used as an MR reporter by increasing the intracellular iron load that can be detected by T_2*-weighted sequences (Haacke et al., 1999; Cohen et al., 2004). Moreover, with advancement and sophistication of molecular targets for visualization of molecular dynamics comes the need for detection of fewer molecules in a smaller space or imaging at the quantum level (Modo & Bulte, 2007).

From the foregoing, we see that hardware systems used for visualization of a specific biological event need to attain sufficient sensitivity to detect the effect of these MR reporters. Although increasing the field strength of the MR scanner can improve some of these detection drawbacks, the different physical and chemical properties of metal particles can facilitate the detection of sparse elements even within a relatively low magnetic field. The commonly used MR contrast agents in current practice are gadolinium, ferric iron, and manganese (Aime et al., 2005; Laurent et al., 2008). From the agents, only iron particles are ferrimagnetic and are able to produce a blooming magnetic effect that involves an area substantially greater than its localization (Bonnemain, 1998; Bjornerud & Johansson, 2004; Reichert et al., 1999). Iron oxides are super magnetic in nature with incredibly high magnetic moment (Bulte & Kraitchman, 2004) such that small quantities of iron oxide particles can be detected on MR scans. In some cases, a single particle or cell can be detected (Shapiro et al., 2004; Shapiro et al., 2006; Heyn et al., 2006).

Despite the success of iron oxides for molecular imaging, there are situations in which iron oxide particles might not be desirable because the distribution of the molecule may be too wide within the organ and the use of ferumoxides may create a large signal void that no longer allows the localization of these events (Reichert et al., 1999). Gadolinium or manganese-based contrast agents give rise to hyperintensities in T_1-weighted images and are hence, able to act as alternatives in these situations. MR contrast agents that give hyperintense signals are highly desirable and preferable to those giving hypointense signals because the task of interpreting a positive signal is quite an easier one than the loss of a signal (Shapiro et al., 2004).

Meanwhile, T_1 agents usually cause hypo-intensities on T_2-weighted scans and consequently may be unable to solve the entire problem. Normally, contrast agents use alternative atoms like fluorine for the detection of the molecular target in a scan such that the anatomical hydrogen-based MR image is unaffected (Schwarz et al., 1999). However, in current experimental practice, iron oxide particles are the preferred agents because they provide sufficient relaxivity that are reliable even when the concentrations of contrast agent are very small (Reichert et al., 1999). We see then that design of contrast agents based on their physicochemical properties greatly influences *in-vivo* MR signal detection. Careful moderation and control of these characteristics will lead to unprecedented improvements in the application of molecular imaging contrast agents.

Although contrast agent design and engineering are developing rapidly and advancing molecular imaging, there is a need to ensure that these exciting advances are translated into clinical applications. This is important because before implementation in human subjects, preclinical studies need to ensure the safety of the newly designed agents are well evaluated and determine the feasibility and reliability of these novel contrast agents (Lorusso et al., 2005).

Despite MR contrast agents not often presenting safety concerns (Runge, 2000), the procedures for translation of molecular imaging contrast agents should follow rigorous tests and analyses that vouch for the agents' biocompatibility. For example, contrast agents to visualize intracellular targets or cellular MR imaging needs to be thoroughly assessed because they will remain localized to the cellular compartments for considerably longer periods than agents used for current conventional MR imaging (Modo et al., 2005). Biocompatibility testing is an important consideration in advanced design and complex functionalization of MR contrast agents.

As more specific elements are incorporated into the design and engineering of MR contrast agents, some of these compounds may expose an immune response. These specific effects are difficult to assess in preclinical experiments with fine details (Li et al., 2006; Chu et al., 2000). Hopefully, extensive testing for biocompatibility of MR contrast agents in human subjects will become a more rigorous procedure for the implementation of novel compounds prior to the testing of their specificity for particular molecules as associated research methods become more advanced (Li et al., 2006; Chu et al., 2000). The present study aims to contribute to the already abundant theoretical studies addressing these issues. However, with theoretical results helping with problems in design and providing new/better ways of contrast agent engineering, clinical translations progress is expected to lead to further refinement as well as improve the use of these agents (Modo & Bulte, 2007).

The potential of MR imaging as an analytical platform increases with improved specificity of imaging contrast agents and more targeted application. This study is important because it falls within the realm of preclinical studies. The possibility of developing new MR imaging methods in preclinical models and easily translating such resulting methods to clinical applications is important to this swift progress.

The versatility of MR imaging, supplementary methods and integration with other imaging techniques such as PET, single photon emission computer tomography (SPECT) and optical imaging (Marsden et al., 2002; Jacobs & Cherry, 2001) will further enhance MR imaging as a fundamental integrative technique for molecular imaging. Multimodal agents have been developed and they hold the promise of expanding the utility and versatility of MR imaging (Roberts et al., 2000). In fact, this study will attempt to develop a method of inferring optical properties of molecules from MR imaging. This may offer new ways of understanding magnetic resonance microscopy. This formulation may prove to be very important in the processing of particles, cells, and droplets for reactions, analyses, labelling, and coating for microfluidic workflows. The ability to control the movement of particles and cells, or the fluidic environment around them, is a crucial factor in various microfluidic applications (Tarn et al., 2014).

One application of this is the single beam optical trap (optical tweezers), a highly focused beam which is presently revolutionizing the fields of colloidal physics, biology, materials science and engineering (Mohanty, 2012). Recently, spatially-extended three-dimensional light patterns are gaining considerable usage for exerting force to alter, manipulate, organize and characterize materials. However, optical tweezers have the challenge of being limited in the depth at which the objects can be manipulated and the achievable throughput (Mohanty, 2012). Optical magnetic resonance imaging may be able to address this problem by providing relaxation properties in addition to optical features.

Molecular MR imaging is an interdisciplinary research field that requires highly specialized expertise in MR physics, contrast agent chemistry, image analysis, and biological disciplines. A successful collaboration between different specialties is required for MR imaging to become a molecular imaging tool fundamental to disease diagnosis and prognosis decision making. This sort of collaboration will definitely expand the frontiers of molecular medicine (Modo & Bulte, 2007).

2.3 Principles of Optics in Magnetic Resonance

Magnetic resonance is governed by the Bloch equation, an equation which relates a macroscopic model of magnetization to the applied radiofrequency, gradient and static magnetic fields. The dynamics of the changes in systems containing NMR-sensitive nuclei, chemical changes are captured by the Bloch equations.

The Bloch equations as originally applied to NMR have a wide application in a variety of physical effects that do not necessarily involve gyromagnetic spin systems (Hahn, 1980). In particular, they are useful in predicting effects involving electric dipole laser resonance phenomena, which in many cases are analogous to the effects observed in NMR.

2.3.1 Bloch Equations in Optics

Among the many contributions of Felix Bloch to physics, the Bloch equations of nuclear induction (the basis for nuclear magnetic resonance), have played a major role in guiding researchers in the interpretation of resonance experiments (Hahn, 1980). The equations apply directly to any two-quantum level or equally spaced quantum level system (Cohen-Tannoudji et al., 1998), and are particularly useful in predicting effects involving electric dipole laser resonance phenomena. This has particularly proved to be important and successful in the design of laser machines.

Bloch-type equations are generally descriptive of systems that contain bistable states. For example, the phase transition in ferroelectric KH_2PO_4, associated with the two potential well model of the hydrogen bond, can be deduced (Blinc & Zeks, 1974) from a set of Bloch equations. Free electron laser action from a relativistic electron beam passing through a spatially periodic magnetic field can be analysed (Hopf et al., 1976) in terms of coupled Maxwell-Bloch equations. In special situations of resonance radiation interacting with unequally spaced multilevel systems, particularly three level ones; such systems can be formulated in terms of the density matrix to give an equivalent two-level Bloch equation description.

The Bloch equations have a historical root in the mechanical torque equation for the motion of a physical top. The torque equation for the description of gyromagnetic resonance phenomena as presented by Bloch provides a powerful phenomenological means for connecting the linear and nonlinear regimes of resonance response in one formulation (Hahn, 1980). Bloch equations introduced the idea that the otherwise coherent time dependence of quantum mechanical macroscopic moment expectation values can be coupled to phenomenological damping terms.

The equations describe resonance and dispersion in classical terms that connect with the quantum mechanical interpretation through the Liouville equation for the density matrix from which they are derived. Bloch's famous damping time constants T_1 and T_2 are common physics expressions among contemporary resonance researchers. Although the original Bloch equations work very well for fluids, they are often not rigorous enough for all systems, particularly solids. Nevertheless, the Bloch formulation has stimulated new statistical investigations with the density matrix that are more rigorous for the particular system under investigation. In addition to these, Bloch equations also enable the predictions of nonlinear quantum macroscopic phenomena (Pintar, 1976).

2.4 Computational Science

In this study, we are attempting to develop computational models for molecular imaging of tissues necessary for diagnosis and prognosis of diseases and it is quite important to explain the science of computational models.

2.4.1   Computational Science and Problem Solving

Computer technology, particularly in the areas of increased speed of calculations and more efficient memory storage devices, has improved at a whirlwind pace over the past years. Many of the improvements in computer hardware and in the algorithms (software) that control computers have presented new tools for investigating and modelling scientific problems. Scientific research can be categorized into four (http://www.shodor.org/chemviz/overview/compsci.html):

1. observational science
2. experimental science
3. theoretical science
4. computational science

The fourth area of scientific research is comparatively new, having emerged in the past two or three decades. Computational science is the application of computational and numerical techniques to solving large and complex problems. It has revolutionized how scientists work and conduct scientific research. Computational science takes advantage of not only the improvements in computer hardware, but also the improvements in computer algorithms and mathematical techniques.

Computational science also enables the construction of models which allow predictions to be made about what may happen in the laboratory, thus preparing scientists to make better observations or to better understand what they see. Computational techniques can also be used to perform experiments that are too expensive or too dangerous to perform in the laboratory. For example, computational techniques may be used to predict how a new drug would behave in the body, thus reducing but not eliminating the number of animal tests that would have been necessary prior to the development of computational pharmacology techniques. Although computational models cannot replace the experimental laboratory science, they have certainly become an intricate part of the overall search for scientific knowledge.

Computational science is not computer science *per se* but a methodology that allows the study of various phenomena. Various fundamental questions in science (especially those with potentially broad social, political, and scientific impact) are sometimes referred to as "grand challenge" problems. Many grand challenge problems can only be solved computationally. Certainly, chemistry problems are considered by computational scientists to be in the major grand challenge category. In chemistry, the argument has been that the theoretical mathematics needed to solve every chemical problem was known. However, it is only since the birth of computational science that the tools and technologies needed to solve these complex mathematical equations borne from the theories have become available.

There are several ways of looking at computational science. Some scientists will describe it as the intersection of three disciplines as shown in Fig. 2.1.

Figure 2.1 shows that computational science is a scientific application that is supported by the concepts and skills of mathematics (algorithms) and computer

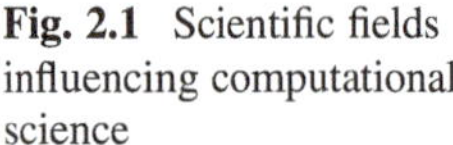

Fig. 2.1 Scientific fields influencing computational science

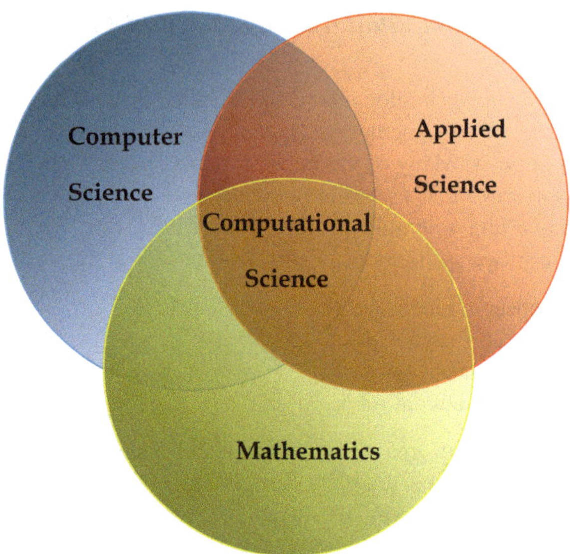

science (architecture). Problem solving in any computational science problem requires an appropriate definition of the scientific event or problem of interest, a perspective provided by applied science. Other crucial issues to be considered are the problem boundaries, components or factors of the system under consideration, assumptions consistent with system behaviour and extent of knowledge of other systems that have similar behaviour to that being studied.

Once these key decisions have been made, then the search begins for a suitable algorithm. This suitable algorithm is a mathematical model that can be created to represent the behaviour described by the parameters of the problem. One or several analytical and numerical methods may be needed to begin the solution of the mathematical model generated. Analytical methods are good for making models because they are flexible and easy to control; but they are very difficult to prosecute. Although much easier to handle, many numerical methods are too complex to calculate by hand and/or require repetitive calculations (iterations) to get close to solutions.

At this point, technologies of computer science can be used to implement the developed algorithm or mathematical model on suitable sized computer using computational software tools. The whole process is iterative; that is, solutions to preliminary algorithmic approaches to the problem generate a better algorithm, perhaps with the need for increasing computational power and precision. The processes involved in this task are illustrated in Fig. 2.2.

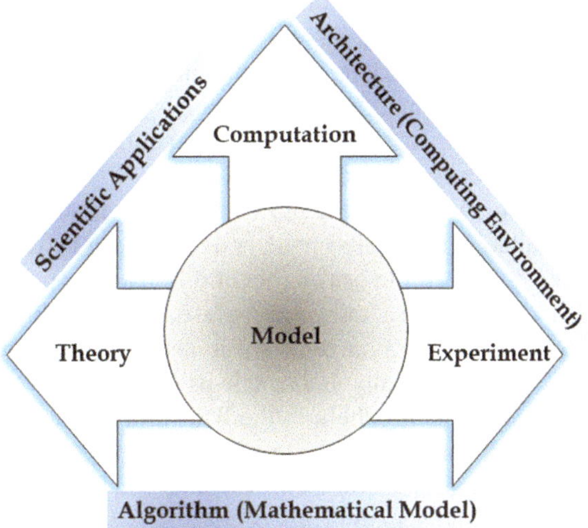

Fig. 2.2 Processes involved in mathematical model generation and their inter-relationships (http://www.shodor.org/chemviz/overview/compsci.html)

2.4.2 Advantages of Computational Model

The process involved in developing a computational model for systems have been explained. This section on the advantages of computational models is from O'Reilly (2000), (http://psych.colorado.edu/~oreilly/cecn/node16.html).

1. Computational models help us understand phenomena by providing novel sources of insight into system behaviour. This might be by providing counter intuitive explanations of phenomena, reconciling seemingly contradictory phenomena (such as complex interactions), and relating seemingly different phenomena in non-obvious ways through a common set of computational mechanisms. Computational models can also be lessened and tested to provide insight into behaviour following specific types of brain damage, and in turn, into normal functioning. Computational models, by being able to translate between functional requirements and the biological mechanisms associated with them, provide an understanding of how the brain is structured and the processes responsible for it.

2. Computational models deal with complexity better than verbal arguments, and can handle complexity across multiple levels of analysis, allowing data across these levels to be integrated and related to each other.

3. The process of computational modelling forces scientists to be explicit about assumptions and exactly how the relevant processes actually work. Such explicitness can be advantageous on many levels. First, it can help in deconstructing concepts such as psychological concepts that may rely on little men; for example, the executive theory of prefrontal cortex function. Then, an explicitly specified computational model can be run to generate novel predictions, which, if necessary, may be modified to account for new data. Any such modification would be

clear and be easy to evaluate by the scientific community. Predictions from verbal theories tend to be more general and couched in vague language, lending the constructs to flexible interpretations. In addition, explicitness can highlight the complexities of otherwise seemingly simple processes. Finally, computational modelling can force scientists to confront aspects of the problem that were hitherto considered irrelevant.

4. Computational models allow the precise control of many more variables than is possible with a real system, and results can be precisely replicated, enabling the exploration of the causal role of different components in unique ways.

5. Models provide a unified framework, so that a single computational framework is used to explain a range of phenomena. They provide a rigorous test of a theory, encourage economy and enable us to relate two seemingly disparate phenomena by understanding them in the light of a common set of basic principles. Inconsistencies may be difficult to detect in a purely verbal theory but a computational model reveals inconsistencies quite readily, because everything has to hang together and actually work.

2.5 Compressed Sensing MRI and Computational Science

Compressed sensing (CS) is aimed at reconstructing signals and images from significantly fewer measurements than were traditionally thought necessary. MRI is an essential medical imaging tool burdened by an inherently slow data acquisition process. The application of CS to MRI has the potential for significant scan time reductions, with benefits for patients and health care economics (Lustig et al., 2008). MRI obeys two key requirements for successful application of CS:

1. medical imagery is naturally compressible by sparse coding in an appropriate transform domain (for example, by wavelet transform)

2. MRI scanners naturally acquire samples of the encoded image in spatial frequency, rather than direct pixel samples.

Imaging speed is important in many MRI applications. However, the speed at which data can be collected in MRI is fundamentally limited by physical (gradient amplitude and slew-rate) and physiological (nerve stimulation) constraints. Therefore, a lot of research seeks methods of reducing the amount of acquired data without degrading the image quality. When k-space is under-sampled, the Nyquist criterion is violated and Fourier reconstructions exhibit aliasing distortions.

Many previous proposals for reduced data imaging try to mitigate under-sampling artifacts. They are classified into the following groups (Lustig et al., 2007):

(a) Methods generating artifacts that are incoherent or less visually apparent, at the expense of reduced apparent signal-to-noise ration

(b) Methods exploiting redundancy in k-space, such as partial-Fourier, parallel imaging, and so on.

(c) Methods exploiting either spatial or temporal redundancy or both.

These methods and other techniques developed for image reconstruction with few data are known as compressed sensing or compressed sampling. Recent theoretical advances in this field hold considerable promise for practical applications in MRI, but the fundamental condition of sparsity required in the CS framework is usually not fulfilled in MR images (Gamper et al., 2008). However, in dynamic imaging, data sparsity can readily be introduced by applying the Fourier transformation along the temporal dimension, assuming that only parts of the field-of-view change at a high temporal rate while other parts remain stationary or change slowly. The second condition for CS is that random sampling can easily be realized by randomly skipping phase-encoding lines in each dynamic frame. In this work, the feasibility of the CS framework for accelerated dynamic MRI is assessed. As can be seen, compressed sensing depends mainly on the computational science of magnetic resonance imaging. In other words, this study also provides an application of compressed sensing to MR molecular imaging.

2.6   MRI in Molecular Imaging

MRI has become a major tool in medical diagnostics, allowing high resolution, non-destructive imaging of the interior of opaque objects. There have been many attempts to use MRI to image small objects such as living cells, because the resolution can be well below the optical diffraction limit (Hemmer, 2013).

However, the detection sensitivity of conventional MRI falls rapidly for smaller feature sizes, making it impossible to resolve features smaller than a few micrometres. In conventional MRI and the related chemical diagnostic nuclear magnetic resonance (NMR), magnetic nuclei such as protons are excited by a radiofrequency field, plus a large magnetic field gradient in the case of MRI. The resulting magnetic signal emitted by the excited nuclei is collected by a magnetic induction coil, much like the antenna loop used in ultrahigh-frequency TV reception, and digitally processed to produce a spectrum or three-dimensional image. However, as the image voxel size is shrunk to allow imaging of ever smaller objects, the magnetic signal becomes too weak to be extracted from the background magnetic noise in the environment. To mitigate this noise, the antenna loop can be shrunk, but so far this approach has not allowed MRI to achieve a resolution beyond a few micrometres (Hemmer, 2013).

Nevertheless, recent works (Mamin et al., 2013; Staudacher et al., 2013) have demonstrated the ability to detect volumes of a few cubic nanometres, comparable to the size of large protein molecules. These independent studies are a crucial step toward molecular-scale MRI. In order to overcome the problem of poor image resolution, nonconventional detection techniques are needed. The first of these was magnetic resonance force microscopy (MRFM) which was developed about two decades ago (Rugar et al., 1994). This approach takes advantage of the force between two magnets: the nuclear or electron spins being imaged on the one hand and a powerful nanomagnet attached to the tip of a scanning probe on the other. This approach allowed detection and imaging of single electrons (Rugar et al., 2004) and imaging of ensembles of nuclei in a virus particle (Degen et al., 2009). However, it required ultralow temperatures and therefore cannot be used to image dynamic samples in ambient environments, like living cells.

Another approach uses a diamond-based nanomagnetometer to sense the magnetic fields generated by small ensembles of nuclear spins. The magnetometer consists of a single nitrogen-vacancy (NV) colour centre in a diamond; it can operate at room temperature in almost any environment (Balasubramanian et al., 2008; Maze et al., 2008). Previously, NVs implanted just below the surface of a bulk diamond have been used to demonstrate near-single-electron spin detection and crude one-dimensional imaging for electron spins outside of the diamond crystal (Grotz et al., 2011). However, to observe the structure and conformations of individual molecules under ambient conditions, imaging of nuclei will be required.

By implementing different advanced noise suppression techniques, Mamin et al. (2013) and Staudacher et al. (2013) have succeeded in using near-surface NVs to detect small volumes of proton spins outside of the diamond crystal. These authors conclude that their observed signals are consistent with a detection volume on the order of 5 nm^3 or less. This sensitivity is comparable to that of the cryogenic MRFM technique and should be adequate for detecting large individual protein molecules. Both groups also project much smaller detection volumes in the future by using NVs closer to the diamond surface. Mamin et al. (2013) project that sensitivity may eventually approach the level of single proton if the NV coherence time can be kept long enough.

To confirm that the observed signals indeed arose from spins outside of the diamond, both groups cleaned the surface to verify that the magnetic proton signals disappeared. This was necessary because single nuclear spins were already detected up to a few nanometres distance inside the diamond lattice (Maurer et al., 2012), but most of the envisioned applications require probing spins in the external environment. Mamin et al. (2013) also observed temporal evolution of the probed nuclear spins. This is only a small step from a fully functional nano-MRI because in conventional MRI, images are acquired by recording the time evolution in the presence of applied magnetic field gradients.

The two approaches have different strengths. Mamin et al. (2013) manipulate the sample spins with an additional radiofrequency field, thus allowing complex protocols of classical NMR to be implemented easily. In contrast, Staudacher et al. (2013) used passive spectroscopy, which does not require an additional radiofrequency field. If magnetic impurities on the diamond surface can be sufficiently controlled, it should be possible to image sub-nanometre volumes of nuclear spins. In this case, single molecule dynamics like protein folding or the action of catalysts in chemical reactions might be observable in real time. Furthermore, if these techniques can be extended to NVs in nano-diamonds, such processes might be observable in more interesting ambient environments like living cells. In contrast to competing techniques like fluorescence resonance energy transfer, the NV can accomplish this task with a larger dynamic range and without the need for separate labels at each site to be monitored. Such capability would substantially enhance our understanding of structural biology and biomolecular processes, as well as numerous additional applications.

In this study, analytical models of NMR Bloch Equations for molecular imaging, *in-vivo* dynamics of contrast agents and modalities for theranostics, the practice of combining two radioactive drugs for diagnosis and therapy, simultaneously or sequentially, are developed.

2.7 Theoretical Treatment of Magnetic Resonance

Magnetic resonance models have been developed to address many biological problems (Awojoyogbe, 1997, 2002, 2003, 2004, 2007). These models were based on treatment of spin dynamics from the fundamental level of the Bloch NMR Equation. The systematic development of this approach to MR dynamics is given in the following subsections.

2.7.1 Theoretical Background of Magnetic Resonance

The energy levels in NMR are associated with the different orientations of the nuclear magnetic moment of an atom in an applied magnetic field. As explained in Chapter One, the spacing between these energy states corresponds to radiation in the RF range. Thus, transitions between the levels will be induced by an applied RF magnetic field (Cowan, 1997; Slichter, 2013). Particles are magnetized within a static magnetic field because of their magnetic nature.

2.7.1.1 Paramagnetism and Curie's Law

When a substance is placed in a static magnetic field B_o, it acquires a magnetization M_o. If the substance is isotropic (exhibiting uniform magnetic properties in all directions), equilibrium magnetization (M_o) will be parallel to B_o. If B_o is reasonably small, the magnetization will be proportional to the applied magnetic field according to the following equation (Cowan, 1997):

$$\vec{M}_0 = \frac{\chi_0}{\mu_0}\vec{B}_0 \qquad (2.10)$$

where χ_0 is the static magnetic susceptibility of the system and μ_0 is the permeability of free space. The size of the NMR free precession signal is proportional to the nuclear spin magnetic susceptibility, and knowing the M_0 of a specimen in a magnetic field, we know the state towards which it will relax following a disturbance.

In a magnetic field, it is energetically favourable for a magnetic dipole to orient itself along the field (the state of minimum energy). However, thermal agitation of the atoms is a challenge to this alignment because it tends to randomize the orientations (the law of entropy increase). The equilibrium state which thus results is a balance between the ordering maximization. The magnetization which results from this mechanism is known as paramagnetism and the magnetization of a paramagnetic substance increases as its temperature decreases (Abragam, 1961; Cowan, 1997).

To determine magnetization, a collection of particles of spin $I = \frac{1}{2}$ is considered because they are easier to handle (Cowan, 1997). Each particle has two spin states, $m = +\frac{1}{2}$ and $m = -\frac{1}{2}$. The number of particles per unit volume is denoted with spin up or $m = +\frac{1}{2}$ by $N_{+1/2}$ and the number per unit volume with spin down or $m = -\frac{1}{2}$ by $N_{-1/2}$. The total number of particles per unit volume may then be given as follows (Abragam, 1961; Haacke et al., 1999):

$$N_v = N_{+1/2} + N_{-1/2}. \tag{2.11}$$

Within a magnetic field B_o the up states and the down states are separated by an energy defined by Eq. (2.11). Hence, the Boltzmann's principle according to Eq. (2.3) may be re-written as:

$$\frac{N_{+1/2}}{N_{-1/2}} = \exp\left(\frac{\Delta E}{k_B T}\right) = \exp\left(\frac{\hbar \gamma B_0}{k_B T}\right), \tag{2.12}$$

where k is Boltzmann's constant and T is the absolute temperature. Given that μ is the magnetic moment of a single particle, the net magnetization of the population, M_o is proportional to the excess of up spins over down spins and is given as (Slichter, 2013):

$$M_0 = \mu\left(N_{+1/2} - N_{-1/2}\right). \tag{2.13}$$

From Eqs. (2.11) and (2.12), we write:

$$N_{+1/2} = \frac{N_v}{\exp\left(-\frac{\hbar \gamma B_0}{k_B T}\right) + 1}, \quad N_{-1/2} = \frac{N_v}{\exp\left(\frac{\hbar \gamma B_0}{k_B T}\right) + 1}. \tag{2.14}$$

Using Eq. (2.14) in (2.13) gives

$$M_0 = N_v\mu\left\{\frac{1 - \exp\left(-\frac{\hbar\gamma B_0}{k_B T}\right)}{1 + \exp\left(-\frac{\hbar\gamma B_0}{k_B T}\right)}\right\}, \tag{2.15}$$

or

$$M_0 = N_v\mu\tanh\left(\frac{\hbar\gamma B_0}{2k_B T}\right). \tag{2.16}$$

Equation (2.16) shows that in general, the magnetization is not proportional to the applied field. Although for low fields, the magnetization response will be linear, at high fields the magnetization will saturate (that is, all dipoles will be oriented along the magnetic field). At this point, no further magnetization can be achieved (Cowan, 1997).

Curie's Law states that when the magnetic field is significantly small and/or the temperature is sufficiently high, then the magnetization is proportional to the applied field. In this situation, this linear response may be obtained by expanding Eq. (2.16) and to first-order approximation, the following expression may be written

$$M_0 = N_v\mu\frac{\hbar\gamma B_0}{2k_B T}. \tag{2.17}$$

Since the magnetic moment μ is related to the spin I, we have (I $= \,^1\!/_2$)

$$\mu = \gamma\hbar I = \frac{\gamma\hbar}{2}, \tag{2.18}$$

$$M_0 = N_v\frac{\hbar^2\gamma^2 B_0}{4k_B T}. \tag{2.19}$$

Equation (2.19) is known as the Curie's Law, which may be stated as follows: for a paramagnet, the magnetization is proportional to the applied field and inversely proportional to the absolute temperature. Eq. (2.19) is otherwise written as:

$$M_0 = C\frac{B_0}{T}, \tag{2.20}$$

where the constant C is referred to as the Curie constant.

2.7.1.2 Magnetic Susceptibility

In the linear region of paramagnetism, it is quite convenient to use the concept of magnetic susceptibility. This is the constant of proportionality between the magnetization M_o and $\frac{B_0}{\mu_0}$. It follows then from Eqs. (2.10) and (2.20) that the magnetic susceptibility is given as (Slichter, 2013; Haacke et al., 1999):

$$\chi_0 = \mu_0 N_v \frac{\hbar^2 \gamma^2}{4k_B T}. \tag{2.21}$$

However, for spin I greater than $^1/_2$, the expression becomes a little more complicated. Hence, for generalized I, χ_0 becomes (Brix et al., 2008):

$$\chi_0 = \mu_0 N_v \frac{\hbar^2 \gamma^2 I(I+1)}{3k_B T}. \tag{2.22}$$

2.7.1.3 Larmor Precession

The net magnetization vector in its equilibrium state is static and does not induce a current in a receiver coil according to the Faraday's law of induction. In order to obtain information from the spins, they must be perturbed and excited. This is normally achieved by irradiating the spins with an RF pulse or a short burst of radiofrequency; a frequency matching the Larmor frequency of the nuclei of interest. This creates a time-dependent deflection of the net magnetization away from its equilibrium orientation. After this RF pulse, the net magnetization recesses about the main magnetic field with the Larmor frequency or resonance frequency.

Spin Equation of Motion

The existence of a collection of magnetic moments and the presence of a magnetic field is of utmost importance in MRI. The dynamics of such system at the simplest level may be treated as a classical problem (Slichter, 2013; Cowan, 1997; Haacke et al., 1999).

A magnetic moment $\vec{\mu}$ in a static magnetic field $\vec{B}$ will experience a torque $\vec{\mu} \times \vec{B}$. According to the rotational analogue of Newton's Second Law, torque equals rate of change of angular momentum; hence, the equation of motion for spin angular momentum is:

$$\vec{\mu} \times \vec{B} = \hbar \vec{I}. \tag{2.23}$$

However, since $\vec{\mu}$ is proportional and parallel to the angular momentum,

$$\vec{\mu} = \gamma \hbar \vec{I}. \tag{2.24}$$

Therefore, the equation of motion for $\vec{\mu}$ is given as:

$$\frac{d\vec{\mu}}{dt} = \gamma \vec{\mu} \times \vec{B}. \tag{2.25}$$

Spin Precession

For an assembly of nuclear spins or population, we introduce the magnetization vector $\vec{M}$, the total magnetic moment per unit volume. If the equations of motion for each spin (same species) are added, the equation of motion for magnetization $\vec{M}$ is obtained:

$$\frac{d\vec{M}}{dt} = \gamma \vec{M} \times \vec{B}. \tag{2.26}$$

This can be solved easily once the direction of $\vec{B}$ has been specified. By convention the magnetic field is assumed to be along the z axis. Hence, we have (Cowan, 1997):

$$\vec{B} = B_0 \hat{k}, \tag{2.27}$$

where $\hat{k}$ is the unit vector in the z-direction. In component form, the equations of motion of the spins may be expressed as:

$$\left.\begin{aligned}
\frac{dM_x}{dt} &= \gamma B_0 M_y, \\
\frac{dM_y}{dt} &= -\gamma B_0 M_x, \\
\frac{dM_z}{dt} &= 0.
\end{aligned}\right\} \tag{2.28}$$

Assuming the initial condition of $M_y(0) = 0$, a simple solution to the differential equations above is given as

$$\left.\begin{aligned}
M_x(t) &= M_x(0) \cos(\gamma B_0 t), \\
M_y(t) &= M_y(0) \sin(\gamma B_0 t), \\
M_z(t) &= M_z(0).
\end{aligned}\right\} \tag{2.29}$$

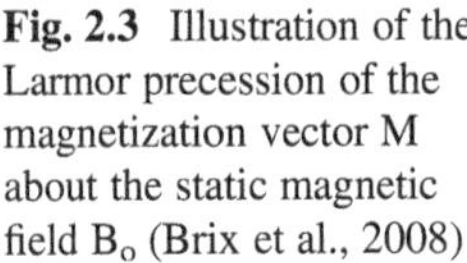

Fig. 2.3 Illustration of the Larmor precession of the magnetization vector M about the static magnetic field B_o (Brix et al., 2008)

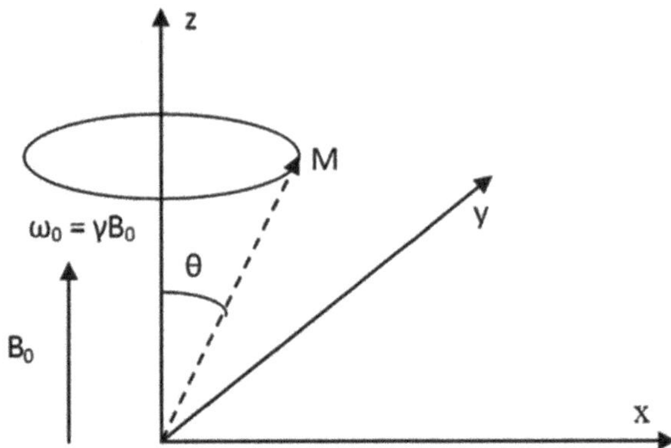

Table 2.1 Abundance of naturally occurring nuclei and their gyromagnetic ratios (Cowan, 1997)

Isotope	Spin (I)	Natural abundance (%)	Gyromagnetic ratio (MHz)
^{1}H	½	100	42.48
^{2}H	1	0.02	6.54
^{3}He	½	0.0001	32.44
^{7}Li	$^3/_2$	92	16.55
^{13}C	½	1.1	10.71
^{19}F	½	100	40.06
^{23}Na	$^3/_2$	100	11.26
^{29}Si	½	4.7	8.46
^{35}Cl	$^3/_2$	75	4.17

These equations describe a rotation in the $x - y$ plane and a constant component in the z direction. The magnetization component in the $x - y$ plane has a constant magnitude $M_x(0)$ and it rotates in the clockwise (negative) direction with an angular velocity of $\omega_0 = \gamma B_0$. This is known as the NMR resonant (angular) frequency.

Figure 2.3 demonstrates the full behaviour of the magnetization vector. The length of the $\vec{M}$ vector remains constant and the angle it makes with the applied magnetic field is constant. The motion of M about B is similar to gyroscope motion in the earth's gravitational field. Such motion is called precession and the behaviour of the magnetization vector in an applied magnetic field is known as Larmor precession. The frequency of the precession is known as the Larmor frequency. Hence, it follows that in Table 2.1, the right-hand column gives the precession frequency for the specified nuclei in a magnetic field of 1 T.

In terms of the magnitude of the magnetization $\vec{M}$ and its inclination θ from the field $\vec{B}$, the evolution of the components of $\vec{M}$ may be expressed as follows (Cowan, 1997):

$$\left. \begin{aligned} M_x(t) &= M \sin\theta \cos(\omega_0 t), \\ M_y(t) &= -M \sin\theta \sin(\omega_0 t), \\ M_z(t) &= M \cos\theta. \end{aligned} \right\} \qquad (2.30)$$

Equation (2.30) demonstrates that the transverse magnetization made up of M_x and M_y, continues to ring indefinitely. This is similar to harmonic oscillator system when the damping constant is zero, which means that the resonance is infinitesimally narrow. The similarity is not complete, however, since now we have two components M_x and M_y rather than the single x. The longitudinal magnetization M_z remains constant, a reflection of the conservation of magnetic energy $-M_z B_o$.

2.7.1.4 NMR Relaxation

After excitation, nuclei making up a population return to equilibrium by losing energy through emission of electromagnetic radiation. This is achieved by transferring energy to the lattice or between them. This process is called relaxation and it commences at the termination of the RF pulse. At this point, both the longitudinal and transverse magnetizations return to their equilibrium values.

Equilibrium States

The existence of equilibrium states is a phenomenon to which the subject of thermodynamics owes its utility; the laws of thermodynamics describe the equilibrium behaviour of complex systems with large numbers of particles in terms of a small number of variables (Cowan, 1997). The calculation of equilibrium properties of a system is in principle a relatively simple procedure involving the Boltzmann distribution function or a similar process. However, the manner in which equilibrium state approached is generally complex and it is not often possible to obtain an explicit expression describing the return of a system to equilibrium following a disturbance. Nuclear spin relaxation is an exception to this general rule.

The Relaxation Process

In a static magnetic field, the equilibrium state of a spin system displays two features of particular scientific importance: a magnetization parallel to the applied magnetic field, and zero magnetization in the plane perpendicular to the field. These features must be compared with the equations of motion for the magnetization vector and their solution given in Eq. (2.29). In NMR experiments, the magnetization achieves its equilibrium value along the B_o direction, but Eq. (2.29) implies that the magnetization in this direction is a constant. In the laboratory, any magnetization in the transverse plane will tend to zero but Eq. (2.29) predicts that the transverse magnetization will continue to precess about B_o indefinitely. This is very contradictory and it means that the mathematical analysis used in the lead up to Eq. (2.29) is inadequate (Cowan, 1997; Callaghan, 1994). This formulation misses the relaxation process itself. The flaw in the model is the assumption that all nuclei see exactly the same magnetic field B_o. In practice, there are other contributions to the local field such that each spin sees a slightly different magnetic field and these fields may change with time as well (Haacke et al., 1999).

2.7.1.5 Spin Interactions

Of the various sources of magnetic fields, two are of particular relevance in spin interactions (Cowan, 1997; Slichter, 2013):

1. the magnet producing the B_o field will not be perfect (there will always be a spread of values of B_o throughout the specimen).
2. the nuclear dipoles produce magnetic fields as well as responding to them.

Other smaller apparent fields may arise from the hyperfine interaction with orbital electrons and possible electric quadrupole interactions with crystal fields. These fields will be experienced by the nuclear spins. When the spins are in motion, the fields seen will fluctuate with time. These different and varying fields cause the spins to precess at different and varying frequencies. The consequent incoherence in the motion of the spins results in dephasing and thus in relaxation. Hence, the relaxation processes observed in the dynamics of a spin system are due to the interactions of the spin dipoles with their environment and with themselves (Haacke et al., 1999; Brix et al., 2008).

2.7.1.6 Spin-Lattice and Spin-Spin Relaxation

It is necessary to make a distinction between the relaxations of magnetization parallel to the B_o field: the approach towards $\vec{M}_0 = \frac{\chi_0}{\mu_0}\vec{B}_0$, and the decay of magnetization in the transverse plane to zero. The former process involves a flow of energy, a change of the spin magnetic energy density given as $-\vec{M}.\vec{B}$. Since the total system energy must be conserved, this process will involve exchange of energy with other degrees of freedom in the system. In terms of NMR, these other degrees of freedom are referred to as the lattice and this is why the processis termed spin-lattice relaxation. Also, since it is the longitudinal component of the magnetization which is varying, this is also referred to as longitudinal relaxation (Haacke et al., 1999; Brix et al., 2008).

Meanwhile, the relaxation of transverse magnetization involves no energy exchange and is therefore an easier process. The mechanism here may be understood simply as the destructive interference in the precessing magnetization as each spin precesses at a slightly different rate. This will be due at least in part to the extra fields seen from other spins or spin-spin relaxation or transverse relaxation (Cowan, 1997).

2.7.2 The Bloch Equations

These relaxation processes can be quantified through the introduction of relaxation times. The spin-lattice relaxation time (T_1) characterizes the relaxation of magnetization parallel to B_o, while the spin-spin relaxation time (T_2) characterizes the relaxation in the plane perpendicular to the B_o field (Cowan, 1997). The relaxation terms may be added to the precession equations for the magnetization which has been outlined earlier on. Hence, Eq. (2.26) becomes:

$$\left.\begin{aligned}
\frac{dM_x}{dt} &= \gamma \left|\vec{M} \times \vec{B}\right|_x - \frac{M_x}{T_2}, \\
\frac{dM_y}{dt} &= \gamma \left|\vec{M} \times \vec{B}\right|_y - \frac{M_y}{T_2}, \\
\frac{dM_z}{dt} &= \gamma \left|\vec{M} \times \vec{B}\right|_z + \frac{(M_0 - M_z)}{T_1}.
\end{aligned}\right\} \tag{2.31}$$

The differential equations in Eq. (2.31) are called the Bloch equations after Felix Bloch who first proposed them in 1946 (Haacke et al., 1999; Clare, 1997). The quantity M_o is the equilibrium value of M_z; and as we have noted in earlier sections, M_x and M_y relax to zero while M_z relaxes to M_0. The relaxation process described by these equations, where the rate of return is proportional to the changes, will follow an exponential law. In the absence of relaxation, the solutions to these equations have been presented already for a constant field B_o, in the z direction by Eqs. (2.30). With the additions of the effects of T_1 and T_2, the magnetization components now evolve according to the following expression (Cowan, 1997):

$$\left.\begin{aligned}
M_x(t) &= M_x(0) \cos(\omega_0 t) \exp\left(-\frac{t}{T_2}\right), \\
M_z(t) &= M_0 - [M_0 - M(0)] \exp\left(-\frac{t}{T_1}\right).
\end{aligned}\right\} \tag{2.32}$$

In a situation where the observed relaxation does not proceed exponentially, the above Bloch behaviour is quantitatively invalid and the relaxation times no longer have the precise definitions as presented. In such cases, T_1 and/or T_2 are usually used simply as a rough measure of the time scale for the relaxation process (Cowan, 1997). Fortunately, it is often observed that the magnetization does indeed relax in an exponential (Haacke et al., 1999) manner (particularly in fluids). T_1 and T_2 are consequently good parameters for characterizing such systems.

2.7.2.1 The Rotating Frame of Reference

The description of resonance phenomenon presented thus far has been in a frame referred to as the laboratory (fixed) reference frame (Slichter, 2013; Callaghan, 1994; Haacke et al., 1999). However, if we are measuring magnetic moment in a reference frame rotating at the Larmor precession frequency, such frames rotate clockwise around the z axis as seen from Fig. 2.4 the origin (z > 0) in a laboratory frame with a constant magnetic field pointing in the positive z direction. This is demonstrated in Fig. 2.4; where the primed frame rotates clockwise around the axis for a static magnetic field pointing in the direction according to a laboratory observer positioned above (z > 0) the x − y plane. Larmor precession frequency is the rotating frequency for the frame.

In a rotating frame, the spin axis is not moving at all and this has been very successful in describing MRI experiments; its mathematical framework has been very impressive also. The basic principles of this approach are presented in the following sections.

2.7.2.2 The Equivalence Principle

As earlier stated, a feature common to almost all NMR events is the application of a static magnetic field B_o, from which a field is formed and leads to Larmor precession at the angular frequency $\omega_0 = \gamma B_0$ (Sprawls, 1987). Another feature which is common to these systems is the use of oscillating magnetic fields. These oscillating magnetic fields are applied in the plane perpendicular to B_o during NMR experiments. In a steady state or continuous wave experiment (frequency domain), a very small oscillating field is applied and the magnetization response is seen as the

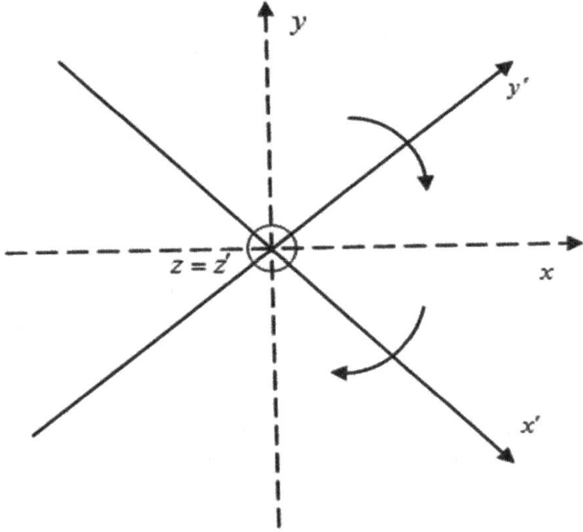

Fig. 2.4 The concept of the rotation of a primed reference frame in which a magnetic moment is at rest (Haacke et al., 1999)

frequency is varied. However, in a pulsed NMR experiment (time domain), a large resonant (at ω_0) oscillating field is applied in a short burst to prepare the system within the required non-equilibrium state (Cowan, 1997).

The concept of the rotating frame proposes a look at the spin system through the eyes of someone who is rotating rather than stationary with respect to the system (Sprawls, 1987; Haacke et al., 1999). A static magnetic field B_o causes a magnetic moment to precess at a frequency $\omega_0 = \gamma B_0$ so that an observer rotating at the same rate ω_0 would observe the magnetic moment to be stationary. The observer is unable to see the effect of the field. This is similar to the equivalence principle of the theory of relativity (that is, if the cables of an elevator are cut the occupants do not know whether they are accelerating to the ground or if gravity has simply been turned off). Hence, we may conveniently say that in a rotating reference frame, a static magnetic field may effectively disappear.

2.7.2.3 Transformation to the Rotating Frame

Within a stationary reference frame, the time derivative of a vector $\vec{\mu}$ is given in terms of its components by (Haacke et al., 1999):

$$\frac{d\vec{\mu}}{dt} = \frac{d\mu_x}{dt}\hat{i} + \frac{d\mu_y}{dt}\hat{j} + \frac{d\mu_z}{dt}\hat{k}. \tag{2.33}$$

However, if the reference is moving, the above basis vectors will not be constant and the time derivative of $\vec{\mu}$ has an extra contribution such that:

$$\frac{d\vec{\mu}}{dt} = \left(\frac{d\mu_x}{dt}\hat{i} + \frac{d\mu_y}{dt}\hat{j} + \frac{d\mu_z}{dt}\hat{k}\right) + \left(\mu_x\frac{d\hat{i}}{dt} + \mu_y\frac{d\hat{j}}{dt} + \mu_z\frac{d\hat{k}}{dt}\right). \tag{2.34}$$

The term within the first brackets corresponds to the rate of change of $\vec{\mu}$ as seen by an observer in the moving frame (the frame from which the unit vectors $\hat{i}, \hat{j}$ and $\hat{k}$ appear stationary). This term may then be written as follows:

$$\frac{\partial\vec{\mu}}{\partial t} = \frac{d\mu_x}{dt}\hat{i} + \frac{d\mu_y}{dt}\hat{j} + \frac{d\mu_z}{dt}\hat{k}. \tag{2.35}$$

If we consider a case in which the reference frame is rotating about the z axis with angular velocity ω, an observer in the fixed frame would see varying unit vectors according to the following expression (Cowan, 1997):

$$\left.\begin{array}{l} \widehat{i}(t) = \widehat{i}(0)\cos\omega t + \widehat{j}(0)\sin\omega t, \\ \widehat{j}(t) = -\widehat{i}(0)\sin\omega t + \widehat{j}(0)\cos\omega t, \\ \widehat{k}(t) = \widehat{k}(0). \end{array}\right\} \tag{2.36}$$

The time derivatives of these basis vectors are given as:

$$\frac{\partial \widehat{i}(t)}{\partial t} = \omega\widehat{j}, \ \frac{\partial \widehat{j}(t)}{\partial t} = -\omega\widehat{i}, \ \frac{\partial \widehat{k}(t)}{\partial t} = 0. \tag{2.37}$$

This expression leads to the total rate of change of the vector u given as follows:

$$\frac{d\vec{\mu}}{dt} = \frac{\partial \vec{\mu}}{\partial t} + \omega\left(\mu_x\widehat{j} - \mu_y\widehat{i}\right). \tag{2.38}$$

Equation (2.38) may be written as:

$$\frac{d\vec{\mu}}{dt} = \frac{\partial \vec{\mu}}{\partial t} + \omega\widehat{k} \times \vec{\mu}. \tag{2.39}$$

Since ω is the magnitude of the angular velocity, the z axis is the direction of the angular velocity and the product $\omega\widehat{k}$ is the angular velocity vector, we may write:

$$\vec{\omega} = \omega\widehat{k}. \tag{2.40}$$

Now if the rotation is about an arbitrary axis, we have:

$$\frac{d\vec{\mu}}{dt} = \frac{\partial \vec{\mu}}{\partial t} + \omega\widehat{k} \times \vec{\mu}. \tag{2.41}$$

2.7.2.4 Equation of Motion in the Fictitious Field

The above results are quite useful in the modification of the behaviour of the equation of motion for the nuclear magnetic moment when placed within a polarizing magnetic field. For the nuclear magnetic moment,

$$\frac{d\vec{\mu}}{dt} = \gamma\vec{\mu} \times \vec{B}. \tag{2.42}$$

In a rotating (at angular velocity ω) frame of reference rotating, the rate of change of $\vec{\mu}$ as seen from the rotating frame $\frac{\partial \vec{\mu}}{\partial t}$ is given by (Cowan, 1997; Haacke et al., 1999):

$$\frac{\partial \vec{\mu}}{\partial t} = \left(-\vec{\omega} \times \vec{\mu}\right) + \frac{d\vec{\mu}}{dt}, \tag{2.43}$$

$$\frac{\partial \vec{\mu}}{\partial t} = \gamma \vec{\mu} \times \left(\frac{\vec{\omega}}{\gamma} + \vec{B}\right). \tag{2.44}$$

Equation (2.44) implies that within a rotating frame, a magnetic moment sees an effective magnetic field $\vec{B} + \frac{\vec{\omega}}{\gamma}$ or

$$\vec{B}_{eff} = \vec{B} + \frac{\vec{\omega}}{\gamma}. \tag{2.45}$$

Hence, the Larmor precession is denoted such that in a frame rotating at angular velocity $\omega_0 = \gamma B_0$, a magnetic moment appears stationary (rotating at angular velocity $-\gamma B_0$). Therefore, in a rotating frame of reference, there appears to be an extra fictitious magnetic field and the rotating frame is chosen to be at the same rate as the applied transverse field such that this frame is then stationary (Haacke et al., 1999).

2.7.2.5   Rotating Field and Oscillating Field

The importance of the rotating field in the description of nuclear resonant processes has been presented. The roles of the oscillating magnetic field and their relationship to the rotating frame of reference will now be presented.

Rotating and Counter-Rotating Components

The rotating frame is obviously best suited for the treatment of rotating magnetic fields because in such a circumstance, the field appears stationary which is much easier to deal with (Cowan, 1997; Haacke et al., 1999). However, in NMR experiments, a linearly polarized field is applied with the aid of a coil system. This normally requires two orthogonally oriented solenoids to produce a circularly polarized field, but in NMR this problem has a simple solution. The linearly polarized field may be regarded as being made up of a rotating and a counter-rotating component (Cowan, 1997):

$$2B_1 \cos\left(\omega t\right)\hat{i} = B_1 \left(\hat{i}\cos \omega t + \hat{j}\sin \omega t\right) + B_1 \left(\hat{i}\cos \omega t - \hat{j}\sin \omega t\right) \tag{2.46}$$

The two components may then be treated separately but this may not always be necessary. The resonant nature of the phenomenon means that the components rotating in the opposite direction to the Larmor precession has negligible effect compared to that moving with the precession; only the fields near resonance are necessary for consideration. Thus, the normal procedure is to use a linearly polarized field while simply ignoring the counter-rotating component.

The Effective Magnetic Field

Considering an assembly of similar magnetic moments placed within a static magnetic field $\vec{B}_0$ and another field $\vec{B}_1$ rotating at angular speed ω in the plane transverse to $\vec{B}_0$, the vector sum of the magnetic field is (Cowan, 1997):

$$\vec{B}(t) = B_1\left(\widehat{i}\cos\omega t + \widehat{j}\sin\omega t\right) + B_0\widehat{k}. \tag{2.47}$$

Within an arbitrary frame rotating about the z-axis at an angular speed of ω, transformation rule as given by Eq. (2.45) implies that the apparent or effective field now becomes:

$$\vec{B}_{eff} = B_1\left(\cos\left[(\omega - \omega')t\right]\widehat{i} + \sin\left[(\omega - \omega')t\right]\widehat{j}\right) + \left(B_0 + \frac{\omega}{\gamma}\right)\widehat{k}. \tag{2.48}$$

This shows that in a reference frame moving at the same rate as the rotating $\vec{B}_1$ field, the apparent field is stationary and we have the following expression (Fig. 2.5)

$$\vec{B}_{eff} = B_1\widehat{i} + \left(B_0 + \frac{\omega}{\gamma}\right)\widehat{k}. \tag{2.49}$$

The description of the motion of a magnetic moment in this combination of magnetic fields is now an easier task. Since we have transformed to a frame from which the total magnetic field appears stationary, the magnetic moment in this frame will simply precess around the effective field at a rate $\omega = \gamma B_{eff}$. The magnetic moment vector $\vec{\mu}$ lies along the surface of a cone and the tip of the vector moves around the base of the cone at the effective Larmor frequency, the effective field lying along the axis of the cone, as illustrated in Fig. 2.6a. Meanwhile in the

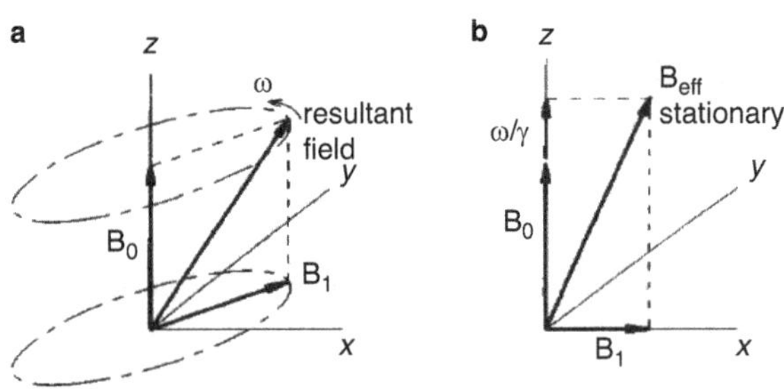

Fig. 2.5 Local magnetic field in (**a**) laboratory frame (**b**) rotating frame (Cowan, 1997)

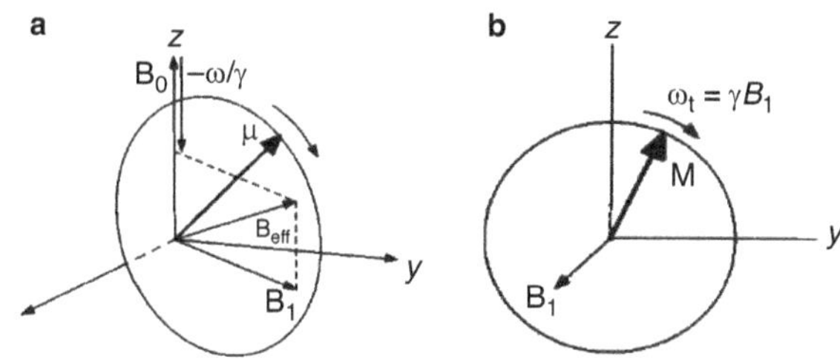

Fig. 2.6 Magnetization vector (**a**) precessing about the effective field (**b**) precessing in the x − y plane (Cowan, 1997)

laboratory frame, the cone is rotating about the z-axis synchronously with the rotating B_1 field.

Resonant Transverse Field

A particularly important example of applied rotating magnetic fields is where the field is on resonance; then it rotates with the same angular velocity as the precessing magnetic moments. In such a case, the effective field

$$\vec{B}_{eff} = B_1\hat{i} + \left(B_0 + \frac{\omega_0}{\gamma}\right)\hat{k},\tag{2.50}$$

becomes:

$$\vec{B}_{eff} = B_1\hat{i},\tag{2.51}$$

because:

$$\omega_0 = -\gamma B_0.\tag{2.52}$$

We see that the effective field is now in the transverse plane and the $\vec{B}_0$ field has been completely eliminated (Cowan, 1997).

2.7.2.6   Excitation Pulses

Excitation pulses are energy pulses which cause the spin within a system to higher energy state. Usually, they are known as 90° ($\pi/2$) and 180° (π) pulses (Hinshaw & Lent, 1983; Haacke et al., 1999). If we now consider an assembly of magnetic moments in a steady magnetic field of polarization $\vec{B}_0$ and if the thermal equilibrium has been established, the magnetization $\vec{M}_0$ is:

$$\vec{M}_0 = \frac{1}{\mu_0}\chi_0\vec{B}_0.$$ (2.53)

However, if we now apply a resonant rotating field in the transverse plane, all that the magnetization in the rotating frame sees is the effective field perpendicular to itself. Hence, $\vec{M}_0$ will simply precess around $\vec{B}_{eff}$, moving in a circle in the vertical plane; the conical hat has become flat as within Fig. 2.6b. The magnetization vector precesses about the vertical plane at an angular velocity of $\omega_1 = \gamma B_1$. Therefore, if the rotating magnetic field is applied for a time t (Cowan, 1997),

$$\gamma B_1 t = \frac{\pi}{2}.$$ (2.54)

Then the magnetization vector will have been rotated through an angle of $\frac{\pi}{2}$ radians or 90°, starting from the z-axis through the plane. Since the vector $\vec{M}$ which is now in the transverse plane will precess at the Larmor frequency while its magnitude decays with a characteristic time T_2, associated burst of resonant rotating field of duration

$$t_{90} = \frac{\pi}{2\gamma B_1},$$ (2.55)

is known as a $\frac{\pi}{2}$ or a 90° pulse. Meanwhile, a pulse of twice this duration would rotate the magnetization through 180°. This pulse usually inverts $\vec{M}$ to $-\vec{M}$ with a length given as follows (Cowan, 1997):

$$t_{180} = \frac{\pi}{\gamma B_1}.$$ (2.56)

That pulse is described by Eq. (2.56) and is known as the π or 180° pulse. Both the 90° pulse and the 180° pulse are used to disturb the equilibrium value of $\vec{M}_z$ and so can be used to initiate longitudinal relaxation. In order to know which parameter must take precedence over others in NMR measurements, the oscillator model which gives a simplified formulation of resonant processes is presented in the next subsection.

2.7.2.7 Harmonic Oscillator Model of NMR and Dynamic Magnetic Susceptibility

All resonance phenomena including NMR share certain features. The one-dimensional harmonic oscillator is usually considered as an embodiment of a resonant system and, denoting the displacement of a unit mass from its equilibrium position by x, the restoring force of the oscillator may be assumed to be proportional to the displacement according to the following expression $-\lambda^2 x$ (Cowan, 1997). If we also consider a friction or damping force proportional to the velocity such that $-2\beta\frac{dx}{dt}$, in the absence of an external driving force, the equation of motion for the system is

$$\frac{d^2x}{dt^2} + 2\beta\frac{dx}{dt} + \lambda^2 x = 0. \tag{2.57}$$

This expression describes the damped harmonic oscillator.

Time Response of Damped Oscillator

The equilibrium state of the harmonic oscillator takes place at the point when the displacement x is zero and the system is stationary. On displacement, the oscillator will return to the equilibrium state according to the equation of motion. Considering the nature of this relaxation from the non-equilibrium to the equilibrium state (free relaxation of the oscillator from some initial condition), we are required to find the time response or the solution to the homogeneous Eq. (2.57). Provided that $\lambda > \beta$, the behaviour of the relaxing harmonic oscillator has the following form (Cowan, 1997):

$$x(t) = [A\cos\omega_0 t + B\cos\omega_0 t]e^{-\beta t}, \tag{2.58}$$

where $\omega_0^2 = \lambda^2 - \beta^2$. An oscillation occurs at the angular frequency ω_0 which is decaying away exponentially with a time constant $1/\beta$. The constants A and B are dependent on the nature of the initial conditions (at time $t = 0$). For example, if the oscillator is released from a displacement of x_0 at $t = 0$, the evolution takes the form

$$x(t) = x_0\cos(\omega_0 t)e^{-\beta t}. \tag{2.59}$$

What is most interesting about this description is that the measurable quantities ω_0 and the damping time $(1/\beta)$ are characterized by restoring force (λ) and frictional force (β).

Frequency Response of Damped Oscillator

Another means of studying the behaviour of the oscillator is by examining how the system responds to irradiation by an oscillatory excitation. If we consider an applied force per unit mass of $F_m\cos(\omega t)$ and noting that in a conventional spectroscopy experiment, the response $x(t)$ is studied as the excitation frequency ω is varied, then the complete equation of motion becomes inhomogeneous. This equation is given as follows (Cowan, 1997; Bracewell, 1986):

$$\frac{d^2x}{dt^2} + 2\beta \frac{dx}{dt} + \lambda^2 x = F_m \cos(\omega t). \qquad (2.60)$$

The solution to this equation is complex and is given as:

$$x(t) = F_m \left\{ \frac{\left(\lambda^2 - \omega^2\right)}{\left(\lambda^2 - \omega^2\right)^2 + (2\beta\omega)^2} \cos(\omega t) + \frac{2\beta\omega}{\left(\lambda^2 - \omega^2\right)^2 + (2\beta\omega)^2} \sin(\omega t) \right\}.$$

$$(2.61)$$

Noting that the excitation is given by $F_0 \cos(\omega t)$, we may now write:

$$\chi'(\omega) = \frac{\left(\lambda^2 - \omega^2\right)}{\left(\lambda^2 - \omega^2\right)^2 + (2\beta\omega)^2}, \qquad (2.62)$$

$$\chi''(\omega) = \frac{2\beta\omega}{\left(\lambda^2 - \omega^2\right)^2 + (2\beta\omega)^2}, \qquad (2.63)$$

where $\chi'(\omega)$ is the in-phase response per unit force while $\chi''(\omega)$ is the quadrature response per unit force. These quantities are the quadrature and in-phase components of the generalized susceptibility. At resonance, $\chi'(\omega)$ goes through zero and these generalized susceptibilities, particularly $\chi''(\omega)$ are the quantities usually investigated in some traditional methods of molecular spectroscopy.

Linearity and Superposition

The parameters which characterize the harmonic oscillator: the frequency and the relaxation time can also be obtained from steady state frequency response or from the time response. In fact, the connection between the frequency domain response and the time domain response is fundamental. Both the exponential decay of Eq. (2.58) and characteristic response of Eq. (2.61) are a consequence of the particular equation of motion for the system (Kreyszig, 1996). A more complex equation of motion would give a different steady state response and different time behaviour, but a simple interrelationship would persist. From Eqs. (2.42) and (2.43), the concept of generalized susceptibility implies that we may now write the steady state response of an arbitrary system to an oscillating force as follows (Slichter, 2013):

$$x(t) = f_m \left[\chi'(\omega) \cos \omega t + \chi''(\omega) \sin \omega t \right]. \qquad (2.64)$$

Implicit in this is the assumption that the system is linear or that the force is sufficiently small that deviations from linearity may be neglected; $x(t)$ is written as being linear in f. At this point, it is important to be able to relate this response to the relaxation following a transient excitation, and it is specifically the linearity condition which helps out since it guarantees that the response to the sum of a number of forces is the sum of the separate responses to the individual forces (Cowan, 1997).

Fourier Duality

Elementary Fourier synthesis may be used to construct an arbitrary force from a distribution of various oscillating waves. The response to this force is constructed using the same distribution from the responses to the various oscillating waves. Now, if the arbitrary force $F(t)$ is given by (Cowan, 1997):

$$F(t) = \int_0^\infty f(\omega) \cos(\omega t) d\omega. \tag{2.65}$$

Since $f(\omega)$ gives the distribution of cosine waves in the force, the corresponding time response becomes:

$$F(t) = \int_0^\infty f(\omega) \left[\chi'(\omega) \cos(\omega t) + \chi''(\omega) \sin(\omega t) \right] d\omega. \tag{2.66}$$

In a situation where there is a unit shock excitation (in which the force is represented by a delta function), the force may be written as follows:

$$F(t) = \delta(t), \tag{2.67}$$

or

$$\delta(t) = \frac{1}{\pi} \int_0^\infty \cos(\omega t) d\omega. \tag{2.68}$$

Equation (2.68) shows that the delta function force is simply the Fourier transform of a constant. Thus, a white distribution may be expressed as follows (Cowan, 1997):

$$f(\omega) = \frac{1}{\pi}. \tag{2.69}$$

The transient response becomes:

$$F(t) = \frac{1}{\pi} \int_0^\infty \left[\chi'(\omega) \cos(\omega t) + \chi''(\omega) \sin(\omega t) \right] d\omega. \tag{2.70}$$

This expression may be inverted to give the quadrature and in-phase components of the generalized susceptibility as the cosine and sine transforms of the transient response (relaxation function), $X(t)$:

$$\left. \begin{aligned} \chi'(\omega) &= \int_0^\infty X(t) \cos \omega t \, dt, \\ \chi''(\omega) &= \int_0^\infty X(t) \sin \omega t \, dt. \end{aligned} \right\} \tag{2.71}$$

The steady state response and the transient response are then related by Fourier transformation. Since the Fourier transform of the delta function gives the cosine function, the response to a cosine excitation can be related directly to a delta function excitation. This demonstrates that the steady state response and the transient response, that is, the resonance and the relaxation, the frequency domain and the time domain, are complementary pairs.

The description above shows that the same information is contained within each member of a pair (within the restriction of linearity). This gives us two ways of looking at resonant systems in general and NMR in particular (Cowan, 1997). Interestingly, in conventional spectroscopy, the focus is on the frequency response, $\chi'(\omega)$ and/or $\chi''(\omega)$. Therefore, it follows that exactly the same information may be obtained from time domain spectroscopy where $X(t)$ which is the relaxation following a transient excitation is studied.

Complex Susceptibility

In the simple harmonic oscillator, we have treated so far, we see that the response $x(t)$ to an oscillatory force $f_0 \cos(\omega t)$ is given by Eq. (2.64). Furthermore, the response X(t) to a transient or delta function force as given by Eq. (2.70) may be re-written as follows (Slichter, 2013; Cowan, 1997):

$$X(t) = \frac{1}{\pi} \int_0^\infty \left[\chi'(\omega) \cos(\omega t) + \chi''(\omega) \sin(\omega t) \right] d\omega. \tag{2.72}$$

This Fourier transform relation may be written in a more compact and symmetrical manner by introducing complex notation. A definition of complex dynamic susceptibility is given as follows (Slichter, 2013):

$$\chi(\omega) = \chi'(\omega) + i\chi''(\omega). \tag{2.73}$$

In terms of the transient response function, we have:

$$\chi(\omega) = \int_0^\infty X(t) \exp(i\omega t)dt. \tag{2.74}$$

The principle of causality implies that $X(t)$ is zero for negative times and there can be no response before the delta function force is applied. Hence, the lower limit of the integral may be extended to minus infinity, giving an expression which is then the usual complex Fourier integral transform (Cowan, 1997):

$$\chi(\omega) = \int_{-\infty}^\infty X(t) \exp(+i\omega t)dt; X(t) = \frac{1}{2\pi} \int_{-\infty}^\infty \chi(\omega) \exp(-i\omega t)d\omega. \tag{2.75}$$

The consequence of this formulation is that the following expressions hold

$$\chi(-\omega) = \chi * (\omega); \quad \chi'(-\omega) = \chi'(\omega); \quad \chi''(-\omega) = -\chi''(\omega). \tag{2.76}$$

According to the method of complex representation, the oscillatory force $f_0 \cos(\omega t)$ may be considered to be real part of a complex exponential $f_0 \exp(-i\omega t)$. From the linearity of the above equations, it follows that the response $x(t)$ may be expressed as the real part of the corresponding complex response:

$$x(t) = \mathrm{Re}\{\, f_0\chi(\omega) \exp(-i\omega t)\}. \tag{2.77}$$

This is analogous to the use of complex exponentials in AC circuit theory (in this case, the response to $f_0 \exp(-i\omega t)$ is $x(t) = f_m\chi(\omega) \exp(-\omega t)$). However, in terms of the harmonic components $x(\omega)$ and $f(\omega)$:

$$x(\omega) = f(\omega)\chi(\omega). \tag{2.78}$$

Dynamic Magnetic Susceptibility

It is very important to point out that the harmonic oscillator phenomenon we have discussed thus far differs from the phenomena associated with NMR (Cowan, 1997). The harmonic oscillator model is one-dimensional while the precession of transverse magnetization takes place in two dimensions. Therefore, an attempt to generalize the resonance views we have considered thus far to the treatment of NMR has been made. Considering a magnetic field of magnitude B_o rotating clockwise in the transverse plane at the angular velocity ω, the transverse components are given as follows:

$$B_x(t) = b_0 \cos \omega t,$$
$$B_y(t) = -b_0 \sin \omega t. \qquad (2.79)$$

A linear response to the fields as expressed in Eq. (2.79) is the rotating magnetization given as follows:

$$M_x(t) = \frac{b_0}{\mu_0} [\chi'(\omega) \cos (\omega t) + \chi''(\omega) \sin (\omega t)], \qquad (2.80)$$

$$M_y(t) = \frac{b_0}{\mu_0} [-\chi'(\omega) \sin (\omega t) + \chi''(\omega) \cos (\omega t)], \qquad (2.81)$$

where μ_0 represents the free space permeability and χ magnetic susceptibility. The expression of Eq. (2.81) follows from the rotational invariance of the system, that is, M_y rotates into M_x in a quarter of the cycle period (this implies a time of t_0). The implication of this is that:

$$M_x(t) = M_y\left(t + \left(t_0 = \frac{\pi}{2\omega}\right)\right). \qquad (2.82)$$

Hence, Eq. (2.80) may be obtained from (2.81) using the expression of Eq. (2.82). Eq. (2.82) is very important because it affords us a simple way of viewing transverse magnetization in the rotating frame. In the two-dimensional/rotational treatment we have considered so far, the hermiticity condition on $\chi(\omega)$ as demonstrated in Eq. (2.76) does not apply (Cowan, 1997). In fact, ω has a physical meaning: the direction of rotation. For an object placed in a magnetic field B_o, the Larmor precession has a well-defined direction such that both $\chi'(\omega)$ and $\chi''(\omega)$ are different from zero only near $\omega = \omega_0$, with no comparable behaviour around $\omega = -\omega_0$. Using complex functions (in which an imaginary part is introduced as an auxiliary quantity solely to simplify the mathematics), the magnetic field and magnetization may be represented as follows (Cowan, 1997):

$$\left.\begin{array}{l} B(t) = B_x(t) + iB_y(t), \\ M(t) = M_x(t) + iM_y(t). \end{array}\right\} \qquad (2.83)$$

From Eq. (2.79), the magnetic field is rotating at a frequency ω and is given as

$$B(t) = b_0 \exp (-i\omega t). \qquad (2.84)$$

Similarly, the resultant magnetization according to Eqs. (2.80) and (2.81) is:

$$M(t) = \frac{b_0}{\mu_0}$$
$$\times \left[\{ \chi'(\omega) \cos \omega t - i\chi'(\omega) \sin \omega t \} + \{ \chi''(\omega) \sin \omega t + i\chi''(\omega) \cos \omega t \} \right],$$
$$(2.85)$$

or

$$M(t) = \frac{b_0}{\mu_0} \left[\chi'(\omega) + i\chi''(\omega) \exp\left(-i\omega t\right) \right]. \tag{2.86}$$

Finally, according to Eq. (2.73), the response to the magnetic field:

$$B(t) = b_0 \exp\left(-i\omega t\right), \tag{2.87}$$

is given by the magnetization

$$M(t) = \frac{b_0}{\mu_0} \chi(\omega) \exp\left(-i\omega t\right), \tag{2.88}$$

The term $\chi(\omega)$ represents the complex (dynamic) magnetic susceptibility. It is important to note that the transverse magnetization of a resonating system can be described with linear response of the complex susceptibility (within some known limitations).

2.7.2.8   The Transverse Relaxation Function

The effects of the complex magnetic susceptibility in the time domain can be studied by examining the response of the system to a transient or delta function burst of magnetic field (Cowan, 1997):

$$\delta(t) = \frac{1}{2\pi} \int_{-\infty}^{\infty} \exp\left(-i\omega t\right) d\omega. \tag{2.89}$$

Hence, for a delta function excitation (superposition of the responses to individual complex exponentials) as defined above, the applied magnetic field and corresponding response are given as follows:

$$B(t) = \frac{\Theta}{2\pi\gamma} \int_{-\infty}^{\infty} \exp\left(-i\omega t\right) d\omega, \tag{2.90}$$

$$M(t) = \frac{\Theta}{2\pi\gamma\mu_0} \int_{-\infty}^{\infty} \chi(\omega)\exp(-i\omega t)d\omega. \tag{2.91}$$

The parameter Θ is related to effect of the applied magnetic field impulse. If this impulse is, for example, that of a magnetic field generated by the solenoid wound around a specimen with its axis in the x direction, a burst of current in the coil produces an impulsive magnetic field (Haacke et al., 1999) along the x direction such that

$$B_x(t) = \frac{\Theta\delta(t)}{\gamma}. \tag{2.92}$$

Applying this magnetic field will make the equilibrium magnetization to rotate around the x axis, producing some magnetization in the y direction. Since γB_x is the instantaneous frequency of precession about the x axis, the angle through which the magnetization is tipped is:

$$\gamma \int B_x(t)dt = \gamma\Theta \int \delta(t)dt. \tag{2.93}$$

The integral of the delta function is unity (Kreyszig, 1996); the tipping angle of tip is ϑ and gives the expression for the magnetic field impulse (Sprawls, 1987). It follows then that on application of the magnetic field impulse, there will be transverse magnetization

$$M_y = M_0 \sin \Theta \tag{2.94}$$

The evolution of the complex M_y as a result of transient excitation is given by:

$$M(t) = \frac{\Theta}{\gamma\mu_0} X(t). \tag{2.95}$$

The physics of the system being described places some restrictions on the form of this transverse relaxation. Normally, the initial transverse magnetization will be perpendicular to the causative magnetic field impulse. Hence, at zero-time, $X(t)$ must be pure imaginary (Cowan, 1997) and hence,

$$\int_{-\infty}^{\infty} \chi'(\omega)d\omega = 0. \tag{2.96}$$

The implication of this expression is that the area under $\chi'(\omega)$ is zero because of equal areas above and below the axis. The parameter $\chi''(\omega)$ is proportional to energy dissipation and this means that the imaginary part of the dynamic susceptibility must always be greater than zero.

The causality condition states that no response occurs before the magnetic field is applied and this condition makes $X(t)$ to be zero for negative times; $X(t)$ is discontinuous at $t = 0$. This will lead to problems if an attempt is made to investigate the behaviour of the transverse magnetization by expanding in powers of time: differentiating M(t) at the origin. This mathematical inconvenience can be overcome by symmetrizing M(t) at the origin. This is achieved by considering a function that is equal to M(t) for positive times with a suitable continuation to negative times.

Just like in the case of the shape of a spin echo, the mirror image may be added since a spin echo grows from zero point to peak value and then returns to zero. The shape is that of two back-to-back free induction decays (Cowan, 1997). The function M(t) and its image M*(t) may then be defined as follows:

$$M(t) = \frac{\Theta}{\gamma\mu_0} \frac{1}{2\pi} \int_{-\infty}^{\infty} \chi(\omega)\exp(-i\omega t)d\omega, \tag{2.97}$$

$$M*(-t) = \frac{\Theta}{\gamma\mu_0} \frac{1}{2\pi} \int_{-\infty}^{\infty} \chi*(\omega)\exp(-i\omega t)d\omega. \tag{2.98}$$

In order to symmetrize these functions by continuation through the origin, we have:

$$M(t) - M*(-t) = \frac{i\Theta}{\gamma\mu_0\pi} \int_{-\infty}^{\infty} \chi''(\omega)\exp(-i\omega t)d\omega. \tag{2.99}$$

Equation (2.99) indicates that the transverse relaxation is the Fourier transform of the imaginary part of the dynamic magnetic susceptibility. That is, the Fourier transform of the energy absorption line shape is related to the transverse relaxation. If the frequency is written as $\omega = \omega_0 + \Delta$ such that Δ is the deviation from the Larmor frequency, then the transverse response as given by Eq. (2.99) become:

$$M(t) - M*(-t) = \frac{\Theta}{\gamma\mu_0} \frac{i}{\pi} \exp(-i\omega_0 t) \int_{-\infty}^{\infty} \chi''(\omega_0 + \Delta)\exp(-i\Delta t)d\Delta. \tag{2.100}$$

That means the envelope of the transverse magnetization decay, F(t) is the Fourier transform of χ'' evaluated about its centre:

$$F(t) \propto \int_{-\infty}^{\infty} \chi''(\omega_0 + \Delta) \exp\left(-i\Delta t\right) d\Delta. \tag{2.101}$$

An implication of this expression is that if χ'' is symmetric about ω_0, it follows that the imaginary sine integral will vanish so that F(t) becomes real. That is, throughout the decay, the magnetization in the rotating frame remains parallel to its initial direction.

Response to Linearly Polarized Magnetic Field

All along, we have treated NMR systems using rotating magnetic fields. However, in most experimental arrangements, it is the linearly polarized field which is actually applied (Hinshaw & Lent, 1983). Since a linearly polarized field may be known as the sum of equal rotating and counter-rotating components, the response will be the sum of the corresponding rotating and counter-rotating magnetizations, which are not equal. Hence, for a magnetic field (Haacke et al., 1999)

$$\vec{B}(t) = \widehat{i}b_0 \cos \omega t = \frac{b_0}{2}\left[\left(\widehat{i}\cos \omega t + \widehat{j}\sin \omega t\right) + \left(\widehat{i}\cos \omega t - \widehat{j}\sin \omega t\right)\right] \tag{2.102}$$

the responses are:

$$M_x(t) = \frac{b_0}{2\mu_0}\left[(\chi'(\omega) + \chi'(-\omega))\cos \omega t + (\chi''(\omega) + \chi''(-\omega))\sin \omega t\right] \tag{2.103}$$

and

$$M_y(t) = \frac{b_0}{2\mu_0}\left[-(\chi'(\omega) + \chi'(-\omega))\sin \omega t + (\chi''(\omega) - \chi''(-\omega))\cos \omega t\right] \tag{2.104}$$

Comparing these results and the response of the one-dimensional oscillator, it can be seen that in the general, each component of the system is equivalent to an oscillator of a real susceptibility component (Cowan, 1997; Haacke et al., 1999)

$$\chi'(\omega) + \chi'(-\omega), \tag{2.105}$$

and an imaginary component

$$\chi''(\omega) - \chi''(-\omega). \tag{2.106}$$

With these expressions, the hermiticity condition shown above has been demonstrated. Where the effects of the counter-rotating magnetization may be ignored (that is, in vicinity of the rotating component), the x-dependent magnetic field becomes

$$B_x(t) = b_0 \cos \omega t, \tag{2.107}$$

and the oscillating magnetization is parallel to this field. This magnetization is given as:

$$M_x(t) = \frac{b_0}{2\mu_0} [\chi'(\omega) \cos \omega t + \chi''(\omega) \sin \omega t]. \tag{2.108}$$

Now, apart from the factor of ½, the excitation and response are identical to those of the one-dimensional oscillator treated earlier as demonstrated in Eqs. (2.79), (2.80), (2.107) and (2.108).

2.7.2.9   Linearity and Saturation

The relationship between NMR excitation and response has been clearly shown. It is now necessary to describe how an NMR system responds to the application of RF pulse. This is very important because many real systems are non-linear, that is, an increase in the input does not translate into an increase in the output.

Conditions for Linearity

At this point, it is necessary to show how small B_o must be to ensure linearity. Linearity is the condition that the hyperbolic tangent in Eq. (2.16) may be approximated by its expansion to first order. Consequently, the following condition must hold

$$\hbar \gamma B_0 << 2k_B T \tag{2.109}$$

In practice, this turns out not to be too restrictive. For the case of protons in water at room temperature, B_o should be less than 7000 T (Cowan, 1997). Therefore, deviations from linearity exist mainly at low temperatures. In realistic fields, the nuclear paramagnetic susceptibility of copper, for instance, obeys Curie's Law well at temperatures as low as 0.001 K. Thus, measurement of magnetization is a common technique of thermometry at such temperatures.

Saturation

The precessing magnetization as exhibited by the transverse component will induce a voltage within an image coil mounted around the specimen to be imaged (Cowan, 1997). The function of the reception side of any MR spectrometer system is fundamentally to detect this voltage signal (Haacke et al., 1999). The size of the absorption signal voltage is seen to be proportional to the applied oscillator voltage amplitude V_o. Essentially, this follows from the linearity assumption that the precessing magnetization has a direct relationship to the oscillating magnetic field B_1 given by the current from the oscillator.

Hence, the NMR absorption signal can be made arbitrarily large through increasing the output level of the oscillator. However, this cannot happen in practical systems. This is because at some stage, non-linearities become inevitable and must be taken into consideration. In fact, the appearance of such non-linearities places an upper limit on the size of the excitation permissible (Cowan, 1997) and consequently on the size of the absorption signal observable.

The framework of the Bloch equations as applied to Continuous Wave (CW) NMR may be useful in elucidating the concept of the non-linearity of the magnetic field response. This treatment considers the Bloch equations for the simple cases of perfect resonance (Cowan, 1997). According to the Bloch equations presented in Eq. (2.31), the field $\vec{B}$ comprises the static component B_o in the z-direction together with the B_1 field rotating at the Larmor frequency. Transforming this expression to the frame rotating at this frequency (in which B_o field vanishes, the B_1 component is zero and the B_x component becomes B_1), the coupled differential equation is

$$\left.\begin{aligned}
\frac{dM_x}{dt} &= -\frac{M_x}{T_2}, \\
\frac{dM_y}{dt} &= \gamma M_z B_1 - \frac{M_y}{T_2}, \\
\frac{dM_z}{dt} &= -\gamma M_y B_1 + \frac{(M_0 - M_z)}{T_1}.
\end{aligned}\right\} \qquad (2.110)$$

Setting the three-time derivatives to zero in order to obtain the equilibrium state of the system, the magnetization components become:

$$M_x = 0, \qquad (2.111)$$

$$M_y = \gamma T_2 M_z B_1, \qquad (2.112)$$

$$M_z = \frac{M_0}{1 + \gamma^2 B_1^2 T_1 T_2}. \qquad (2.113)$$

Equation (2.111) confirms that there is no in-phase response on resonance (Cowan, 1997). The transverse field response is embodied in Eq. (2.112) so that the seeming linearity of M_y in the excitation B_1 neglects the dependence of M_z on B_x. If we then substitute for Eq. (2.113) in Eq. (2.112), we have:

$$M_y = \frac{\gamma B_1 T_2 M_0}{1 + \gamma^2 B_1^2 T_1 T_2}. \qquad (2.114)$$

The linearity of response and excitation then follows when B_x is absent from the denominator of Eq. (2.114) and we may write:

$$\gamma^2 B_1^2 T_1 T_2 \ll 1. \tag{2.115}$$

According to Eq. (2.113), this condition is equivalent to:

$$\frac{M_0 - M_z}{M_z} \ll 1. \tag{2.116}$$

This implies that the z-component should not be reduced significantly from its equilibrium value M_0 (Cowan, 1997). The condition given in Eqs. (2.115) and (2.116) is known as the non-adiabatic condition (Awojoyogbe, 1997, 2002, 2003, 2004). The general solution to the Bloch equations may be separated into real and imaginary components of the dynamic susceptibility in terms of the relaxation times T_1 and T_2:

$$\chi'(\omega) = \frac{\chi_0 \omega_0 T_2^2 (\omega - \omega_0)}{1 + T_2^2 (\omega - \omega_0)^2 + \gamma^2 B_1^2 T_1 T_2}, \tag{2.117}$$

$$\chi''(\omega) = \frac{\chi_0 \omega_0 T_2}{1 + T_2^2 (\omega - \omega_0)^2 + \gamma^2 B_1^2 T_1 T_2}. \tag{2.118}$$

These equations have non-linear behaviour since the magnitude of the susceptibility is dependent on the strength of the applied oscillating field B_1. It is quite important to we note that the effect of a large B_1 is to reduce the z component of magnetization by tending to equalize the populations of the up and the down spin states. This phenomenon is called saturation (Cowan, 1997). Thus, in order to maintain linearity of response, saturation must be avoided. This phenomenon allows us to estimate the typical maximum signal voltage observable in a CW NMR experiment. For a coil of N turns and area A, Faraday's Law states that

$$V = NA\omega B, \tag{2.119}$$

where V is the voltage on the coil at frequency ω and B is the field in the coil (which is assumed to be uniform). The voltage change is now given as (Cowan, 1997; Haacke et al., 1999):

$$\Delta V = NA\omega BQ\eta\chi'' = 2NA\omega B_1 Q\eta\chi''. \tag{2.120}$$

This shows that the signal increases with strength of the oscillating field B_1. However, there is a maximum realistic value for B_1 dictated by the saturation condition:

$$\gamma^2 B_1^2 T_1 T_2 = 1 \tag{2.121}$$

or

$$B_1 = \frac{1}{\gamma\sqrt{T_1 T_2}}.$$ (2.122)

With this degree of saturation, we have:

$$\chi''_{\max} = \chi''(\omega_0) = \frac{T_2 \omega_0 \chi_0}{2}.$$ (2.123)

Therefore, the maximum change in voltage is given as:

$$(\Delta V)_{\max} = \frac{N A \omega_0^2 Q \eta \chi_0}{\gamma} \sqrt{\frac{T_2}{T_1}}.$$ (2.124)

Non-Adiabatic Condition

There is a limit to which the B_1 term may be ignored in the denominator of Eq. (2.114) such that the observed susceptibility is independent of the strength of the excitation. This limit is defined by the condition (Cowan, 1997; Awojoyogbe, 2003):

$$\gamma^2 B_1^2 T_1 T_2 \ll 1.$$ (2.125)

When this condition holds, the frequency response of the system assumes Lorentzian behaviour (Cowan, 1997). This condition is also known as the non-adiabatic condition (Abragam, 1961). Hence, Eqs. (2.117) and (2.118) become:

$$\chi'(\omega) = \frac{\chi_0 \omega_0 T_2^2 (\omega - \omega_0)}{1 + T_2^2 (\omega - \omega_0)^2},$$ (2.126)

and

$$\chi''(\omega) = \frac{\chi_0 \omega_0 T_2}{1 + T_2^2 (\omega - \omega_0)^2}.$$ (2.127)

Adiabatic Condition

In magnetic resonance applications, it is also possible to suppress the applied field very slowly and reversibly, taking care to introduce as little spin relaxation as possible in the process such that the relaxation rate changes very slowly compared to the applied RF field. In such a system, increase in entropy is made as small as possible (Abragam, 1996). The condition under which this situation holds is known as the adiabatic condition (Awojoyogbe, 1997, 2007) given as:

$$\gamma^2 B_1^2 T_1 T_2 \gg 1. \tag{2.128}$$

Then, Eqs. (2.127) and (2.128) become

$$\chi'(\omega) = \frac{\chi_0 \omega_0 T_2^2 (\omega - \omega_0)}{T_2^2 (\omega - \omega_0)^2 + \gamma^2 B_1^2 T_1 T_2}, \tag{2.129}$$

and

$$\chi''(\omega) = \frac{\chi_0 \omega_0 T_2}{T_2^2 (\omega - \omega_0)^2 + \gamma^2 B_1^2 T_1 T_2}. \tag{2.130}$$

2.7.3   *Quantum and Classical Descriptions of Spin Motion*

It is noteworthy that the emergence of molecular imaging as a research field combining *in-vivo* images and molecular biology (Xue et al., 2013) led to the technique being considered as a cellular function visualization method which does not significantly disturb molecular processes in living systems. The theoretical background of MR presented thus far has been based on numerous molecular spins (macroscopic description) (Slichter, 2013; Callaghan, 1994).

Quantum mechanics is the most often used to describe matter on the scale of molecules, atoms or elementary particles. Therefore, proper formulation of NMR application for molecular imaging requires quantum mechanics of magnetic resonance molecular imaging. Classical and quantum views of magnetic resonance have presented a historical challenge. These views will now be examined in some detail in the subsection below.

2.7.3.1   Quantum and Classical Treatment

The Larmor frequency is important in both the quantum and the classical definitions of magnetic resonance (Abragam, 1961; Slichter, 2013). These two views of magnetic resonance appear different but they are somewhat related. From the quantum mechanics perspective, it is the energy equivalent of the resonance frequency that is fundamental (quanta of the requisite energy are absorbed, causing energy level transitions). Meanwhile, in the classical approach, it is the actual frequency that is important: the magnetic moment vectors precess at this particular rate (Cowan, 1997).

This duality of views precisely demonstrates the twin independent discoveries of nuclear magnetic resonance. The team from Harvard University, made up of Purcell, Torrey and Pound were fundamentally spectroscopists and they regarded the effect of a magnetic field on an individual magnetic moment in terms of splitting of its energy levels. From their point of view, the NMR experiment then involved the stimulation of transitions between these levels. The Stanford University group, comprising Bloch, Hansen and Packard, dealt with the gross effect of a magnetic field by considering a collection of magnetic moments and their reorientation. This approach is responsible for their picture in terms of observable quantities and why the approach is called the classical description (Cowan, 1997).

The unification of these apparently different views has roots in the dual role of energy in physics. The elementary interpretation of energy as work done is connected with the internal state of the system and this implies that a bit more work may be done on the system and its internal state will have changed correspondingly. This is the point of view of NMR spectroscopists. However, a more advanced interpretation of energy follows from the Hamiltonian formulation of dynamics (Cowan, 1997). This formulation conceptualizes energy function as being the determinant of the time evolution of the system, giving the Hamilton's equations in classical mechanics and Heisenberg's equation, also known as the time dependent Schrödinger equation, in quantum mechanics. Therefore, the foregoing is responsible for the inextricable connection between work done and time evolution— the spectroscopic approach of Purcell at Edward Harvard and the reorientation approach of Felix Bloch at Stanford.

Unfortunately, this has constituted in a challenge for MRI because these different views have been held differently ever since. Although the view which any scientist holds is usually very sufficient enough for their research, the different perspectives in MRI are better taken as complementary approaches rather than exclusive definitions. This brings out a better understanding of NMR. In fact, NMR may be regarded as a macroscopic quantum mechanical phenomenon (describing processes that show quantum behaviour at macroscopic scale (Leggett, 1980; Chen, 2013). This point is considered in subsequent sections.

2.7.3.2 The Heisenberg Equation

In quantum mechanics, observable quantities are represented by operators (Peleg et al., 1998). For an arbitrary operator $\mathfrak{R}$ representing the observable R, the equation of motion which is known as the Heisenberg equation is given as (Cowan, 1997):

$$\frac{d\mathfrak{R}}{dt} = -\frac{i}{\hbar}[\mathfrak{R}, \aleph],
\qquad (2.131)$$

where $\aleph$ is the Hamiltonian or energy operator and the square bracket denotes the commutator expressed as:

$$[\mathfrak{H}, \aleph] = \mathfrak{H}\aleph - \aleph\mathfrak{H}. \tag{2.132}$$

2.7.3.3   Equation of Motion for Magnetic Moment Operator $\widehat{\mu}$

For this study, the observable operator of interest is the magnetic moment operator $\vec{\mu}$, which, by comparison with the classical case, will be parallel and proportional to the spin angular momentum operator $\widehat{\ell}$ such that we may now write (Haacke et al., 1999):

$$\widehat{\mu} = \hbar\gamma\widehat{\ell}. \tag{2.133}$$

Having denoted the operator, representing the magnetic moment as $\widehat{\mu}$, the angular momentum $\vec{I}$ may as well be represented as $\widehat{\ell}$. Additional information necessary for solving the equation of motion is the form of the Hamiltonian operator representing the energy of the spin magnetic moment. In classical terms, the energy of the nuclear magnetic moment placed inside a magnetic field $\vec{B}$ is:

$$E = -\vec{\mu} \cdot \vec{B}. \tag{2.134}$$

Thus, the quantum expression corresponding to the above expression is given as:

$$\aleph = -\widehat{\mu} \cdot \vec{B} = -\gamma\hbar\widehat{\ell} \cdot \vec{B}. \tag{2.135}$$

If we now substitute for this expression in the Heisenberg equation, we have (Cowan, 1997; Peleg et al., 1998):

$$\frac{d\widehat{\ell}}{dt} = -\frac{i}{\hbar}\left[\widehat{\ell}, -\gamma\hbar\widehat{\ell} \cdot \vec{B}\right] = i\gamma\left[\widehat{\ell}, \widehat{\ell} \cdot \vec{B}\right]. \tag{2.136}$$

In order to use the commutation relations for angular momenta, we chose to work with the (dimensionless) spin angular momentum operator $\widehat{\ell}$ rather than the magnetic moment $\widehat{\mu}$. Hence, the equation of motion for x component of $\widehat{\ell}$ is given as (Cohen-Tannoudji et al., 1977):

$$\frac{d\widehat{\ell}_x}{dt} = i\gamma\left[\widehat{\ell}_x, \widehat{\ell}_x B_x + \widehat{\ell}_y B_y + \widehat{\ell}_z B_z\right]. \tag{2.137}$$

or

$$\frac{d\widehat{\ell}_x}{dt} = i\gamma B_x \left[\widehat{\ell}_x, \widehat{\ell}_x\right] + i\gamma B_y \left[\widehat{\ell}_x, \widehat{\ell}_y\right] + i\gamma B_z \left[\widehat{\ell}_x, \widehat{\ell}_z\right]. \tag{2.138}$$

2.7.3.4 Evaluation of Commutators

Angular momentum commutators may be arranged in cyclic permutations such that we have (Cohen-Tannoudji et al., 1977):

$$\left[\widehat{\ell}_x, \widehat{\ell}_y\right] = i\widehat{\ell}_z. \tag{2.139}$$

Since the commutator of an operator with itself is zero, the first term of the equation of motion vanishes (Cowan, 1997; Cohen-Tannoudji et al., 1977). If we now make use of Eq. (2.139) in Eq. (2.138), we have:

$$\frac{d\widehat{\ell}_x}{dt} = \gamma\left(\widehat{\ell}_y B_z - \widehat{\ell}_z B_y\right). \tag{2.140}$$

Here are some observations in these expressions;

1. Planck's constant has cancelled out and this implies that the resulting equation is not inherently quantum mechanical.
2. Since the equation is linear in $\widehat{\ell}$, the same equation holds for the magnetic moment operator $\widehat{\mu}$ on multiplying through by the gyromagnetic ratio γ.
3. Finally, the right side of Eq. (2.140) is equal to the x-component of the cross product $\widehat{\ell} \times \vec{B}$, or

$$\left(\widehat{\ell}_y B_z - \widehat{\ell}_z B_y\right) = \left|\widehat{\ell} \times \vec{B}\right|_x. \tag{2.141}$$

Consequently, the equation of motion for nuclear magnetic moment operator is given as:

$$\frac{d\widehat{\mu}}{dt} = \gamma\widehat{\mu} \times \vec{B}. \tag{2.142}$$

For the quantum mechanical magnetic moment operator, the equation of motion is identical to the classical equation for the magnetic moment given by Eq. (2.42). This formulation is very remarkable because individual magnetic moments and macroscopic magnetization (due to the magnetic moments) of a sample evolve according to the classical equation of spin motion (Hanson, 2008).

2.7.3.5   Computation of Expectation Values

To further establish the equivalence of the classical and quantum descriptions of the behaviour of nuclear magnetic moment, it is crucial consider a description of the quantum state of the system (Cowan, 1997). If a quantum system is in a state s, described by a wave function $\psi_s(x)$, then the mean or expectation value of an operator $\Re$ is:

$$R = \langle \Re \rangle = \int \psi_s^*(x)\Re\psi_s(x)dx \tag{2.143}$$

where x represents a set of coordinates defining the state of the system. It is worthy of note that in quantum mechanics, to perform the integration, we include summation over discrete variables as well as integration over continuous ones (Cowan, 1997; Hansen, 2008). This is normally accounted for using the Dirac notation for Eq. (2.143):

$$\langle \Re \rangle = \langle b|\Re|b \rangle. \tag{2.144}$$

Performing this operation on the equation of motion for the quantum operator, $\widehat{\mu}$ gives:

$$\frac{d\langle \widehat{\mu} \rangle}{dt} = \gamma \langle \widehat{\mu} \rangle \times \vec{B}. \tag{2.145}$$

It can therefore be seen that the quantum equation for the expectation value of the operator $\widehat{\mu}$ is identical to the classical equation for the quantity $\widehat{\mu}$. Further demonstration relies on the quantum equation of motion being linear in $\widehat{\mu}$ and the fact that there are no other operators multiplying $\widehat{\mu}$. In general, quantum and classical equations do not have equivalent forms. Hence, switching between quantum and classical approach requires the measurement of an operator or its expectation value. For example, this task will require that an observer either takes the measurement of the spin angular momentum operator or the expectation value of spin angular momentum.

2.7.4   Quantum Treatment of NMR Relaxation

A simple description of NMR relaxation as it relates to the classical equations of Bloch has been presented. The following section applies this theoretical formulation to relaxation mechanism in magnetic resonance.

2.7.4.1   Relaxation, Resonance and Equilibrium State

In the simplest pulsed NMR experiments, a system is studied as it returns to equilibrium following some initial disturbance (Cowan, 1997). At this point, it is useful to describe NMR relaxation and the physics of the process. The quantum mechanical state of a system is described by a wave function and the Schrödinger equation for the wave function gives the time evolution of that state in terms of the Hamiltonian (Bransden & Joachain, 2000; Cohen-Tannoudji et al., 1977). The Hamiltonian contains eigenstates that have fixed time energy and, according to statistical mechanics, when a system is in thermal equilibrium, the system may be found in any of its energy eigenstates with a probability given as:

$$P \propto \exp\left(-\frac{E}{k_B T}\right) \tag{2.146}$$

where E represents the eigenstate energy, k_B represents the Boltzmann constant and the system's absolute temperature is given by T. Eq. (2.146) implies that an equilibrium state is made up of a large number of quantum states (Cowan, 1997).

2.7.4.2   Expectation Values of Quantum Operators

As presented in earlier sub-sections, in the scenario where a quantum system is in a state defined by a wave function $\psi_s(x)$, it follows that the expectation value R of an operator $\mathfrak{R}$ is described by Eq. (2.143). This may be considered to be a simple ss matrix element of the operator $\mathfrak{R}$, that is, we have (Cowan, 1997):

$$\mathfrak{R}_{ss} = \langle \mathfrak{R} \rangle = \int \psi*_s(x)\mathfrak{R}\psi_s(x)dx. \tag{2.147}$$

Hence, obtaining the value of an observable quantity requires that the state of the system as well as the wave function must first be known. For simple systems, this may easily be obtained by solving the Schrödinger equation subject to boundary conditions appropriate to the system. For complex systems, this task usually proves to be extremely difficult and since there may be as many as 10^{23} variables to be considered, an appropriate alternative approach is sought in such circumstances.

2.7.4.3   Systems in Equilibrium

For a system that is initially at thermal equilibrium, it is quite impossible to know its precise quantum state but the probability of finding the system in any one of its stationary states (its energy eigenstates) may be known (Cowan, 1997). A system in thermal equilibrium is characterized by a temperature T and the probability that the system is found in the energy state s is given as follows:

$$P_s = P_0 \exp\left(-\frac{E_s}{k_B T}\right) \tag{2.148}$$

where E_s is the energy eigenvalue of the eigenstate s with k_B representing the Boltzmann constant. The constant P_0 may be computed by normalization since

$$\sum_{all\ states} P_s = 1 \tag{2.149}$$

Hence, the probability P_s is given as:

$$P_s = \frac{\exp\left(-\frac{E_s}{k_B T}\right)}{\sum_s \exp\left(-\frac{E_s}{k_B T}\right)} \tag{2.150}$$

where the denominator is the partition function for the system. Consequently, the observed value of an operator $\mathfrak{R}$ will be the average over the set of possible states s so that we now have (Cowan, 1997):

$$R = \sum_s \mathfrak{R}_{ss} P_s. \tag{2.151}$$

If we then set $\zeta = \frac{1}{k_B T}$, this expression becomes

$$R = \frac{\sum_s \mathfrak{R}_{ss} \exp\left(-\zeta E_s\right)}{\sum_s \exp\left(-\zeta E_s\right)}. \tag{2.152}$$

We may consider the energy E_s of the state, s as ss matrix element of the Hamiltonian or energy operator $\aleph$. Noting that $\aleph$ is diagonal in this representation since the states considered are energy eigenstates, we may write:

$$E_s = \aleph_{ss} \tag{2.153}$$

Since $\aleph$ is diagonal, we have the following expression (Cowan, 1997):

$$\exp\left(-\zeta E_s\right) = \left|\exp\left(-\zeta \aleph\right)\right|_{ss} \tag{2.154}$$

where the quantity $|\exp(-\zeta \aleph)|_{ss}$ is the ss matrix element of the operator $\exp(-\zeta \aleph)$. Since $\exp(-\zeta \aleph)$ is also diagonal, we write

$$\mathfrak{R}_{ss}|\exp\left(-\zeta\aleph\right)|_{ss} = |\mathfrak{R}\exp\left(-\zeta\aleph\right)|_{ss} \tag{2.155}$$

The observed value R as given by Eq. (2.152) becomes:

$$R = \frac{\sum_s |\mathfrak{R}\exp\left(-\zeta\aleph\right)|_{ss}}{\sum_s |\exp\left(-\zeta\aleph\right)|_{ss}} \tag{2.156}$$

It is observed that both the numerator and denominator represent the sum of the diagonal matrix elements of operators. A special quantity called trace may be used to represent the diagonal sum of a matrix (Cowan, 1997); the value of which is usually independent of the representation used. Hence, the matrix does not have to be diagonal and we may now write the expression for R as follows:

$$R = \frac{Tr\{\mathfrak{R}\exp\left(-\zeta\aleph\right)\}}{Tr\{\exp\left(-\zeta\aleph\right)\}}. \tag{2.157}$$

Any set of states can be used and so, in order to find the expectation value of an observable we no longer need to find the eigenstates of the system. The symbol Tr represents the trace of its matrix argument. It is extremely difficult to obtain the stationary states of a macroscopic system (Bransden & Joachain, 2000). The expression for the thermal averages in terms of traces of operators obviates the need to find these states (Cowan, 1997). Any convenient basis set of states may be used and the evaluation of expectation values for systems in thermal equilibrium thus becomes a relatively simple problem.

2.7.4.4 Curie's Law for Systems with Interactions

At this point, Curie's Law for complex interacting systems is derived in order to demonstrate the results obtained above. Eq. (2.157) may be used to find the expectation value of the magnetization $\vec{M}$ and hence, it may be assumed that for a polarizing magnetic field $\vec{B}$ applied along the z-direction (Cowan, 1997),

$$M_z = \frac{1}{V}\sum \mu_z = \frac{\gamma\hbar}{V}\sum_i I_z^i. \tag{2.158}$$

And using Eq. (2.157), we write:

$$\langle M_z \rangle = \frac{\gamma\hbar}{V}\frac{\sum_i Tr\{I_z^i\exp\left(-\zeta\aleph\right)\}}{Tr\{1\}}. \tag{2.159}$$

Within high temperature and low polarization limit, the exponential expansion leads to:

$$\langle M_z \rangle = \frac{\gamma\hbar}{V} \frac{\sum_i Tr\{I_z^i(1 - \zeta\aleph)\}}{Tr\{1\}}. \tag{2.160}$$

Since the trace of I_z vanishes when the first term of Eq. (2.160) vanishes, we have

$$\langle M_z \rangle = -\frac{\gamma\hbar\zeta}{V} \frac{\sum_i Tr\{I_z^i\aleph\}}{Tr\{1\}} \tag{2.161}$$

This expression shows why the term ζ introduces 1/T behaviour in the magnetization of the system. The Hamiltonian for the NMR system will be made up of the Zeeman term (because of spin interaction in the presence of an external magnetic field) with inter-spin interactions. Hence, we may write:

$$\aleph = -\gamma\hbar B_z I_z + \aleph'. \tag{2.162}$$

The expectation value of the magnetization then becomes (Cowan, 1997)

$$\langle M_z \rangle = \frac{\gamma^2\hbar^2\zeta B_z}{V} \frac{\sum_i Tr\{I_z^i I_z\}}{Tr\{1\}} - \frac{\gamma\hbar\zeta}{V} \frac{\sum_i Tr\{I_z^i\aleph'\}}{Tr\{1\}}. \tag{2.163}$$

The I_z is made up of an overall sum I_z^i, but only the i $=$ j term will be non-zero. Since all spins are equivalent, the sum is the same as the product of N and the trace over a single spin, the expectation value of the magnetization becomes:

$$\langle M_z \rangle = N_v\gamma^2\hbar^2\zeta B_z \frac{Tr\{(I_z^i)^2\}}{Tr\{1\}}, \tag{2.164}$$

where N_v represents spin density. The trace of the numerator can easily be evaluated for spin $^1/_2$. Since I_z is diagonal, the elements $+^1/_2$ and $-^1/_2$, I_z^2 are also diagonal and the trace is $^1/_2$. The denominator trace is trivial and for a 2×2-unit matrix, the trace must be 2 and consequently, for spin $^1/_2$, we have

$$\langle M_z \rangle = \frac{N_v\gamma^2\hbar^2 B_z}{4kT}. \tag{2.165}$$

To generalize this result to a spin of arbitrary magnitude I, the traces are given as (Cowan, 1997):

$$\left.\begin{array}{l} Tr\{I_z^2\} = \dfrac{I(I+1)(2I+1)}{3}, \\[2mm] Tr\{1\} = I(I+1). \end{array}\right\} \tag{2.166}$$

Hence, the results for high temperature/low polarization limits lead to the standard Curie Law:

$$\langle M_z \rangle = \frac{N_v \gamma^2 \hbar^2 I(I+1)}{3kT} B_z \tag{2.167}$$

2.7.4.5 Behaviour of Magnetization in Pulsed NMR

In order to state the expressions for the relaxation of the magnetization from the non-equilibrium states created by 90° and 180° RF pulses, it is noteworthy that in quantum theory, a system and its evolution are specified by the Hamiltonian operator (Cowan, 1997).

The Hamiltonian and Approximations

Considering the general form of the Hamiltonian for an assembly of nuclear spins, we may write

$$\aleph = \aleph_z + \aleph_m + \aleph_d, \tag{2.168}$$

where $\aleph_z$ is the Zeeman part of the Hamiltonian (Slichter, 2013; Abragam, 1961) which represents the interactions of the individual nuclear magnetic moments placed in the static magnetic field. Since the magnetic energy density may be written as $-B_oM$, the Zeeman Hamiltonian operator becomes:

$$\aleph_z = -B_0 \gamma \hbar I_z = -\hbar \omega_0 I_z, \tag{2.169}$$

where I_z is the operator for the total z component of the spin angular momentum for the system. The Hamiltonian component $\aleph_m$ generates any motion of the particles and comprises the potential and kinetic energy of the particles. In most cases, the associated motion may be treated in a classical fashion such that $\aleph_m$ becomes negligible. The parameter $\aleph_d$ represents the inter-spin interactions and usually accounts for the inter-nuclear dipole-dipole interaction. However, it can also include a quadrupole or a hyperfine interaction (Cowan, 1997).

At this point, an approximation can be made such that the static magnetic field B_o is said to be much greater than the local fields associated with the interaction Hamiltonian $\aleph_d$. This is no restriction for the majority of NMR experiments where

B_o is much greater than the few gauss of the dipolar or other interactions. The consequence of this assumption is that the spin energy states are then essentially the equally spaced eigenstates of the Zeeman Hamiltonian and the effect of $\aleph_d$ is to cause small splitting of these energy levels. In the Boltzmann distribution of energy states for the system $(\exp(-\zeta\aleph))$, it follows that $\aleph_d$ is insignificant compared to $\aleph_z$. We shall also ignore any motion Hamiltonian in the Boltzmann distribution. From earlier sections, it is to be noted that for the high temperature or low polarization limit, the following approximation holds

$$\exp\left(-\zeta\aleph\right) \approx 1 - \zeta\aleph. \tag{2.170}$$

The relaxation of the quantity R may be written as (Cowan, 1997):

$$R(t) = \frac{Tr\{\mathfrak{R}(t)(1 - \zeta\aleph')\}}{Tr\{1\}}, \tag{2.171}$$

where $\aleph'$ is the fictitious Hamiltonian related to the initial non-equilibrium state being treated. The denominator is the trace of the unit operator. It takes this simple form because the trace of $\aleph'$, a spin angular momentum operator, is zero since the trace of I_z is zero. What are interested in the behaviour of magnetization: M_x, M_y and M_z whose traces similarly are zero. The assumption $Tr\{\mathfrak{R}\}$ leads to:

$$R(t) = -\zeta\frac{Tr\{\mathfrak{R}(t)\aleph'\}}{Tr\{1\}}. \tag{2.172}$$

Quantum Responses Due to 90° and 180° Pulses

In pulsed NMR experiments, the non-equilibrium states of interest are created using bursts of RF pulses in the transverse plane (Cowan, 1997). Hence, the equilibrium magnetization, which was pointing in the z-direction, can be rotated to the x direction (90° pulse) or the z-direction (180° pulse). A more realistic view of the creation of non-equilibrium states with the RF pulses may be seen considering these states as the equilibrium configuration of a fictitious Hamiltonian. Hence, in order to create x-plane magnetization, the B_o field needs to be applied along the x-axis and fictitious Zeeman Hamiltonian for a 90° pulse is then given by:

$$\aleph_{90} = -B_0\gamma\hbar I_x = -\hbar\omega_0 I_x. \tag{2.173}$$

Similarly, in order to create the magnetization along the z direction (magnetization reversal), we require that the B_o field points in the z axis such that the fictitious Hamiltonian for a 180° pulse is written as:

$$\aleph_{180} = +B_0\gamma\hbar I_z = +\hbar\omega_0 I_x. \tag{2.174}$$

Equations (2.173) and (2.174) are the Hamiltonians to be employed in Eq. (2.172) in order to describe the relaxation of spin magnetization following 90° and 180° pulses. Fortunately, with this formulation, the spin system would have remained unchanged by these fictitious magnetic fields even if the motion Hamiltonian in the Boltzmann distribution was not neglected.

Relaxation of Longitudinal and Transverse Magnetization

In NMR experiments, the behaviour of magnetization functions is usually studied. To study transverse relaxations, the transverse component of these magnetization functions is created by applying a 90° pulse such that the transverse magnetization such as M_x may be observed (note that the transverse magnetization functions M_x and M_y may be taken to be equivalent in the rotational frame of reference (Haacke et al., 1999)). Then according to Eq. (2.172), M_x becomes (Cowan, 1997):

$$M_x(t) = -\zeta \frac{Tr\{M_x(t)\aleph_{90}\}}{Tr\{1\}} = -\zeta\hbar\omega_0 \frac{Tr\{M_x(t)I_x\}}{Tr\{1\}}. \tag{2.175}$$

Since M_x is the magnetic moment per unit volume, Eq. (2.175) becomes:

$$M_x(t) = \frac{1}{V}\sum_i \mu_x^i = \frac{\gamma\hbar}{V} I_x, \tag{2.176}$$

$$M_x(t) = \frac{\gamma\zeta\hbar^2\omega_0}{V} \frac{Tr\{I_x(t)I_x\}}{Tr\{1\}}. \tag{2.177}$$

An application of the 180° pulse leads to longitudinal relaxation, thereby reversing the magnetization. When M_z returns to its equilibrium value (M_0), we have:

$$M_z(t) = -\zeta \frac{Tr\{M_z(t)\aleph_{180}\}}{Tr\{1\}} = -\frac{\gamma\zeta\hbar^2\omega_0}{V} \frac{Tr\{I_z(t)I_z\}}{Tr\{1\}}. \tag{2.178}$$

It is worthy of note that the magnetization in the longitudinal direction cannot be easily observed (Cowan, 1997); it needs to be tipped into the transverse plane with a 90° pulse in order to map it and reconstruct images from it.

The operator in the trace $I_x(t)$ is what is usually observed (Cowan, 1997). The second operator I_x may be written as $I_x(0)$. This is the zero-time value of the Heisenberg operator $I_x(t)$ and we have decided to write out this term $I_x(0)$ to show this explicitly and to enhance the symmetry of the expression. The term (I_x) is written in this way because the 90° pulse rotated the equilibrium I_z into I_x. The expression for the behaviour of the y component of the magnetization function after the application of same 90° pulse may then be written as:

$$M_y(t) = \frac{\gamma \zeta \hbar^2 \omega_0}{V} \frac{Tr\{I_y(t)I_x(0)\}}{Tr\{1\}}$$

(2.179)

Therefore, Eqs. (2.177) and (2.179) further confirm that M_y and M_x interchange orientations in a rotating frame of reference since I_x and I_y rotate interchangeably. The form of the magnetization function in a rotating frame of reference has been shown and now in a Laboratory frame of reference, the magnetization function may be written as:

$$M = M_x + iM_y.$$

(2.180)

From Eqs. (2.177) and (2.179), the complex magnetization now becomes:

$$M(t) = \frac{\gamma \zeta \hbar^2 \omega_0}{V} \frac{Tr\{I_+(t)I_x(0)\}}{Tr\{1\}}$$

(2.181)

or more conveniently,

$$M(t) = \frac{\gamma \zeta \hbar^2 \omega_0}{V} \frac{Tr\{I_+(t)I_-(0)\}}{Tr\{1\}}$$

(2.182)

where

$$I_+ = I_x + iI_y, \quad I_- = I_x - iI_y$$

(2.183)

2.7.4.6 Quantum Mechanical Description of Spin in a Static Field B_o

Quantum mechanical description of a spin in the uniform magnetic field B_o leads to energies in terms of the magnetic quantum number, m (Slichter, 2013). The parameter m is the eigenvalue of the component of spin I_z that is in a direction parallel to the uniform magnetic field B_o. The energies of the spin are then given as follows:

$$E_m = -m\gamma\hbar B_0.$$

(2.184)

This leads to eigen functions of the time-independent Schrödinger equation with the representation $\psi^0_{I,m}$ and is related to the wave function presented in Sect. 2.7.4.2. Hence, the time-dependent solution to the Schrödinger equation is:

$$\psi_{I,m}(t) = \psi_{I,m}^0 e^{-\frac{iE_m t}{\hbar}} \tag{2.185}$$

In general terms, Eq. (2.185) may be written as follows:

$$\psi(t) = \sum_{m=-I}^{I} A_m \psi_{I,m}^0 e^{-\frac{iE_m t}{\hbar}} \tag{2.186}$$

where A_m represents complex constants (Slichter, 2013). Using Eq. (2.186), the expectation value of an observable such as the x-component of magnetic moment is:

$$\langle \mu_x(t) \rangle = \int \psi^*(t) \mu_x \psi(t) d\tau. \tag{2.187}$$

The x-component of magnetic moment as given in Eq. (2.133) may be given as

$$\mu_x = \gamma \hbar I_x. \tag{2.188}$$

Therefore, from Eqs. (2.186) and (2.188),

$$\langle \mu_x(t) \rangle = \sum_{m,\,m'} \gamma \hbar A_{m'}^* A_m (m'|I_x|m) e^{-\frac{i\left(E_{m'}-E_m\right)t}{\hbar}}, \tag{2.189}$$

where

$$(m'|I_x|m) = \int A_{m'}^* I_x A_m d\tau. \tag{2.190}$$

Equation (2.190) is a time-dependent matrix and both Eqs. (2.189) and (2.190) hold for any operator. For general description, the expectation value will constitute a number of terms which are oscillating harmonically. The possible frequencies $f_{m'm}$ are:

$$f_{m'm} = \frac{E_{m'} - E_m}{\hbar}, \tag{2.191}$$

They correspond to the frequency of absorption or emission between the states, m and m'. The observable properties of any quantum system are usually given in terms of Eq. (2.189). This is an assumption which is the basis of Heisenberg and Born's formulations of quantum theory in matrix form (Slichter, 2013). In this study, attention has been focused on Heisenberg and Born's formulations because they make room for quantitative analysis in molecular imaging of a single particle. However, since the matrix elements in Eq. (2.190) vanish unless $m' = m \pm 1$, all the terms of Eq. (2.189) possess an angular frequency of either $+\gamma B_0$ or $-\gamma B_0$. Their sum must also contain just the term γB_0 and the expectation value $\langle \mu_x(t) \rangle$ oscillates in time at the classical precession frequency.

Equation of Motion for the System

Assuming there exists a pair of wave functions $\Psi(t)$ and $\Phi(t)$, both of which are solutions of the same Schrödinger equation. That is, for a Hamiltonian operator $\aleph$, these equations are written as follows:

$$-\frac{\hbar}{i}\frac{\partial \Psi}{\partial t} = \aleph\Psi, \tag{2.192}$$

$$-\frac{\hbar}{i}\frac{\partial \Phi}{\partial t} = \aleph\Phi. \tag{2.193}$$

If some operator $\widehat{O}$ with no explicit time dependence is introduced,

$$\frac{d}{dt}\int \Phi^*\widehat{O}\Psi d\tau = \int \frac{\partial \Phi^*}{\partial t}\widehat{O}\Psi d\tau + \int \Phi^*\widehat{O}\frac{\partial \Psi}{\partial t}d\tau. \tag{2.194}$$

From Eqs. (2.192) and (2.193),

$$\left.\begin{array}{l} \dfrac{\partial \Psi}{\partial t} = -\dfrac{i}{\hbar}\aleph\Psi, \\[2mm] \dfrac{\partial \Phi^*}{\partial t} = -\dfrac{i}{\hbar}\aleph\Phi^*. \end{array}\right\} \tag{2.195}$$

Hence, Eq. (2.194) becomes:

$$\frac{d}{dt}\int \Phi^*\widehat{O}\Psi d\tau = \int \left(-\frac{i}{\hbar}\aleph\Phi^*\right)\widehat{O}\Psi d\tau + \int \Phi^*\widehat{O}\left(-\frac{i}{\hbar}\aleph\Psi\right)d\tau$$

$$= \frac{i}{\hbar}\int \Phi^*\left(\aleph\widehat{O} - \widehat{O}\aleph\right)\Psi d\tau = \frac{i}{\hbar}\int \Phi^*\left[\aleph, \widehat{O}\right]\Psi d\tau \tag{2.196}$$

Now, if a symbolic derivative of the operator $\widehat{O}$ is given as follows (Slichter, 2013)

$$\frac{d\widehat{O}}{dt} = \widehat{O} \tag{2.197}$$

introducing Eq. (2.197) into Eq. (2.196) leads to:

$$\frac{d\widehat{O}}{dt} = \frac{i}{\hbar}\left[\aleph, \widehat{O}\right] \tag{2.198}$$

where $\left[\aleph, \widehat{O}\right]$ represents the commutator $\aleph\widehat{O} - \aleph\widehat{H}$ (Cowan, 1997).

Applying this formulation to the expectation values of the magnetic moment (μ_x, μ_y, μ_z), the x-, y-, z-axes may be fixed in space such that the z-axis lies in the same direction as the magnetic field at any point in time. Thus, this gives room for the inclusion of static and time varying fields. Hence, in the laboratory frame of reference, the Hamiltonian operator may be written from Eq. (2.169):

$$\aleph = -\gamma\hbar\vec{B}I_z \tag{2.199}$$

where $\vec{B}$ is the total magnetic field. Before applying any transverse field, the total magnetic field may be given as (Haacke et al., 1999):

$$\left|\vec{B}\right| = B_0 \tag{2.200}$$

Consequently, a form of Eq. (2.198) for this problem becomes:

$$\frac{dI_x}{dt} = \frac{i}{\hbar}\left[-\gamma\hbar B_0 I_z, I_x\right] = -\gamma\hbar B_0\frac{i}{\hbar}\left[I_z, I_x\right] \tag{2.201}$$

According to Eq. (2.139),

$$[I_z, I_x] = iI_y \tag{2.202}$$

Hence, Eq. (2.201) becomes:

$$\frac{dI_x}{dt} = -\gamma\hbar B_0\frac{i}{\hbar}\left(iI_y\right) = \gamma B_0 I_y \tag{2.203}$$

Evaluation of commutation for the other spin components gives

$$\frac{dI_y}{dt} = -\gamma B_0 I_x \tag{2.204}$$

$$\frac{dI_z}{dt} = 0 \tag{2.205}$$

Generalizing these equations gives the spin equation of motion:

$$\frac{d\vec{I}}{dt} = \vec{I} \times \gamma \vec{B} \tag{2.206}$$

where

$$\frac{d\vec{I}}{dt} = \hat{i}\frac{dI_x}{dt} + \hat{j}\frac{dI_y}{dt} + \hat{k}\frac{dI_z}{dt}$$

Therefore, the equation for the expectation value of magnetization is given as:

$$\frac{d\langle\vec{\mu}\rangle}{dt} = \langle\vec{\mu}\rangle \times \gamma\vec{B} \tag{2.207}$$

where $\vec{\mu} = \gamma\hbar\vec{I}$. Eq. (2.207) has been derived for the expectation value of magnetic moment of a single spin. For a group of spins with moments μ_n, for the nth spin such that the total magnetic moment is given as:

$$\vec{\mu} = \sum_n \vec{\mu}_n \tag{2.208}$$

Where spins do not interact with one another, Eq. (2.207) may be proved to be true for the expectation value of the total magnetization. In experimental measurements, a number of spins are considered such that the expectation values of the various components of magnetization are measured. This means that the measured bulk magnetization is simply the expectation value of the total magnetic moment. The implication of this is that the classical equation can adequately describe the dynamics of the magnetization as long as the spins are not interacting with each other.

2.7.4.7   Quantum Mechanical Description of Spin in a Rotating Magnetic Field

Since the emphasis of this study is on the rotating frame, it is important to state the quantum mechanical formulation for a system in which the magnetic field B_1 is rotating at an angular velocity ω_z, in addition to the static field $\hat{k}B_0$. The means that the total magnetic field may be written as (Abragam, 1961; Slichter, 2013):

$$\vec{B}(t) = \hat{i}B_1 \cos \omega_z t + \hat{j}B_1 \sin \omega_z t + \hat{k}B_0. \tag{2.209}$$

Hence, the Schrödinger equation may be written as:

$$\frac{\hbar}{i}\frac{\partial \Psi}{\partial t} = -\vec{\mu}\cdot\vec{B}\Psi = -\gamma\hbar\left[B_0 I_z + B_1\left(I_x \cos\omega_z t + I_y \sin\omega_z t\right)\right]\Psi. \qquad (2.210)$$

By close observation, the Hamiltonian of the system is

$$\aleph = -\gamma\hbar\left(B_0 I_z + B_1 e^{-i\omega_z t I_z} I_x e^{i\omega_z t I_z}\right). \qquad (2.211)$$

Meanwhile, the wave function may be written as follows:

$$\Psi' = e^{i\omega_z t I_z}\Psi \quad \text{or} \quad \Psi = e^{-i\omega_z t I_z}\Psi'. \qquad (2.212)$$

The meaning of Eq. (2.212) is that Ψ and Ψ' differ by a rotation of axes through an angle $\omega_z t$ (a rotating coordinate transformation). Hence, with the use of Eq. (2.212) and rotating coordinate transformation, Eq. (2.210) becomes:

$$\frac{\partial \Psi}{\partial t} = -i\omega_z I_z e^{-i\omega_z t I_z}\Psi' + e^{-i\omega_z t I_z}\frac{\partial \Psi'}{\partial t}. \qquad (2.213)$$

Substituting for Eqs. (2.212) and (2.213) in Eq. (2.210) and multiplying all through by $e^{i\omega_z t I_z}$ gives:

$$-\frac{\hbar}{i}\frac{\partial \Psi'}{\partial t} = -\left[\hbar(\omega_z + \gamma B_0) + \gamma\hbar B_1 I_x\right]\Psi'. \qquad (2.214)$$

It is noteworthy that time dependence of the transverse applied field has been eliminated. This is because time dependence represents the coupling of the spins with an effective field given by Eq. (2.49). That is, for the classical equation,

$$\vec{B}_{eff} = \left(B_0 + \frac{\omega}{\gamma}\right)\hat{k} + B_1\hat{i}$$

It follows that the spins are quantized along the effective field in the rotating coordinate system such that the spacing between the energy levels is given as:

$$\gamma\hbar B_{eff} \qquad (2.215)$$

The wave function Ψ' is therefore related to the function Ψ by coordinate rotation; that is the forward motion of I_x relative to a stationary Ψ being replaced by a stationary I_x and backward rotating the function Ψ'. Since resonance conditions exist at $\omega_z \approx -\gamma B_0$, the transformed Hamiltonian may be written as:

$$\aleph' = -\left[\hbar(\omega_z + \gamma B_0)I_z + \gamma\hbar B_1 I_x\right]. \qquad (2.216)$$

Therefore, Eq. (2.214) becomes:

$$\frac{\hbar}{i}\frac{\partial \Psi'}{\partial t} = \aleph' \Psi'. \tag{2.217}$$

The solution to Eq. (2.217) is given as (Slichter, 2013)

$$\Psi'(t) = e^{-(i/\hbar)\aleph' t}\Psi'(0). \tag{2.218}$$

Using Eq. (2.212), the wave function becomes:

$$\Psi(t) = e^{-i\omega_z t I_z}\Psi' e^{-(i/\hbar)\aleph' t}\Psi(0); \quad \Psi(0) = \Psi'(0). \tag{2.219}$$

Equation (2.219) is a compact manner in which the solution of the Schrödinger equation when oscillating magnetic field is present. If the B_1 field is applied exactly at resonance ($\omega_z = -\gamma B_o$), the transformed Hamiltonian is

$$\aleph' = -\gamma \hbar B_1 I_x. \tag{2.220}$$

Then the effects of Eqs. (2.219) and (2.220) on the expectation value of the magnetic moment is given as:

$$\langle \mu_z(t) \rangle = \int \Psi^* \mu_z \Psi d\tau$$

$$= \int \left[e^{-i\omega_z t I_z} e^{(i/\hbar)\gamma \hbar B_1 I_x t}\Psi(0) \right]^* \gamma \hbar I_z \left[e^{-i\omega_z t I_z} e^{(i/\hbar)\gamma \hbar B_1 I_x t}\Psi(0) \right] d\tau,$$

$$\langle \mu_z(t) \rangle = \int \Psi^* \mu_z \Psi d\tau = \gamma \hbar \int \left[e^{-i\omega_z t I_z} e^{i\gamma B_1 I_x t}\Psi(0) \right]^* I_z \left[e^{-i\omega_z t I_z} e^{i\gamma B_1 I_x t}\Psi(0) \right] d\tau.$$

If the applied field is defined as:

$$\omega_1 = \gamma B_1. \tag{2.221}$$

Since I_x and I_z are Hermitian, it follows that

$$\langle \mu_z(t) \rangle = \int \Psi^*(t)\mu_z \Psi(t)d\tau$$

$$= \gamma \hbar \int e^{i\omega_z t I_z} e^{-i\omega_1 I_x t}\Psi^*(0) I_z e^{i\omega_z t I_z} e^{-i\omega_1 I_x t}\Psi(0)d\tau$$

$$= \gamma \hbar \int \Psi^*(0) e^{-i\omega_1 t I_x} I_z e^{i\omega_1 t I_x}\Psi(0)d\tau. \tag{2.222}$$

Since the terms on the right of Equation (2.222) may be expressed as:

$$e^{-i\omega_1 tI_x} I_z e^{i\omega_1 tI_x} = -I_y \sin \omega_1 t + I_z \cos \omega_1 t, \tag{2.223}$$

Equation (2.222) then becomes:

$$\langle \mu_z(t) \rangle = -\langle \mu_y(0) \rangle \sin \omega_1 t + \langle \mu_z(0) \rangle \cos \omega_1 t. \tag{2.224}$$

For a magnetization lying on the z-axis, the components along the transverse direction at t = 0, become zero (that is, $\langle \mu_y(0) \rangle = 0$)

$$\langle \mu_z(t) \rangle = \langle \mu_z(0) \rangle \cos \omega_1 t. \tag{2.225}$$

Equation (2.225) shows that M_z oscillates in time at ω_1 and this is the same as the precession of $\langle \vec{\mu} \rangle$ about the rotating field B_1.

2.7.4.8 Quantum Mechanical Pictures for Description of Physical Systems

Quantum mechanical pictures are formulations or representations of quantum dynamic motion based on the behaviour of state and quantum operators (observables and other parameters) with respect to time. The difference of these formulations is either in the state vector or the operators having time-dependence or both carrying part of the time dependence of observables. A brief outline of these pictures is given in this section.

The Heisenberg Picture

This is a formulation of quantum mechanics in which the operators incorporate time-dependency while the state vectors remain time-independent. This formulation is attributed to Werner Heisenberg's work of 1925 (Bransden & Joachain, 2000; Burkhardt & Leventhal, 2008). Considering a system in a quantum state described by the ket$|\psi(0)\rangle$, it follows that the expectation value R of an arbitrary $\mathfrak{R}$ is (Cowan, 1997):

$$R = \langle \psi(0) | \mathfrak{R} | \psi(0) \rangle. \tag{2.226}$$

Hence, considering a system whose Hamiltonian is given by $\aleph$, the time evolution Eq. (2.226) is given as

$$R(t) = \left\langle \psi(0) \left| e^{\frac{i\aleph t}{\hbar}} \mathfrak{R} e^{-\frac{i\aleph t}{\hbar}} \right| \psi(0) \right\rangle. \tag{2.227}$$

The implication of Eq. (2.227) is that the expectation value of the time-dependent operator $e^{i\frac{\aleph t}{\hbar}}$ in the state $|\psi(0)\rangle$ can be modified to:

$$R = \langle \psi(0)|\mathfrak{R}(t)|\psi(0)\rangle, \tag{2.228}$$

where

$$\mathfrak{R}(t) = e^{i\frac{\aleph t}{\hbar}}\mathfrak{R}e^{-i\frac{\aleph t}{\hbar}}. \tag{2.229}$$

Differentiating Eq. (2.229) with respect to time gives

$$\frac{d\mathfrak{R}(t)}{dt} = \frac{i}{\hbar}\aleph e^{i\frac{\aleph t}{\hbar}}\mathfrak{R}e^{-i\frac{\aleph t}{\hbar}} + e^{i\frac{\aleph t}{\hbar}}\left(\frac{\partial \mathfrak{R}}{\partial t}\right)e^{-i\frac{\aleph t}{\hbar}} + \frac{i}{\hbar}e^{i\frac{\aleph t}{\hbar}}\mathfrak{R}(-\aleph)e^{-i\frac{\aleph t}{\hbar}},$$

$$\frac{d\mathfrak{R}(t)}{dt} = \frac{i}{\hbar}e^{i\frac{\aleph t}{\hbar}}(\aleph\mathfrak{R} - \mathfrak{R}\aleph)e^{-i\frac{\aleph t}{\hbar}} + e^{i\frac{\aleph t}{\hbar}}\left(\frac{\partial \mathfrak{R}}{\partial t}\right)e^{-i\frac{\aleph t}{\hbar}},$$

$$\frac{d\mathfrak{R}(t)}{dt} = \frac{i}{\hbar}(\aleph\mathfrak{R}(t) - \mathfrak{R}(t)\aleph) + e^{i\frac{\aleph t}{\hbar}}\left(\frac{\partial \mathfrak{R}}{\partial t}\right)e^{-i\frac{\aleph t}{\hbar}}. \tag{2.230}$$

Since $e^{-i\frac{\aleph t}{\hbar}}$ commutes with $\aleph$, Eq. (2.230) becomes

$$\frac{d\mathfrak{R}(t)}{dt} = \frac{i}{\hbar}[\aleph, \mathfrak{R}(t)] + e^{i\frac{\aleph t}{\hbar}}\left(\frac{\partial \mathfrak{R}}{\partial t}\right)e^{-i\frac{\aleph t}{\hbar}} = \frac{i}{\hbar}[\aleph, \mathfrak{R}(t)] + \left(\frac{\partial \mathfrak{R}}{\partial t}\right). \tag{2.231}$$

If the operator $\mathfrak{R}$ is not dependent on time,

$$\frac{d\mathfrak{R}(t)}{dt} = \frac{i}{\hbar}[\aleph, \mathfrak{R}(t)]. \tag{2.232}$$

This is the Heisenberg equation of motion and it has been derived for NMR systems in Eqs. (2.195) to (2.201). Eq. (2.232) represents the Heisenberg picture.

The Schrödinger Picture

In this formulation, quantum state vectors evolve in time while the observables or operators remain time-independent. An alternative way of viewing Eq. (2.227) is to consider a quantum state where the required description is in terms of the expected value computed for the operator $\mathfrak{R}$ in the time-dependent state $|\psi(t)\rangle$ such that (Cowan, 1997)

$$R(t) = \langle \psi(t)|\mathfrak{R}|\psi(t)\rangle, \tag{2.233}$$

where the wave function is given as:

$$\psi(t) = e^{-i\frac{\aleph t}{\hbar}}\psi(0).\tag{2.234}$$

Equations (2.233) and (2.234) imply that the quantum state evolves with time while the operator remains constant. If Eq. (2.234) is differentiated with respect to time,

$$\frac{d\psi(t)}{dt} = -i\frac{\aleph}{\hbar}e^{-i\frac{\aleph t}{\hbar}}\psi(0),$$

$$\frac{d\psi(t)}{dt} = -\frac{i}{\hbar}\aleph\psi(t).\tag{2.235}$$

from a mathematical point of view, these two formulations of quantum mechanics are equivalent (Cowan, 1997).

The Interaction (Dirac) Picture

The interaction picture has application in situations where the wave functions and observables are changing because of the interactions (Sakurai, 1994). This formulation is used for a whole range of intermediate scenarios in which the evolution is separated into two parts such that the time evolution is shared between the quantum states and the operators. In this representation, one term is associated with the state vector so that it has a Schrödinger-like evolution. This formulation is particularly useful in situations where the Hamiltonian separates naturally into two parts such that one partial has a trivial effect on the system (for example, the Zeeman Hamiltonian). It is interesting to note that this picture makes perturbation computations; in magnetic resonance for instance, the applied RF field may be considered a small perturbation on the NMR system and the interaction Hamiltonian is a sum of the Hamiltonian due to the uniform magnetic field B_o and the trivial part due to the B_1 field.

Considering a system Hamiltonian given by the following sum:

$$\aleph = \aleph_a + \aleph_b.\tag{2.236}$$

the time evolution operator is:

$$e^{i\frac{\aleph t}{\hbar}} = e^{i(\aleph_a + \aleph_b)\frac{t}{\hbar}}.\tag{2.237}$$

If the Hamiltonian $\aleph_a$ and $\aleph_b$ commute, the following factorization is allowed.

$$e^{i(\aleph_a + \aleph_b)\frac{t}{\hbar}} = e^{i\aleph_b\frac{t}{\hbar}}.e^{i\aleph_a\frac{t}{\hbar}}\tag{2.238}$$

However, if the Hamiltonians do not commute, Eq. (2.238) is not possible (Cowan, 1997). In the circumstance where Eq. (2.238) is possible, the expectation value for the operator $\mathfrak{R}$ becomes:

$$R(t) = \left\langle \psi(0) \left| e^{i\aleph_a \frac{t}{\hbar}} e^{i\aleph_b \frac{t}{\hbar}} \mathfrak{R} e^{-i\aleph_b \frac{t}{\hbar}} e^{-i\aleph_a \frac{t}{\hbar}} \right| \psi(0) \right\rangle. \tag{2.239}$$

Symbolically, Eq. (2.239) may be written as

$$R(t) = \langle \psi(t) | \mathfrak{R}(t) | \psi(t) \rangle. \tag{2.240}$$

It is then clear from Eq. (2.240) that the operator $\aleph_a$ leads to evolution of the state:

$$|\psi(t)\rangle = e^{-i\aleph_a \frac{t}{\hbar}} |\psi(0)\rangle, \tag{2.241}$$

leading to the Schrödinger formulation

$$\frac{d|\psi(t)\rangle}{dt} = -\frac{i}{\hbar} \aleph_a |\psi(0)\rangle. \tag{2.242}$$

Meanwhile, $\aleph_b$ leads to evolution of the operator

$$\mathfrak{R}(t) = e^{i\aleph_b \frac{t}{\hbar}} \mathfrak{R} e^{-i\aleph_b \frac{t}{\hbar}}, \tag{2.243}$$

which is connected to the Heisenberg formulation given as

$$\frac{d\mathfrak{R}}{dt} = -\frac{i}{\hbar} [\mathfrak{R}, \aleph].$$

However, in the general scenario where the operators $\aleph_a$ and $\aleph_b$ do not commute; that is, where Eq. (2.238) is impossible, the equation is re-written as follows (Cowan, 1997):

$$e^{i(\aleph_a + \aleph_b)\frac{t}{\hbar}} = f(t) e^{i\aleph_b \frac{t}{\hbar}}. \tag{2.244}$$

Differentiating this equation with respect to time gives

$$\frac{i(\aleph_b + \aleph_a)}{\hbar} e^{i(\aleph_b + \aleph_a)\frac{t}{\hbar}} = \frac{df(t)}{dt} e^{i\aleph_b \frac{t}{\hbar}} + f(t) \frac{i\aleph_b}{\hbar} e^{i\aleph_b \frac{t}{\hbar}}, \tag{2.245}$$

$$\frac{i\aleph_a}{\hbar} e^{i(\aleph_b + \aleph_a)\frac{t}{\hbar}} + \frac{i\aleph_b}{\hbar} e^{i(\aleph_b + \aleph_a)\frac{t}{\hbar}} = \frac{df(t)}{dt} e^{i\aleph_b \frac{t}{\hbar}} + f(t) \frac{i\aleph_b}{\hbar} e^{i\aleph_b \frac{t}{\hbar}},$$

$$\frac{i\aleph_a}{\hbar}e^{i\aleph_a\frac{t}{\hbar}}e^{i\aleph_b\frac{t}{\hbar}} + \frac{i\aleph_b}{\hbar}e^{i\aleph_a\frac{t}{\hbar}}e^{i\aleph_b\frac{t}{\hbar}} = \frac{df(t)}{dt}e^{i\aleph_b\frac{t}{\hbar}} + f(t)\frac{i\aleph_b}{\hbar}e^{i\aleph_b\frac{t}{\hbar}}. \tag{2.246}$$

From Eq. (2.244), the function $f(t) = e^{i\aleph_a\frac{t}{\hbar}}$ and Equation (2.246) becomes:

$$\frac{df(t)}{dt} = \frac{i}{\hbar}f(t)e^{i\aleph_b\frac{t}{\hbar}}\aleph_a e^{-i\aleph_b\frac{t}{\hbar}}. \tag{2.247}$$

The expression $e^{i\aleph_b\frac{t}{\hbar}}\aleph_a e^{-i\aleph_b\frac{t}{\hbar}}$ represents the Hamiltonian $\aleph_a$ in which the Heisenberg time evolution is generated by $\aleph_b$. Therefore

$$\frac{df(t)}{dt} = \frac{i}{\hbar}f(t)\aleph_a, \tag{2.248}$$

which can be integrated to give the following expression

$$f(t) = f(0)e^{\frac{i}{\hbar}\int_0^t \aleph_a(\delta)d\delta}, \tag{2.249}$$

where $f(0)$ is the unit operator.

2.7.5   Modelling and Differential Equations in Computational MRI

Modelling is a very important general process in physics, medical imaging, engineering, financial economics, clinical medicine, computer science, laboratory biology, analytical chemistry, environmental science and other fields in which physical processes can be mimicked with mathematical expressions (Kreyszig, 1996). Engineering mixing problem, Newton's cooling, Gompertz model, population control, carbon dating are a few examples of how mathematical models can help solve physical problems. Whenever these models are developed, the problem to be solved must be correctly set up in terms of differential equations.

In order to solve a scientific problem of physical significance, the problem needs to be formulated in terms of a mathematical expression where variables, functions and equations form the elements of the expression. The resulting expressions are known as mathematical models of the physical problems already defined. The process of solving scientific problems in terms of mathematical models involves the following steps (Kreyszig, 1996):

1. setting up a model in terms of differential equations
2. seeking mathematical/computational solutions to the equations
3. interpreting the result in physical, chemical or NMR relaxation terms.

Because many physical concepts are made up of derivatives, mathematical models are often equations with derivatives whose functions are unknown. The mathematical model obtained in this manner is normally referred to as a differential equation. Once models have been set up, solutions or functions, satisfying the differential equations must now be sought. Such a solution can then be explored (in terms of its properties), plotted, evaluated for its values, making it possible to explore how their behaviour can be used to unravel physical systems. In fact, this exploration may lead to the discovery of new features that were previously hidden.

The differential equations in the developed models can be either ordinary differential equations (ODEs) or partial differential equations (PDEs). ODEs are reduced forms of PDEs. PDEs are quite crucial to most problems in science and they have wider applications to physical problems than ODEs because of their ability to accommodate more complexities. Consequently, PDEs shall be widely used in this study. Fortunately, some of the PDEs required for the computational models in this study have been presented in earlier sections. Meanwhile, it is quite important to note that modelling with PDEs is more demanding than modelling with ODEs (Kreyszig, 1996; Polyanin et al., 2008).

A mathematical equation that contains two or more independent variables, an unknown function and partial derivatives of the unknown function with respect to the independent variables is known as a partial differential equation (Polyanin et al., 2008). The order of a PDE is the order of the highest derivative involved. PDEs govern numerous qualitatively comparable processes or phenomena (Camazine, 2003; Polyanin et al., 2008). Just as PDEs have a large number of applications, so do they have numerous solutions, which are results obtained after initial and boundary conditions must have been applied to the general solutions (Polyanin et al., 2008).

2.7.6 Formulation of the Bloch Equations for Treatment of Fluids in Motion

Thus far, the Bloch formulation for systems in which the spin magnetic moments were immobile, located at fixed positions has been presented. In the treatment, the transverse relaxation which is measured in such cases could have a complex relaxation profile. It is equally important to consider nuclear resonance behaviour in fluid systems, that is, systems where the spins are in motion. This is because the molecular process, the imaging of which this study is attempting to describe, is always dynamic (Nelson & Cox, 2008) and their spins are dynamic as well. This means that the Bloch equations which have been described in earlier sections need to be modified to account for spins in fluid motion (Awojoyogbe, 1997, 2002).

There are two ways in which flow may be studied using NMR. Using selective pulses, it is possible to label the magnetization (Axel & Dougherty, 1989a, b) in a given region. This is known as NMR tagging. For instance, a 90° pulse will destroy the equilibrium z-component of magnetization. We may then use an imaging procedure to follow this labelled collection of spins as it moves within the specimen (Cowan, 1997). It is expected that the magnetization will recover over the order of T_1 so that the mapped magnetization is uniquely synonymous to the nature and manner of fluid motion. Alternatively, a field gradient may be used for encoding the spatial information. The phase angle accumulated by a spin as it moves in this field gradient is regarded as the phase of the spin. The modification that has been made to the Bloch equation for fluids in motion is based on an unconventional method of NMR studies of flow (Awojoyogbe, 1997, 2002, 2003, 2004, 2007). The development of this method of NMR flow study is from the kinematics of the fluids and it is presented in the following subsection.

2.7.6.1 Kinematic of Fluids in Motion

In order to describe the motion of fluids, the individual points within the fluids may be explored by assigning coordinates x, y, z to each constituent particle of the fluid such that their motion is specified as a time-dependent function. For example, a fluid particle may be described by the point, x_o, y_o, z_o, at an initial time $t = t_o$ such that the motion of the fluid may now be described by means of functions x (x_o, y_o, z_o, t), y (x_o, y_o, z_o, t), z (x_o, y_o, z_o, t). This demonstrate that the particle initially at x_o, y_o, z_o, t_o may now be found at the x, y, z at a time t. Although this description is a generalization of the idea of particle mechanics, Euler's method, which is more convenient, may be employed in explaining this concept. This method works by neglecting each fluid particle while placing emphasis on the velocity $\vec{v}(x, y, z, t)$ and density $\rho(x, y, z, t)$ at time-specific points in space (Symon, 1971).

Any quantity describing the state of a fluid, such as density ρ may be represented with the function [ρ (x, y, z, t)] of the spatial coordinates x, y, z and t. This means that the density has a definite value at the specified points in space. The manner of description used in this section emphasizes points in space rather than fluid particles so that it would be very difficult to avoid monitoring the motion of the fluid particles even for a short period of time dt. This is because the laws of mechanics apply only to the particles and not to their spatial locations. It is therefore now of great interest to consider a given quantity ρ (density) in two-time rates. The rate at which the density is changing with respect to fixed spatial points is given by the partial derivative $\left(\frac{\partial \rho}{\partial t}\right)$. The total derivative is then given as (Symon, 1971):

$$\frac{d\rho}{dt} = \frac{\partial \rho}{\partial t} + \frac{\partial \rho}{\partial x}\frac{dx}{dt} + \frac{\partial \rho}{\partial y}\frac{dy}{dt} + \frac{\partial \rho}{\partial z}\frac{dz}{dt}, \qquad (2.250)$$

where the fluid velocity $\vec{v}$ has the components dx/dt, dy/dt, dz/dt. Hence, rate of change of the density for a fluid particle moving from x, y, z to x + dx, y + dy, z + dz in a time period of t is:

$$dp = \rho(x + dx, y + dy, z + dz, t + dt) - \rho(x, y, z, t)$$

$$= \frac{\partial \rho}{\partial x} dx + \frac{\partial \rho}{\partial y} dy + \frac{\partial \rho}{\partial z} dz + \frac{\partial \rho}{\partial t} dt. \tag{2.251}$$

and if $dt \to 0$, we have the following expression:

$$\frac{d\rho}{dt} = \frac{\partial \rho}{\partial t} + v_x \frac{\partial \rho}{\partial x} + v_y \frac{\partial \rho}{\partial y} + v_z \frac{\partial \rho}{\partial z}. \tag{2.252}$$

This expression may be written in a compact form as follows:

$$\frac{d\rho}{dt} = \frac{\partial \rho}{\partial t} + \vec{v} \cdot \nabla \rho, \tag{2.253}$$

Equation (2.253) is a shorter form of Eq. (2.252). Since the total derivative dp/dt is also a function of the spatial points x, y, z, and time t, a relationship between the total and partial derivatives for any physical quantity may be given as (Symon, 1971):

$$\frac{d}{dt} = \frac{\partial}{\partial t} + \vec{v} \cdot \nabla, \tag{2.254}$$

This equation is a derivative taken with respect to a moving coordinate system; it known as the Lagrangian derivative, the substantive derivative (Tritton, 1988) or the Stokes derivative (Kaplan, 1991). Deriving the equation of continuity is now a simple task using this method.

2.7.6.2   The Bloch NMR Flow Equations

Since Equation (2.254) holds for any quantity (Symon, 1971; Kaplan, 1991), the rule is applicable to the magnetization function M such that:

$$\frac{dM}{dt} = \frac{\partial M}{\partial t} + \vec{v} \cdot \nabla M \tag{2.255}$$

where $\vec{v}$ is the fluid velocity of fluid motion and the Del operator $\nabla = \hat{i} \frac{\partial}{\partial x} + \hat{j} \frac{\partial}{\partial y} + \hat{k} \frac{\partial}{\partial z}$ (Wylie & Barrett, 1982; Kreyszig, 1996). Using Eq. (2.255), the Bloch Equation given in Eq. (2.110) can be re-written as follows (Awojoyogbe, 2002, 2003, 2004):

$$\frac{dM_x}{dt} = \frac{\partial M_x}{\partial t} + \vec{v} \cdot \nabla M_x = -\frac{M_x}{T_2}, \tag{2.256}$$

$$\frac{dM_y}{dt} = \frac{\partial M_y}{\partial t} + \vec{v} \cdot \nabla M_y = \gamma M_z B_1 - \frac{M_y}{T_2}, \tag{2.257}$$

$$\frac{dM_z}{dt} = \frac{\partial M_z}{\partial t} + \vec{v} \cdot \nabla M_z = -\gamma M_y B_1 + \frac{(M_0 - M_z)}{T_1}. \tag{2.258}$$

2.7.6.3 NMR Flow at Constant Fluid Velocity

Fluid flow is at the heart of chemical and biochemical reactivity (Price, 2009; Nelson & Cox, 2008) and as has been noted earlier on, is very important to molecular imaging (Modo & Bulte, 2007). The treatment of the Bloch flow phenomenon when the fluid velocity if fairly constant is presented in the following derivations.

The General Bloch NMR Flow Equation for Constant Fluid Velocity

Considering fluid flow along a horizontal x-direction, partial derivatives along the y and z directions are zero (Awojoyogbe, 2002, 2003, 2004). Therefore,

$$\vec{v} \cdot \nabla M_x = v\frac{\partial M_x}{\partial x}; \quad \vec{v} \cdot \nabla M_y = v\frac{\partial M_y}{\partial x}; \quad \vec{v} \cdot \nabla M_z = v\frac{\partial M_z}{\partial x}. \tag{2.259}$$

It can be seen that the fluid velocity and the magnetization are assumed to take the following forms:

$$\vec{v} = \hat{i}v + \hat{j}v + \hat{k}v, \tag{2.260}$$

$$M = \hat{i}M_x + \hat{j}M_y + \hat{k}M_z. \tag{2.261}$$

For the one-dimensional case of Eq. (2.259), Eqs. (2.256)–(2.258) then become:

$$\frac{dM_x}{dt} = \frac{\partial M_x}{\partial t} + v\frac{\partial M_x}{\partial x} = -\frac{M_x}{T_2}, \tag{2.262}$$

$$\frac{dM_y}{dt} = \frac{\partial M_y}{\partial t} + v\frac{\partial M_y}{\partial x} = \gamma M_z B_1 - \frac{M_y}{T_2}, \tag{2.263}$$

and

$$\frac{dM_z}{dt} = \frac{\partial M_z}{\partial t} + v\frac{\partial M_z}{\partial x} = -\gamma M_y B_1 + \frac{(M_0 - M_z)}{T_1}. \tag{2.264}$$

Equations (2.262) to (2.264) may be used to derive a second-order partial differential equation (Awojoyogbe, 2002, 2003, 2004; Awojoyogbe et al., 2011a, b) which is very useful in elucidating the spatial and temporal NMR transverse magnetization. From the above Eqs. (2.262) to (2.264), the Bloch NMR flow equations for constant fluid velocity (Gupta et al., 2014) are:

$$\frac{\partial M_x}{\partial t} + v\frac{\partial M_x}{\partial x} = -\frac{M_x}{T_2}, \tag{2.265}$$

$$\frac{\partial M_y}{\partial t} + v\frac{\partial M_y}{\partial x} = \gamma M_z B_1 - \frac{M_y}{T_2}, \tag{2.266}$$

and

$$\frac{\partial M_z}{\partial t} + v\frac{\partial M_z}{\partial x} = -\gamma M_y B_1 + \frac{(M_o - M_z)}{T_1}. \tag{2.267}$$

From Eq. (2.267), we write:

$$v\frac{\partial M_z}{\partial x} + \frac{\partial M_z}{\partial t} + \frac{M_z}{T_1} = -\gamma M_y B_1 + \frac{M_o}{T_1},$$

or

$$\left(v\frac{\partial}{\partial x} + \frac{\partial}{\partial t} + \frac{1}{T_1} \right) M_z = -\gamma M_y B_1 + \frac{M_o}{T_1},$$

$$M_z = \left(-\gamma M_y B_1 + \frac{M_o}{T_1} \right) \frac{1}{\left(v\frac{\partial}{\partial x} + \frac{\partial}{\partial t} + \frac{1}{T_1} \right)}. \tag{2.268}$$

Substituting for Eq. (2.268) in Eq. (2.266), we write:

$$\frac{\partial M_y}{\partial t} + v\frac{\partial M_y}{\partial x} = \gamma \left(-\gamma M_y B_1 + \frac{M_o}{T_1} \right) \frac{B_1}{\left(v\frac{\partial}{\partial x} + \frac{\partial}{\partial t} + \frac{1}{T_1} \right)} - \frac{M_y}{T_2},$$

$$v\frac{\partial M_y}{\partial x}\left(v\frac{\partial}{\partial x} + \frac{\partial}{\partial t} + \frac{1}{T_1} \right) + \frac{\partial M_y}{\partial t}\left(v\frac{\partial}{\partial x} + \frac{\partial}{\partial t} + \frac{1}{T_1} \right) + \frac{M_y}{T_2}\left(v\frac{\partial}{\partial x} + \frac{\partial}{\partial t} + \frac{1}{T_1} \right)$$

$$= \gamma \left(-\gamma M_y B_1(x) + \frac{M_o}{T_1} \right) B_1,$$

$$v^2 \frac{\partial^2 M_y}{\partial x^2} + v\frac{\partial^2 M_y}{\partial x \partial t} + \frac{v}{T_1}\frac{\partial M_y}{\partial x} + v\frac{\partial^2 M_y}{\partial x \partial t} + \frac{\partial^2 M_y}{\partial t^2} + \frac{1}{T_1}\frac{\partial M_y}{\partial t} + \frac{v}{T_2}\frac{\partial M_y}{\partial x} + \frac{1}{T_2}\frac{\partial M_y}{\partial t}$$

$$+ \frac{1}{T_1 T_2} M_y$$

$$= -\gamma^2 B_1{}^2 M_y + \frac{\gamma B_1 M_o}{T_1}$$

$$v^2 \frac{\partial M_y}{\partial x^2} + 2v\frac{\partial^2 M_y}{\partial x \partial t} + v\left(\frac{1}{T_1} + \frac{1}{T_2}\right)\frac{\partial M_y}{\partial x} + \left(\frac{1}{T_1} + \frac{1}{T_2}\right)\frac{\partial M_y}{\partial t} + \frac{\partial^2 M_y}{\partial t^2}$$

$$+ \left(\frac{1}{T_1 T_2} + \gamma^2 B_1{}^2\right) M_y$$

$$= \frac{\gamma B_1 M_o}{T_1}.$$

This equation becomes:

$$v^2 \frac{\partial^2 M_y}{\partial x^2} + 2v\frac{\partial^2 M_y}{\partial x \partial t} + T_o \frac{\partial M_y}{\partial t} + vT_o \frac{\partial M_y}{\partial x} + \frac{\partial^2 M_y}{\partial t^2} + \left(T_g + \gamma^2 B_1{}^2\right) M_y$$

$$= F_o \gamma B_1, \tag{2.269}$$

where $T_o = \frac{1}{T_1} + \frac{1}{T_2}$, $T_g = \frac{1}{T_1 T_2}$ and $F_o = \frac{M_o}{T_1}$.

Equation (2.269) is a general second-order differential equation applicable to most fluid flow problems (Awojoyogbe, 2002, 2003, 2004; Awojoyogbe et al., 2011a, b). At any point in time, physical system information may be obtained after appropriate initial and boundary conditions have been applied (Gupta et al., 2014). From such differential equations, process equations such the wave equation, diffusion equation and diffusion-advection equations can be derived quite easily. These equations are suitable for modelling various NMR applications and they can be resolved into functions that are dependent on NMR parameters but subject to reasonable restraints on the system. Hence, obtaining crucial dynamical system information becomes quite easy with these equations. Since the parameter $F_o \gamma B_1(x, t)$ represents the forcing function, the NMR system can be studied while undergoing free vibration ($F_o \gamma B_1 = 0$) or forced vibration ($F_o \gamma B_1 \neq 0$).

Time-Independent Bloch NMR Flow Equation for Constant Fluid Velocity

For molecular fluids with a relatively steady flow, the time-independent partial derivatives are set to zero. Consequently, Eqs. (2.262) to (2.264) become (Gupta et al., 2014; Awojoyogbe et al., 2011a, b):

$$v\frac{dM_x}{dx} = -\frac{M_x}{T_2}, \tag{2.270}$$

$$v\frac{dM_y}{dx} = \gamma M_z B_1 - \frac{M_y}{T_2}, \tag{2.271}$$

and

$$v\frac{dM_z}{dx} = -\gamma M_z B_1 + \frac{(M_0 - M_z)}{T_1}. \tag{2.272}$$

From Eq. (2.271) we write:

$$v\frac{dM_z}{dx} = -\gamma M_y B_1 + \frac{M_o}{T_1} - \frac{M_z}{T_1},$$

$$\left(v\frac{d}{dx} + \frac{1}{T_1}\right)M_z = -\gamma M_y B_1 + \frac{M_o}{T_1},$$

or

$$M_z = \left(-\gamma M_y B_1 + \frac{M_o}{T_1}\right)\frac{1}{\left(v\frac{d}{dx} + \frac{1}{T_1}\right)}. \tag{2.273}$$

Substituting for M_z in Eq. (2.271),

$$v\frac{dM_y}{dx} = \gamma\left(-\gamma B_1 M_y + \frac{M_o}{T_1}\right)\frac{1}{\left(v\frac{d}{dx} + \frac{1}{T_1}\right)}B_1 - \frac{M_y}{T_2},$$

$$v\frac{dM_y}{dx} + \frac{M_y}{T_2} = \left(-\gamma^2 M_y B_1{}^2 + \frac{\gamma B_1 M_o}{T_1}\right)\frac{1}{\left(v\frac{d}{dx} + \frac{1}{T_1}\right)},$$

$$v\frac{dM_y}{dx}\left(v\frac{d}{dx} + \frac{1}{T_1}\right) + \frac{M_y}{T_2}\left(v\frac{d}{dx} + \frac{1}{T_1}\right) = -\gamma^2 M_y B_1{}^2(x) + \frac{\gamma B_1 M_o}{T_1},$$

$$v^2\frac{d^2 M_y}{dx^2} + v\left(\frac{1}{T_1} + \frac{1}{T_2}\right)\frac{dM_y}{dx} + \left(\frac{1}{T_1 T_2} + \gamma^2 B_1{}^2\right)M_y = \frac{\gamma B_1 M_o}{T_1},$$

$$v^2\frac{d^2 M_y}{dx^2} + vT_o\frac{dM_y}{dx} + \left(T_g + \gamma^2 B_1{}^2\right)M_y = \frac{M_o}{T_1}\gamma B_1. \tag{2.274}$$

Equation (2.274) is known as the time-independent Bloch NMR flow equation.

Time-Dependent Bloch NMR Flow Equation for Constant Fluid Velocity

In the event that the fluid flow is not spatially dependent and the nuclear magnetization does not change appreciably with x, then all spatially dependent partial derivatives are set to zero. Hence, Eqs. (2.262) to (2.264) become:

$$\frac{dM_x}{dt} = -\frac{M_x}{T_2}, \tag{2.275}$$

$$\frac{dM_y}{dt} = \gamma M_z B_1(t) - \frac{M_y}{T_2}, \tag{2.276}$$

and

$$\frac{dM_z}{dt} = -\gamma B_1(t)M_y + \frac{(M_0 - M_z)}{T_1}. \tag{2.277}$$

From Eq. (2.277),

$$\frac{dM_z}{dt} = -\gamma M_y B_1(t) + \frac{M_o}{T_1} - \frac{M_z}{T_1},$$

$$\left(\frac{d}{dt} + \frac{1}{T_1}\right) M_z = -\gamma M_y B_1(t) + \frac{M_o}{T_1},$$

$$M_z = \left(-\gamma M_y B_1(t) + \frac{M_o}{T_1}\right) \frac{1}{\left(\frac{d}{dt} + \frac{1}{T_1}\right)}. \tag{2.278}$$

Substituting for M_z in Eq. (2.210),

$$\frac{dM_y}{dt} = \gamma \left(-\gamma M_y B_1(t) + \frac{M_o}{T_1}\right) \frac{1}{\left(\frac{d}{dt} + \frac{1}{T_1}\right)} B_1(t) - \frac{M_y}{T_2},$$

$$\left(\frac{d}{dt} + \frac{1}{T_1}\right)\frac{dM_y}{dt} + \left(\frac{d}{dt} + \frac{1}{T_1}\right)\frac{M_y}{T_2} = -\gamma^2 B_1{}^2(t)M_y + \frac{\gamma B_1(t)M_o}{T_1},$$

$$\frac{d^2 M_y}{dt^2} + \frac{1}{T_1}\frac{dM_y}{dt} + \frac{1}{T_2}\frac{dM_y}{dt} + \frac{1}{T_1 T_2}M_y + \gamma^2 M_y B_1{}^2(t) = \frac{\gamma B_1(t)M_o}{T_1},$$

$$\frac{d^2 M_y}{dt^2} + T_o\frac{dM_y}{dt} + \left(T_g + \gamma^2 B_1{}^2(t)\right)M_y = \frac{M_o}{T_1}\gamma B_1(t). \tag{2.279}$$

Equation (2.279) is called the time-dependent Bloch NMR flow equation. Now, any of the equations can be tailored into appropriate physical situations.

2.7.7   Diffusion Process in Magnetic Resonance

The molecular motions observed in fluid spins are usually random and haphazard such that these processes directly influence NMR signal attenuation. Nuclear spins are always in motion while travelling significantly short distances such that the spins are considered as particles diffusing over small distances (Cowan, 1997). This results in the emergence of irreversible damping of the magnitude of the nuclear magnetization and an eventual disappearance of the information contained in the spatial frequency. The implication of this information is that molecular diffusion is directly impressed on the NMR signal given that the information obtained can be interpreted appropriately. In order to account for the motion of spins in magnetic resonance, a diffusion term is often added to the Bloch NMR Equation. This modification to the fundamental Bloch NMR equation is known as the Bloch-Torrey equation (Cowan, 1997; Torrey, 1956).

The Bloch-Torrey equation generalizes the Bloch NMR equations to include parameters quantifying magnetization transfer caused by molecular diffusion (Torrey, 1956). This equation has various forms, one of which is given as follows (Cowan, 1997; Haacke et al., 1999):

$$\frac{d\vec{M}}{dt} = \gamma \cdot \left(\vec{M} \times \vec{B}_0\right)_{B_0\ \text{excitation}} + \begin{pmatrix} \dfrac{M_x}{T_2} \\ \dfrac{M_y}{T_2} \\ \dfrac{M_0 - M_z}{T_1} \end{pmatrix}_{\text{longitudinal,transverse,relaxation}} \tag{2.280}$$

Adding the diffusion term to this expression gives:

$$\frac{d\vec{M}}{dt} = \gamma \cdot \left(\vec{M} \times \vec{B}_0\right)_{B_0\ \text{excitation}} + \begin{pmatrix} \dfrac{M_x}{T_2} \\ \dfrac{M_y}{T_2} \\ \dfrac{M_0 - M_z}{T_1} \end{pmatrix}_{\text{longitudinal,transverse,relaxation}}$$

$$+ D\nabla^2 \vec{M}_{\text{diffusion}}. \tag{2.281}$$

The meaning of the Bloch-Torrey equation as stated in Eq. (2.281) is that the NMR signal can be obtained in the absence of molecular diffusion. Although this idea is quite plausible, contemporary experimental experiences in molecular dynamics have shown that there is no molecular scenario where diffusion can be ruled out. The immediate consequence of this is that the resonance signals measured from molecular dynamical processes are dependent on all system processes that is observed in molecular dynamics and are a consequence of all possible processes occurring in the sample. Hence, attempts have been made to derive NMR diffusion from the Bloch NMR flow equations (Gupta et al., 2014; Awojoyogbe et al., 2011a, b). The derivation was achieved by setting:

$$\frac{\partial^2 M_y}{\partial t^2} + vT_o \frac{\partial M_y}{\partial x} + 2v \frac{\partial^2 M_y}{\partial x \partial t} + \left(\gamma^2 B_1{}^2 + T_g\right) M_y = 0. \tag{2.282}$$

Equation (2.269) becomes:

$$v^2 \frac{\partial^2 M_y}{\partial x^2} + T_o \frac{\partial M_y}{\partial t} = F_o \gamma B_1, \tag{2.283}$$

$$\frac{\partial M_y}{\partial t} = \frac{-v^2}{T_o} \frac{\partial^2 M_y}{\partial x^2} + \frac{F_o}{T_o} \gamma B_1. \tag{2.284}$$

If the diffusion coefficient takes the following form

$$D = \frac{-v^2}{T_o}. \tag{2.285}$$

then, Eq. (2.218) becomes:

$$\frac{\partial M_y}{\partial t} = D \frac{\partial^2 M_y}{\partial x^2} + \frac{F_0}{T_0} \gamma B_1. \tag{2.286}$$

Equation (2.220) represents the equation of diffusion of magnetization according to the motion of nuclear spins (Awojoyogbe et al., 2011a, b). The right side of Eq. (2.286) implies that the application of the *RF* B_1 field modifies the behaviour of diffused magnetization inside a voxel.

2.7.7.1 Multidimensional Diffusion with Constant Diffusion Coefficient

It is possible for the transverse magnetization to vary with more than one spatial coordinate even when the diffusion coefficient is constant. This happens when the fluid molecules are moving in different directions without a significant change in the diffusion coefficient. In such circumstances, Eq. (2.256) can be reworked to (2.257) so that a modified form of Eq. (2.286) is obtained such that the following expression holds (Awojoyogbe et al., 2011a, b):

$$D \nabla^2 M_y + \frac{F_o}{T_o} \gamma B_1 \left(\vec{r}, t\right) = \frac{\partial M_y}{\partial t}, \tag{2.287}$$

where ∇^2 represents the Laplacian in the coordinate (Wylie & Barrett, 1982; Kreyszig, 1996) appropriate to the voxel. In common geometries, Eq. (2.287) has the following forms (Awojoyogbe et al., 2011a, b; Gupta et al., 2014):

Cartesian Geometry:

$$D\left(\frac{\partial^2 M_y}{\partial x^2} + \frac{\partial^2 M_y}{\partial y^2} + \frac{\partial^2 M_y}{\partial z^2}\right) + \frac{F_o}{T_o}\gamma B_1\left(\vec{r},t\right) = \frac{\partial M_y}{\partial t} \tag{2.288}$$

Cylindrical Geometry:

$$D\left(\frac{1}{r}\frac{\partial}{\partial r}\left(r\frac{\partial M_y}{\partial r}\right) + \frac{1}{r^2}\frac{\partial^2 M_y}{\partial \varphi^2} + \frac{\partial^2 M_y}{\partial z^2}\right) + \frac{F_o}{T_o}\gamma B_1(x,t) = \frac{\partial M_y}{\partial t} \tag{2.289}$$

Spherical Geometry:

$$\frac{D}{r^2}\left(\frac{\partial}{\partial r}\left(r^2\frac{\partial M_y}{\partial r}\right) + \frac{1}{\sin^2\theta}\frac{\partial^2 M_y}{\partial \varphi^2} + \frac{1}{\sin\theta}\frac{\partial}{\partial\theta}\left(\sin\theta\frac{\partial M_y}{\partial\theta}\right)\right)$$
$$+ \frac{F_o}{T_o}\gamma B_1\left(\vec{r},t\right)$$
$$= \frac{\partial M_y}{\partial t} \tag{2.290}$$

2.7.7.2 Multidimensional Diffusion with Variable Diffusion Coefficient

Generally, variable-coefficient diffusion processes are more prominent in multi-dimensional molecular problems and magnetic resonance should be able to account for every small change observed (Tahayori et al., 2009; Awojoyogbe et al., 2010; Awojoyogbe & Dada, 2011a, b). It is therefore crucial to demonstrate the directional changes of the NMR diffusion coefficient as they affect the NMR signal. So, Eq. (2.286) may be generalized into the following expression (Awojoyogbe et al., 2011a, b):

$$\nabla\cdot\left(D\nabla M_y\right) + \frac{F_o}{T_o}\gamma B_1\left(\vec{r},t\right) = \frac{\partial M_y}{\partial t} \tag{2.291}$$

The forms of Eq. (2.291) in various geometries are now presented (Gupta et al., 2014; Awojoyogbe et al., 2011a, b).

Rectangular Geometry:

$$\frac{\partial M_y}{\partial t} = \frac{\partial}{\partial x}\left(D\frac{\partial M_y}{\partial x}\right) + \frac{\partial}{\partial y}\left(D\frac{\partial M_y}{\partial y}\right) + \frac{\partial}{\partial z}\left(D\frac{\partial M_y}{\partial z}\right) + \frac{F_o}{T_o}\gamma B_1\left(\vec{r},t\right), \quad (2.292)$$

or

$$\frac{\partial M_y}{\partial t} = \frac{\partial}{\partial x}\left(D_x\frac{\partial M_y}{\partial x}\right) + \frac{\partial}{\partial y}\left(D_y\frac{\partial M_y}{\partial y}\right) + \frac{\partial}{\partial z}\left(D_z\frac{\partial M_y}{\partial z}\right)$$
$$+ \frac{F_o}{T_o}\gamma B_1\left(\vec{r},t\right). \quad (2.293)$$

Cylindrical Geometry:

$$\frac{\partial M_y}{\partial t} = \frac{1}{r}\frac{\partial}{\partial r}\left(Dr\frac{\partial M_y}{\partial r}\right) + \frac{1}{r^2}\frac{\partial}{\partial \varphi}\left(D\frac{\partial M_y}{\partial \varphi}\right) + \frac{\partial}{\partial z}\left(D\frac{\partial M_y}{\partial z}\right)$$
$$+ \frac{F_o}{T_o}\gamma B_1\left(\vec{r},t\right), \quad (2.294)$$

or

$$\frac{\partial M_y}{\partial t} = \frac{1}{r}\frac{\partial}{\partial r}\left(D_r r\frac{\partial M_y}{\partial r}\right) + \frac{1}{r^2}\frac{\partial}{\partial \varphi}\left(D_\varphi\frac{\partial M_y}{\partial \varphi}\right) + \frac{\partial}{\partial z}\left(D_z\frac{\partial M_y}{\partial z}\right)$$
$$+ \frac{F_o}{T_o}\gamma B_1\left(\vec{r},t\right). \quad (2.295)$$

Spherical Geometry:

$$\frac{\partial M_y}{\partial t} = \frac{1}{r^2}\frac{\partial}{\partial r}\left(Dr^2\frac{\partial M_y}{\partial r}\right) + \frac{1}{r^2\sin\theta}\frac{\partial}{\partial \theta}\left(D\sin\theta\frac{\partial M_y}{\partial \theta}\right) + \frac{1}{r^2\sin^2\theta}\frac{\partial}{\partial \varphi}$$
$$\times\left(D\frac{\partial M_y}{\partial \varphi}\right) + \frac{F_o}{T_o}\gamma B_1\left(\vec{r},t\right), \quad (2.296)$$

or

$$\frac{\partial M_y}{\partial t} = \frac{1}{r^2}\frac{\partial}{\partial r}\left(D_r r^2\frac{\partial M_y}{\partial r}\right) + \frac{1}{r^2\sin\theta}\frac{\partial}{\partial \theta}\left(D_\theta\sin\theta\frac{\partial M_y}{\partial \theta}\right)$$
$$+ \frac{1}{r^2\sin^2\theta}\frac{\partial}{\partial \varphi}\left(D_\varphi\frac{\partial M_y}{\partial \varphi}\right) + \frac{F_o}{T_o}\gamma B_1\left(\vec{r},t\right). \quad (2.297)$$

2.7.8 Advection–Diffusion in Nuclear Magnetic Resonance

In natural transport processes, convection (advection)plays a very important role such that unlike in diffusion, there are molecular mixings and mass transport in the absence of bulk molecular motion. Thus, actual transport of molecules is known as convection or advection (http://en.wikipedia.org/wiki/Advection). Consequently, within image voxel, the measured NMR signal damping is not a consequence of motion of spins across the voxel but rather, a result of small molecular interactions and motions within the region under investigation. In order to describe these processes, Advection-Diffusion equation may be employed.

In addition to this, if molecular diffusion is taking place in the presence of mass transport, the diffusion-advection equation is used. The addition of the advection operator to the diffusion equation leads to the diffusion-advection equation. The advection operator (in the Cartesian coordinate) is given as

$$\vec{v} \cdot \nabla = v_x \frac{\partial}{\partial x} + v_y \frac{\partial}{\partial y} + v_z \frac{\partial}{\partial z},$$

where $\vec{v} = v_x\widehat{i} + v_y\widehat{j} + v_z\widehat{k}$. Therefore, from Eq. (2.269),

$$\frac{\partial^2 M_y}{\partial t^2} + 2v \frac{\partial^2 M_y}{\partial x \partial t} + T_g M_y = 0. \tag{2.298}$$

If the non-adiabatic condition $T_g > > \gamma^2 B_1{}^2(x, t)$ holds, it follows that

$$v^2 \frac{\partial^2 M_y}{\partial x^2} + vT_o \frac{\partial M_y}{\partial x} + T_o \frac{\partial M_y}{\partial t} = F_o \gamma B_1(x, t), \tag{2.299}$$

$$vT_o \frac{\partial M_y}{\partial x} + T_o \frac{\partial M_y}{\partial t} = -v^2 \frac{\partial^2 M_y}{\partial x^2} + F_o \gamma B_1(x, t). \tag{2.300}$$

From earlier assumptions, setting

$$D = -\frac{v^2}{T_o}. \tag{2.301}$$

Equation (2.300) becomes:

$$\frac{\partial M_y}{\partial t} + v \frac{\partial M_y}{\partial x} = D \frac{\partial^2 M_y}{\partial x^2} + \frac{F_o}{T_o} \gamma B_1(x, t). \tag{2.302}$$

Given that D is the molecular diffusion coefficient, Eq. (2.302) is the diffusion-advection equation of the magnetization (Awojoyogbe et al., 2011a, b).

2.7.8.1 One-Dimensional Diffusion-Advection with Constant Diffusion Coefficient

Equation (2.302) is exactly the one-dimensional, constant-coefficient NMR diffusion-advection equation (Gupta et al., 2014). The implication of this is that the diffusion coefficient and fluid velocity are dependent on x.

2.7.8.2 Multidimensional Diffusion-Advection with Constant Diffusion Coefficient

If the transverse magnetization varies significantly in various spatial directions, Eq. (2.302) may be generalized into (Awojoyogbe et al., 2011a, b):

$$v\nabla M_y + \frac{\partial M_y}{\partial t} = D\nabla^2 M_y + \frac{F_o}{T_o}\gamma B_1\left(\overrightarrow{r},t\right), \tag{2.303}$$

In the common coordinates, this equation has the following forms:
Cartesian Geometry:

$$v\left(\frac{\partial M_y}{\partial x} + \frac{\partial M_y}{\partial y} + \frac{\partial M_y}{\partial z}\right) + \frac{\partial M_y}{\partial t} = D\left(\frac{\partial^2 M_y}{\partial x^2} + \frac{\partial^2 M_y}{\partial y^2} + \frac{\partial^2 M_y}{\partial z^2}\right)$$
$$+ \frac{F_o}{T_o}\gamma B_1\left(\overrightarrow{r},t\right). \tag{2.304}$$

and
Cylindrical Geometry:

$$v\left(\frac{\partial M_y}{\partial r} + \frac{1}{r}\frac{\partial M_y}{\partial \varphi} + \frac{\partial M_y}{\partial z}\right) + \frac{\partial M_y}{\partial t}$$
$$= D\left(\frac{\partial^2 M_y}{\partial r^2} + \frac{1}{r}\frac{\partial M_y}{\partial r} + \frac{1}{r^2}\frac{\partial^2 M_y}{\partial \varphi^2} + \frac{\partial^2 M_y}{\partial z^2}\right) + \frac{F_o}{T_o}\gamma B_1\left(\overrightarrow{r},t\right). \tag{2.305}$$

Since the advection operator in cylindrical coordinate is defined as $\overrightarrow{v}\cdot\nabla = v\left(\frac{\partial}{\partial r} + \frac{1}{r}\frac{\partial}{\partial \varphi} + \frac{\partial}{\partial z}\right)$ for constant fluid velocity v.

Spherical Geometry:

$$v\left(\frac{\partial M_y}{\partial r} + \frac{1}{r}\frac{\partial M_y}{\partial \theta} + \frac{1}{r\sin\theta}\frac{\partial M_y}{\partial \varphi}\right) + \frac{\partial M_y}{\partial t}$$

$$= \frac{D}{r^2}\left(\frac{\partial}{\partial r}\left(r^2\frac{\partial M_y}{\partial r}\right) + \frac{1}{\sin^2\theta}\frac{\partial^2 M_y}{\partial \varphi^2} + \frac{1}{\sin\theta}\left(\sin\theta\frac{\partial M_y}{\partial \theta}\right)\right)$$

$$+ \frac{F_o}{T_o}\gamma B_1\left(\vec{r},t\right), \tag{2.306}$$

since $\vec{v}\cdot\nabla = v\left(\frac{\partial}{\partial r} + \frac{1}{r}\frac{\partial}{\partial \theta} + \frac{1}{r\sin\theta}\frac{\partial}{\partial \varphi}\right)$.

2.7.8.3   Multidimensional Diffusion-Advection with Variable Diffusion Coefficient

Equation (2.303) can be generalized into the NMR diffusion-advection equation with variable coefficient such that:

$$\nabla\cdot\left(vM_y\right) + \frac{\partial M_y}{\partial t} = \nabla\cdot\left(D\nabla M_y\right) + \frac{F_o}{T_o}\gamma B_1\left(\vec{r},t\right), \tag{2.307}$$

The advection operator needs to be re-written since the fluid velocity is now a factor the various variables of the Del operator. An expansion of the advection term for the transverse magnetization is $\nabla\cdot(vM_y)$ is:

$$\nabla\cdot\left(vM_y\right) = (\nabla\cdot v)M_y + v\cdot\nabla M_y.$$

For constant fluid velocity, $\nabla. v = 0$ and $\nabla\cdot(vM_y) = v\nabla M_y$. This is the situation for incompressible fluids (Batchelor, 1967) and the different forms of the diffusion-advection equation for this case are given as (Awojoyogbe et al., 2011a, b):
Rectangular Geometry:

$$\frac{\partial\left(vM_y\right)}{\partial x} + \frac{\partial\left(vM_y\right)}{\partial y} + \frac{\partial\left(vM_y\right)}{\partial z} + \frac{\partial M_y}{\partial t}$$

$$= \frac{\partial}{\partial x}\left(D\frac{\partial M_y}{\partial x}\right) + \frac{\partial}{\partial y}\left(D\frac{\partial M_y}{\partial y}\right) + \frac{\partial}{\partial z}\left(D\frac{\partial M_y}{\partial z}\right) + \frac{F_o}{T_o}\gamma B_1\left(\vec{r},t\right), \tag{2.308}$$

or

$$\frac{\partial\left(v_x M_y\right)}{\partial x}+\frac{\partial\left(v_y M_y\right)}{\partial y}+\frac{\partial\left(v_z M_y\right)}{\partial z}+\frac{\partial M_y}{\partial t}$$

$$=\frac{\partial}{\partial x}\left(D_x\frac{\partial M_y}{\partial x}\right)+\frac{\partial}{\partial y}\left(D_y\frac{\partial M_y}{\partial y}\right)+\frac{\partial}{\partial z}\left(D_z\frac{\partial M_y}{\partial z}\right)+\frac{F_o}{T_o}\gamma B_1\left(\vec{r},t\right). \tag{2.309}$$

Cylindrical Geometry:

$$\frac{\partial\left(v M_y\right)}{\partial r}+\frac{1}{r}\frac{\partial\left(v M_y\right)}{\partial\varphi}+\frac{\partial\left(v M_y\right)}{\partial z}+\frac{\partial M_y}{\partial t}$$

$$=\frac{1}{r}\frac{\partial}{\partial r}\left(Dr\frac{\partial M_y}{\partial r}\right)+\frac{1}{r^2}\frac{\partial}{\partial\varphi}\left(D\frac{\partial M_y}{\partial\varphi}\right)+\frac{\partial}{\partial z}\left(D\frac{\partial M_y}{\partial z}\right)$$

$$+\frac{F_o}{T_o}\gamma B_1\left(\vec{r},t\right), \tag{2.310}$$

or

$$\frac{\partial\left(v_r M_y\right)}{\partial r}+\frac{1}{r}\frac{\partial\left(v_\varphi M_y\right)}{\partial\varphi}+\frac{\partial\left(v_z M_y\right)}{\partial z}+\frac{\partial M_y}{\partial t}$$

$$=\frac{1}{r}\frac{\partial}{\partial r}\left(D_r r\frac{\partial M_y}{\partial r}\right)+\frac{1}{r^2}\frac{\partial}{\partial\varphi}\left(D_\varphi\frac{\partial M_y}{\partial\varphi}\right)+\frac{\partial}{\partial z}\left(D_z\frac{\partial M_y}{\partial z}\right)$$

$$+\frac{F_o}{T_o}\gamma B_1\left(\vec{r},t\right). \tag{2.311}$$

where v_r, v_ϕ, and v_z are the fluid velocities along the r, ϕ, and z directions respectively.

Spherical Geometry:

$$\frac{\partial}{\partial r}\left(v M_y\right)+\frac{1}{r}\frac{\partial}{\partial\theta}\left(v M_y\right)+\frac{1}{r\sin\theta}\frac{\partial}{\partial\varphi}\left(v M_y\right)+\frac{\partial M_y}{\partial t}$$

$$=\frac{1}{r^2}\frac{\partial}{\partial r}\left(Dr^2\frac{\partial M_y}{\partial r}\right)+\frac{1}{r^2\sin\theta}\frac{\partial}{\partial\theta}\left(D\sin\theta\frac{\partial M_y}{\partial\theta}\right)+\frac{1}{r^2\sin^2\theta}\frac{\partial}{\partial\varphi}$$

$$\times\left(D\frac{\partial M_y}{\partial\varphi}\right)+\frac{F_o}{T_o}\gamma B_1\left(\vec{r},t\right), \tag{2.312}$$

or

$$\frac{\partial}{\partial r}\left(v_r M_y\right) + \frac{1}{r}\frac{\partial}{\partial \theta}\left(v_\theta M_y\right) + \frac{1}{r\sin\theta}\frac{\partial}{\partial \varphi}\left(v_\varphi M_y\right) + \frac{\partial M_y}{\partial t}$$

$$= \frac{1}{r^2}\frac{\partial}{\partial r}\left(D_r r^2 \frac{\partial M_y}{\partial r}\right) + \frac{1}{r^2\sin\theta}\frac{\partial}{\partial \theta}\left(D_\theta \sin\theta \frac{\partial M_y}{\partial \theta}\right) + \frac{1}{r^2\sin^2\theta}\frac{\partial}{\partial \varphi}$$

$$\times \left(D_\varphi \frac{\partial M_y}{\partial \varphi}\right) + \frac{F_o}{T_o}\gamma B_1\left(\vec{r},t\right). \tag{2.313}$$

where v_r, v_ϕ, and v_θ are the fluid velocities along the r, ϕ, and θ directions respectively.

2.7.9   Justification for Assumption of the Nature of Transverse Magnetization

The formulations presented in Sects. 2.7.6–2.7.8 are only possible if the assumptions made in Eqs. (2.282) and (2.298) hold. These assumptions are indications of the possible nature of the NMR signal as given by the transverse magnetization. The most straight forward solution to Eqs. (2.282) and (2.298) is the plane wave solution. Consequently, if the NMR signal is a plane wave, all formulations based on these equations are justified such that if we set

$$\Omega = \frac{1}{T_1 T_2} + \gamma^2 B_1{}^2(x,t). \tag{2.314}$$

The following combinations are possible (Awojoyogbe et al., 2011a, b):

1. $\Omega = \frac{1}{T_1 T_2} + \gamma^2 B_1{}^2(x,t)$.
2. $\Omega = \frac{1}{T_1 T_2} + \gamma^2 B_1{}^2$, where B_1 is constant-independent of x and t.
3. $\Omega = \frac{1}{T_1 T_2} + \gamma^2 B_1{}^2(x)$, where $B_1(x)$ is independent of t.
4. $\Omega = \frac{1}{T_1 T_2} + \gamma^2 B_1^2(t)$, where $B_1(t)$ is independent of x.
5. $\Omega \approx \frac{1}{T_1 T_2}$, where $\frac{1}{T_1 T_2} >> \gamma^2 B_1{}^2(x,t)$ holds (non-adiabatic condition).
6. $\Omega \approx \gamma^2 B_1{}^2$, where $\frac{1}{T_1 T_2} << \gamma^2 B_1{}^2$ and B_1 is constant-independent of x and t.
7. $\Omega \approx \gamma^2 B_1{}^2(x,t)$, where $\frac{1}{T_1 T_2} << \gamma^2 B_1{}^2(x,t)$ holds.
8. $\Omega \approx \gamma^2 B_1{}^2(x)$, where $\frac{1}{T_1 T_2} << \gamma^2 B_1{}^2(x)$ and B_1 is independent t.
9. $\Omega \approx \gamma^2 B_1{}^2(t)$, where $\frac{1}{T_1 T_2} << \gamma^2 B_1{}^2(t)$ and B_1 is independent of x.

It is worthy of note that in multiple dimensions (3D), the RF frequency $\gamma B_1(x,t)$ is replaced with $\gamma B_1\left(\vec{r},t\right)$. Furthermore, for the cases where the diffusion coefficient is constant, the transverse magnetization can take the form of a plane wave such that (Awojoyogbe et al., 2011a, b)

$$M_y(x, t) = Ae^{px+qt}, \tag{2.315}$$

where p and q represent constants, which are dependent on the NMR parameters. This form of transverse magnetization is only possible for differential equations with constant coefficients (that is, Ω is independent of x and t) (Wylie & Barrett, 1982; Spiegel, 1983). Therefore, the implication of the assumptions is that only cases (2), (5) and (6) justify the formulations of diffusion as presented in this section (Awojoyogbe et al., 2011a, b).

2.7.10 Variable Fluid Flow Corrections to the Bloch NMR Flow Equation

In earlier studies, constant velocity approximations have been used. It can be observed as well that this led to the constant diffusion coefficient. However, in real biological systems, molecular dynamics problems generally involve non-uniform velocity fields and variable diffusion coefficients (Dahlem & Müller, 2004; Siepmann & Siepmann, 2012) and since this study deals with molecular imaging, the sizes of the flowing molecular species are small enough such that the effect of surrounding geometric structure is no longer negligible. Hence, this challenge demands a modification of the Bloch NMR flow equation and derivation of the coupled differential equations for flow problems with spatial dependence.

2.7.10.1 NMR Flow with Variable Fluid Velocity

In biological systems, fluid flow is not always as has been described so far. Fluid velocity shows complex behaviour such they change their behaviour from point to point (Price, 2009; Nelson & Cox, 2008; Awojoyogbe & Dada, 2013). This is especially very conspicuous in blood flow (Dada et al., 2013a, b) and hence, it is very crucial for to outline how the Bloch NMR flow formulation will look when the fluid velocity has spatial dependence.

The General Bloch NMR Flow Equation for Variable Fluid Velocity

If it is assumed that the variable fluid velocity in the NMR system is dependent only on x, Eq. (2.193) may now be written as follows:

$$\vec{v} \cdot \nabla M_x = v(x)\frac{\partial M_x}{\partial x}, \quad \vec{v} \cdot \nabla M_y = v(x)\frac{\partial M_y}{\partial x}, \quad \vec{v} \cdot \nabla M_z$$

$$= v(x)\frac{\partial M_z}{\partial x}, \tag{2.316}$$

The earlier assumption made in Eqs. (2.194) and (2.260) may now be modified as follows:

$$\vec{v} = \widehat{i}v(x) + \widehat{j}v(x) + \widehat{k}v(x), \tag{2.317}$$

Since Eq. (2.261) still holds, Eqs. (2.262)–(2.264) then become:

$$\frac{dM_x}{dt} = \frac{\partial M_x}{\partial t} + v(x)\frac{\partial M_x}{\partial x} = -\frac{M_x}{T_2}, \tag{2.318}$$

$$\frac{dM_y}{dt} = \frac{\partial M_y}{\partial t} + v(x)\frac{\partial M_y}{\partial x} = \gamma M_z B_1 - \frac{M_y}{T_2}, \tag{2.319}$$

and

$$\frac{dM_z}{dt} = \frac{\partial M_z}{\partial t} + v(x)\frac{\partial M_z}{\partial x} = -\gamma M_y B_1 + \frac{(M_0 - M_z)}{T_1}, \tag{2.320}$$

or

$$\frac{\partial M_x}{\partial t} + v(x)\frac{\partial M_x}{\partial x} = -\frac{M_x}{T_2}, \tag{2.321}$$

$$\frac{\partial M_y}{\partial t} + v(x)\frac{\partial M_y}{\partial x} = \gamma M_z B_1 - \frac{M_y}{T_2}, \tag{2.322}$$

and

$$\frac{\partial M_z}{\partial t} + v(x)\frac{\partial M_z}{\partial x} = -\gamma M_y B_1 + \frac{(M_o - M_z)}{T_1}. \tag{2.323}$$

From Eq. (2.323), the following expression can be written:

$$\left(v(x)\frac{\partial}{\partial x} + \frac{\partial}{\partial t} + \frac{1}{T_1}\right)M_z = -\gamma M_y B_1 + \frac{M_0}{T_1},$$

$$M_z = \left(-\gamma M_y B_1 + \frac{M_o}{T_1}\right)\frac{1}{\left(v(x)\frac{\partial}{\partial x} + \frac{\partial}{\partial t} + \frac{1}{T_1}\right)}. \tag{2.324}$$

Substituting for Eq. (2.324) in Eq. (2.322) gives:

$$\frac{\partial M_y}{\partial t} + v(x)\frac{\partial M_y}{\partial x} = \gamma\left(-\gamma M_y B_1 + \frac{M_0}{T_1}\right)\frac{B_1}{\left(v(x)\frac{\partial}{\partial x} + \frac{\partial}{\partial t} + \frac{1}{T_1}\right)} - \frac{M_y}{T_2}$$

$$\left(v(x)\frac{\partial}{\partial x} + \frac{\partial}{\partial t} + \frac{1}{T_1}\right)\left\{v(x)\frac{\partial M_y}{\partial x} + \frac{\partial M_y}{\partial t} + \frac{M_y}{T_2}\right\} = \gamma\left(-\gamma M_y B_1 + \frac{M_o}{T_1}\right)B_1,$$

$$v^2(x)\frac{\partial^2 M_y}{\partial x^2} + v(x)\frac{dv}{dx}\frac{\partial M_y}{\partial x} + v(x)\frac{\partial^2 M_y}{\partial x \partial t} + \frac{v(x)}{T_1}\frac{\partial M_y}{\partial x} + v(x)\frac{\partial^2 M_y}{\partial x \partial t} + \frac{\partial^2 M_y}{\partial t^2}$$

$$+ \frac{1}{T_1}\frac{\partial M_y}{\partial t} + \frac{v(x)}{T_2}\frac{\partial M_y}{\partial x} + \frac{1}{T_2}\frac{\partial M_y}{\partial t} + \frac{1}{T_1 T_2}M_y$$

$$= -\gamma^2 B_1{}^2 M_y + \frac{\gamma B_1 M_o}{T_1},$$

$$v^2(x)\frac{\partial M_y}{\partial x^2} + v(x)\frac{dv(x)}{dx}\frac{\partial M_y}{\partial x} + 2v\frac{\partial^2 M_y}{\partial x \partial t} + v(x)\left(\frac{1}{T_1} + \frac{1}{T_2}\right)\frac{\partial M_y}{\partial x} + \left(\frac{1}{T_1} + \frac{1}{T_2}\right)$$

$$\times \frac{\partial M_y}{\partial t} + \frac{\partial^2 M_y}{\partial t^2} + \left(\frac{1}{T_1 T_2} + \gamma^2 B_1^2\right)M_y$$

$$= \frac{\gamma B_1 M_o}{T_1}.$$

This expression may be written as:

$$v^2(x)\frac{\partial^2 M_y}{\partial x^2} + 2v(x)\frac{\partial^2 M_y}{\partial x \partial t} + v(x)\left(T_0 + \frac{dv(x)}{dx}\right)\frac{\partial M_y}{\partial x} + T_0\frac{\partial M_y}{\partial t}$$

$$+ \frac{\partial^2 M_y}{\partial t^2} + (T_g + \gamma^2 B_1{}^2)M_y = F_o \gamma B_1 \tag{2.325}$$

where

$$T_o = \frac{1}{T_1} + \frac{1}{T_2}, \quad T_g = \frac{1}{T_1 T_2} \quad \text{and} \quad F_o = \frac{M_o}{T_1}.$$

Equation (2.325) is a general second-order differential equation which can be applied to NMR flow systems with spatial fluid velocity dependence.

The Time-Independent Bloch NMR Flow Equation for Variable Fluid Velocity

Within steady molecular fluid flow situations and spatial fluid velocity, Eqs. (2.270)–(2.272) can be modified as follows:

$$v(x)\frac{dM_x}{dx} = -\frac{M_x}{T_2} \tag{2.326}$$

$$v(x)\frac{dM_y}{dx} = \gamma M_z B_1 - \frac{M_y}{T_2} \tag{2.327}$$

and

$$v(x)\frac{dM_z}{dx} = -\gamma M_z B_1 + \frac{(M_0 - M_z)}{T_1} \tag{2.328}$$

From Eq. (2.328), the following equation is obtained:

$$M_z = \left(-\gamma M_y B_1 + \frac{M_o}{T_1}\right)\frac{1}{\left(v(x)\frac{d}{dx} + \frac{1}{T_1}\right)}. \tag{2.329}$$

Substituting for Eq. (2.329) into Eq. (2.327), we have:

$$v(x)\frac{dM_y}{dx} = \gamma\left(-\gamma M_y B_1 + \frac{M_0}{T_1}\right)\frac{1}{\left(v(x)\frac{d}{dx} + \frac{1}{T_1}\right)}B_1 - \frac{M_y}{T_2}$$

$$v(x)\frac{dM_y}{dx} + \frac{M_y}{T_2} = \left(-\gamma^2 M_y B_1{}^2 + \frac{\gamma B_1 M_0}{T_1}\right)\frac{1}{\left(v(x)\frac{d}{dx} + \frac{1}{T_1}\right)}$$

$$\left(v(x)\frac{d}{dx} + \frac{1}{T_1}\right)v(x)\frac{dM_y}{dx} + \left(v(x)\frac{d}{dx} + \frac{1}{T_1}\right)\frac{M_y}{T_2} = -\gamma^2 M_y B_1{}^2(x) + \frac{\gamma B_1 M_0}{T_1},$$

$$v^2(x)\frac{d^2 M_y}{dx^2} + v(x)\frac{dv(x)}{dx}\frac{dM_y}{dx} + v\left(\frac{1}{T_1} + \frac{1}{T_2}\right)\frac{dM_y}{dx} + \left(\frac{1}{T_1 T_2} + \gamma^2 B_1{}^2(x)\right)M_y$$

$$= \frac{\gamma B_1(x)M_0}{T_1},$$

$$v^2(x)\frac{d^2 M_y}{dx^2} + v(x)\left(T_0 + \frac{dv(x)}{dx}\right)\frac{dM_y}{dx} + \left(T_g + \gamma^2 B_1{}^2(x)\right)M_y$$

$$= \frac{M_0}{T_1}\gamma B_1(x). \tag{2.330}$$

Equation (2.330) is a variable time-independent Bloch NMR flow equation.

2.7.10.2 Derivation of Diffusion Equation with Variable Diffusion Coefficient

Biological systems usually have compartments occupied by fluids with different hydrodynamic properties (Nelson & Cox, 2008) such that the system's diffusion coefficient varies in different locations. As earlier noted, this is consistent with fluid flow in real biological systems (Dahlem & Müller, 2004; Siepmann & Siepmann, 2012). Therefore, quantifying the spatial dependence may hold important information about the system and associated molecular dynamics at any instant. Meanwhile, this problem has been stated in earlier studies (Awojoyogbe et al., 2011a, b; Gupta et al., 2014) but at this point, it is important to state the conditions under which the variable diffusion equation holds. Consequently, assuming:

$$\frac{\partial^2 M_y}{\partial t^2} + 2v(x)\frac{\partial^2 M_y}{\partial x \partial t} = 0. \tag{2.331}$$

Then, the following differential equation is obtained:

$$v^2(x)\frac{\partial^2 M_y}{\partial x^2} + v(x)\left(T_o + \frac{dv(x)}{dx}\right)\frac{\partial M_y}{\partial x} + T_0\frac{\partial M_y}{\partial t} + \left(T_g + \gamma^2 B_1{}^2\right)M_y$$

$$= F_0 \gamma B_1. \tag{2.332}$$

If the applied RF field is then designed such that

$$\gamma B_1 = \sqrt{\frac{v(x)}{M_y}\left(T_0 + \frac{dv(x)}{dx}\right)\frac{\partial M_y}{\partial x} - T_g}. \tag{2.333}$$

If the velocity gradient changes very slowly compared to the relaxation rate of molecular dynamics, we may assume that

$$T_0 << \frac{dv(x)}{dx}. \tag{2.334}$$

Equations (2.333) and (2.332) then become:

$$\gamma B_1(x) = \sqrt{\frac{v(x)}{M_y}\frac{dv(x)}{dx}\frac{\partial M_y}{\partial x} - T_g}, \tag{2.335}$$

$$v^2(x)\frac{\partial^2 M_y}{\partial x^2} + 2v(x)\frac{dv(x)}{dx}\frac{\partial M_y}{\partial x} + T_0\frac{\partial M_y}{\partial t} = F_0 \gamma B_1(x),$$

or

$$\frac{\partial}{\partial x}\left(v^2(x)\frac{\partial M_y}{\partial x}\right) + \frac{1}{T_0}\frac{\partial M_y}{\partial t} = F_0\gamma B_1. \tag{2.336}$$

This equation may be written as

$$\frac{\partial M_y}{\partial t} = \frac{\partial}{\partial x}\left(-\frac{v^2(x)}{T_0}\frac{\partial M_y}{\partial x}\right) + F_0\gamma B_1. \tag{2.337}$$

If the following term holds

$$D(x) = -\frac{v^2(x)}{T_0}. \tag{2.338}$$

Provided that D(x) represents the spatial diffusion coefficient, we finally have:

$$\frac{\partial M_y}{\partial t} = \frac{\partial}{\partial x}\left(D(x)\frac{\partial M_y}{\partial x}\right) + \frac{F_o}{T_o}\gamma B_1\left(\vec{r},t\right). \tag{2.339}$$

Equation (2.339) is the one-dimensional NMR diffusion equation with variable coefficient.

2.7.10.3 One-Dimensional Diffusion-Advection with Variable Diffusion Coefficient

If the fluid velocity and the diffusion coefficient D are both spatially dependent on x and making an assumption (Eq. (2.325)) of the following form:

$$\frac{\partial^2 M_y}{\partial t^2} + 2v(x)\frac{\partial^2 M_y}{\partial x\partial t} = 0. \tag{2.340}$$

The equation becomes

$$v^2(x)\frac{\partial^2 M_y}{\partial x^2} + v(x)\left(T_o + \frac{dv(x)}{dx}\right)\frac{\partial M_y}{\partial x} + T_o\frac{\partial M_y}{\partial t} + \left(T_g + \gamma^2 B_1{}^2\right)M_y$$

$$= F_o\gamma B_1. \tag{2.341}$$

If the applied RF field is then designed such that

$$\gamma B_1 = \sqrt{\frac{v(x)}{M_y}\frac{dv(x)}{dx}\frac{\partial M_y}{\partial x} + T_o\frac{dv(x)}{dx} - T_g}. \qquad (2.342)$$

Equation (2.341) becomes:

$$v^2(x)\frac{\partial^2 M_y}{\partial x^2} + v(x)\left(T_0 + \frac{dv(x)}{dx}\right)\frac{\partial M_y}{\partial x} + T_0\frac{\partial M_y}{\partial t} + v(x)\frac{dv(x)}{dx}\frac{\partial M_y}{\partial x} + T_0\frac{dv(x)}{dx}M_y$$

$$= F_0\gamma B_1,$$

$$v^2(x)\frac{\partial^2 M_y}{\partial x^2} + 2v(x)\frac{dv(x)}{dx}\frac{\partial M_y}{\partial x} + T_0\frac{\partial M_y}{\partial t} + v(x)T_0\frac{\partial M_y}{\partial x} + T_0\frac{dv(x)}{dx}M_y = F_0\gamma B_1,$$

$$\frac{\partial}{\partial x}\left(v^2(x)\frac{\partial M_y}{\partial x}\right) + T_0\frac{\partial M_y}{\partial t} + \frac{\partial}{\partial x}\left(v(x)T_0 M_y\right) = F_0\gamma B_1. \qquad (2.343)$$

Multiplying all through by $1/T_0$ and re-arranging the terms leads to:

$$\frac{\partial}{\partial x}\left(v M_y\right) + \frac{\partial M_y}{\partial t} = \frac{\partial}{\partial x}\left(D(x)\frac{\partial M_y}{\partial x}\right) + \frac{F_0}{T_0}\gamma B_1(x,t), \qquad (2.344)$$

where

$$D(x) = -\frac{v^2(x)}{T_0}.$$

2.7.10.4 Multidimensional Diffusion and Diffusion-Advection with Variable Diffusion Coefficient

The multidimensional diffusion and diffusion-advection (diffusion with transport) differential equation have been presented in Eqs. (2.303) and (2.307) respectively. Specifying the space vector $\vec{r}$, these equations are written as follows:

$$\frac{\partial M_y}{\partial t} = \nabla \cdot \left(D\left(\vec{r}\right)\nabla M_y\right) + \frac{F_o}{T_o}\gamma B_1\left(\vec{r},t\right), \qquad (2.345)$$

$$\nabla \cdot \left(v\left(\vec{r}\right)M_y\right) + \frac{\partial M_y}{\partial t} = \nabla \cdot \left(D\left(\vec{r}\right)\nabla M_y\right) + \frac{F_o}{T_o}\gamma B_1\left(\vec{r},t\right). \qquad (2.346)$$

Solving this problem is very tasking but possible with minimum assumptions. These assumptions are made by either setting the fluid velocity and diffusion coefficient equal to functions of only radial distance or eliminating the effects of the RF B_1 pulse.

Chapter 3
Radiofrequency Identification System for Computational Diffusion Magnetic Resonance Imaging Based on Bloch's NMR Flow Equation and Hermite Functions

Abstract In this Chapter, a mathematical analysis for Radiofrequency Identification System (RFID) systems for computational diffusion MRI based on Bloch's NMR flow equation and Hermite functions has been developed for detailed studies of processes taking place at the molecular level in living tissues. The goal is to explore the fundamental physics of RFID in MRI so that they can be further developed for individuals or hospitals to benefit from novel situation-aware technologies in real-world settings through the exploration of big data technologies. This will occur in the context of smart healthcare-wearable sensors, body area sensors, IoT, machine learning and artificial intelligence. Based on this theoretical innovation, we may be able to develop training software to stimulate MRI experiments and provide visual training tools to help understand diffusion MRI technology for future generations.

Keywords Bloch NMR flow equation · Diffusion magnetic resonance imaging · Radiofrequency identification system (RFID) systems · Hermite functions

3.1 Introduction

Diffusion Magnetic Resonance Imaging (diffusion MRI) has become popular in its numerous clinical and research applications which span brain stroke evaluation to fiber tracking. With a few exceptions, these methods are rooted in the classic Stejskal-Tanner formula for the diffusion-attenuated signal, usually obtained by solving the Bloch-Torrey partial differential equations (Price, 1997; Tanner, 1978; Price & Kuchel, 1991; Taylor & Bushell, 1985). The basic derivation guiding most diffusion MRI is:

$$S = S_0 e^{-bD} \tag{3.1}$$

$$\ln\left(\frac{S}{S_0}\right) = -bD \tag{3.2}$$

where

$$b = \gamma^2 G^2 \delta^2 \left(\tau - \frac{\delta}{3}\right) \tag{3.3}$$

$$b = \gamma^2 G^2 \delta^2 \tau \tag{3.4}$$

In Eq. (3.1), S is the signal strength in a pulse sequence with a pair of balanced diffusion-sensitizing gradients of strength G, each of a duration δ and with a delay τ between them. S_o is the signal strength in an identical experiment but without the diffusion gradient pair. Where it can be safely assumed that $\delta \ll \tau$, the expression for b (usually called b-value) simplifies to Eq. (3.4). In Eq. (3.1), the smallest value of τ (the time between pulses) will be limited by the performance of the gradient system. In practice, τ is normally between 1 ms and 1 s. The time of pulse δ must be smaller than τ and is typically in the range of 0–10 ms. The magnitude of the magnetic field gradient pulse G is machine dependent.

Diffusion sequences based on Eq. (3.1) are very demanding on the scanner. The gradient set is required to perform very accurately and achieve perfect symmetry of the diffusion gradient pairs. The b-value depends on a square of the gradient; therefore, nonlinearity over the subject head will translate into an error in diffusion coefficient. Furthermore, because efficient signal detection requires that the exponent bD does not become too small, larger and slower metabolites require even larger b-values to compensate for their lower diffusion coefficient and are therefore particularly affected. Larger anisotropy of brain tissues may require more sampling directions. In general, given a finite available imaging time, the trade-off is made by spending it on more signal averages, more gradient directions, more different b-values, or more different diffusion times. There is no simple and universal answer to this puzzle.

Depending on the exact sequence design, such experiments can become essentially qualitative in nature and difficult to compare with those obtained by another sequence. The results acquired with such sequences would be truly quantitative only if the diffusion was strictly Gaussian, that is, if the diffusion constant (or tensor) did not depend on the diffusion time for the whole range of the diffusion times involved.

However, despite all the problems, much valuable and relevant information has been extracted by their employment. Clearly, a good understanding of the Stejskal-Tanner formulae is crucial to avoid suboptimal experimental designs and misinterpretations of results (Chang et al., 1972, 1973; Rorschach et al., 1973; Cooper et al., 1974; Hazlewood et al., 1991).

Dimensionless parameter and scaling laws can be framed in order to predict the prototype behaviour from Eq. (3.1) and the dimensionless variable can be defined as (Price, 1997):

$$\frac{D\tau}{R^2} = \eta \tag{3.5}$$

Equation (3.5) is useful in characterizing restricted diffusion. The length R denotes the measured displacement in the direction of the gradient. We consider three relevant time scales for the measurement of the effects of the restricted diffusion:

$$\frac{D\tau}{R^2} = \eta << 1 \tag{3.6}$$

Equation (3.6) represents the short time limit. The particle does not diffuse far enough during τ to feel the effects of restriction. Measurement performed within this time scale leads to the true diffusion coefficient D.

$$\frac{D\tau}{R^2} = \eta \simeq 1 \tag{3.7}$$

In Eq. (3.7), some of the particles feel the effects of restriction and the diffusion coefficient measured within this time scale will be apparent D_{app}. The fraction of particles that feel the effects of the boundary will be dependent on the surface-to-volume ratio.

$$\frac{aD}{R^2} = \eta > 1 \tag{3.8}$$

Equation (3.8) describes the long-time limit where all particles feel the effects of restriction. In this time scale, the displacement of the particle is independent of τ and depends only on R. Thus, restriction causes measuring-time-dependent diffusion coefficient in which at τ the displacement is limited by the embedding geometry.

We study the flow properties of the modified time-independent Bloch NMR flow equations which describe the dynamics of the hydrogen atom under the influence of a radio frequency (RF) magnetic field (Hinshaw & Lent, 1983; Harris, 1986; Brix et al., 2008; Tung et al., 2013). The RF system provides the communication link with the patient's body for the purpose of producing an image. All medical imaging modalities use some form of radiation (e.g., X-ray, gamma-ray, etc.) or energy (e.g., ultrasound) to extract information in the form of images from the patient. The MRI process uses RF signals to transmit the information from the patient's body to the scanner. The RF energy used is a form of non-ionizing radiation in which excitation pulses that are applied to the patient are absorbed by the tissue and converted to heat. A small amount of the energy is emitted as radio signals used to produce images. Actually, the image itself is not formed within and transmitted from the body but the RF signals provide information (data) from which the image is reconstructed by the computer. However, the resulting image is a display of RF signal intensities produced by different tissues.

Radio-frequency identification (RFID) employs electromagnetic fields to automatically identify and track tags attached to or placed on objects. A specific RFID tag is made up of a very small radio transponder, a radio receiver and a transmitter. On excitation by an electromagnetic pulse emanating from the RFID reader device, the tag transmits digital information, such as an identifying inventory number, back to the reader (Chiagozie & Nwaji, 2012). A typical RFID system is incredibly similar to a typical MRI system. For example, RF-based tagging has been used extensively in harmonic phase-magnetic resonance imaging (Parthasarathy, 2006; Osman et al., 1999, 2000; Dada et al., 2017). Since T_1 and T_2 relaxation times are unique to physiological states of tissues and the applied magnetic fields, they can be used as tags as long as they can be tuned to the applied RF pulse in MRI experiments. This book presents the theoretical background for MR spin-based RFID system which could prove to be useful in disease diagnosis, treatment monitoring, drug discovery and materials science investigations.

Consequently, a mathematical analysis for RFID systems for computational diffusion MRI based on Bloch's NMR flow equation and Hermite functions has been developed for detailed studies of processes taking place at molecular level in living tissues. The goal is to explore the fundamental physics of RFID in MRI so that they can be further developed for individuals or hospitals to benefit from novel, situation-aware technologies in real-world settings through the exploitation of big data technologies. This will occur in the context of smart healthcare-wearable sensors, body area sensors, IoT, machine learning and artificial intelligence (Firouzi et al., 2018). Based on this theoretical innovation, we may be able to develop training software to stimulate MRI experiments and provide visual training tools to help understand MRI technology for future generations.

3.2 Mathematical Analysis

We study the flow properties of the modified time independent Bloch NMR flow equation which describes the dynamics of fluid flow under the influence of RF field (Awojoyogbe et al., 2011a, b, 2016a, b; Dada et al., 2017) is given by Eq. (2.269). That is,

$$v^2 \frac{\partial^2 M_y}{\partial x^2} + 2v \frac{\partial^2 M_y}{\partial x \partial t} + \frac{\partial^2 M_y}{\partial t^2} + \left(\frac{1}{T_1} + \frac{1}{T_2} \right) v \frac{\partial M_y}{\partial x} + \left(\frac{1}{T_1} + \frac{1}{T_2} \right) \frac{\partial M_y}{\partial t}$$

$$+ S(x,t)M_y = \frac{M_0 \gamma B_1(x,t)}{T_1} \tag{3.9}$$

subject to the conditions as highlighted in earlier studies (Awojoyogbe et al., 2011a, b). T_1 and T_2 are the spin-lattice and spin-spin relaxation times respectively, the reciprocals of T_1 and T_2 are defined as relaxation rates. RF B_1 is the spatially varying magnetic field and v is the fluid flow velocity. When the RF $B_1(x)$ field is applied at $t = 0$, M_y has a maximum value when RF $B_1(x)$ is maximum and $M_0 = 0$. At the points when maximum NMR signal is received (maximum values of M_y and $B_1(x)$ respectively), Eq. (3.9) becomes:

$$\frac{d^2 M}{dx^2} + \frac{1}{vT_A}\frac{dM_y}{dx} + \frac{S(x)}{v^2}M_y = 0 \tag{3.10}$$

where

$$S(x,t) = \gamma^2 B_1^2(x,t) + \frac{1}{T_1 T_2}, \quad \frac{1}{T_A} = \frac{1}{T_1} + \frac{1}{T_2} \tag{3.11}$$

We shall assume a stream function of the following form:

$$M_y(x) = \psi(x)e^{\lambda x} \tag{3.12}$$

Using Eq. (3.12), we have:

$$\frac{dM_y}{dx} = e^{\lambda x}\frac{d\psi(x)}{dx} + \lambda e^{\lambda x}\psi(x), \quad \frac{d^2 M_y}{dx^2} = e^{\lambda x}\frac{d^2\psi(x)}{dx^2} + 2\lambda e^{\lambda x}\frac{d\psi(x)}{sx} + \lambda^2 e^{\lambda x}\psi(x)$$

Equation (3.10) then becomes:

$$e^{\lambda x}\left\{\frac{d^2\psi}{dx^2} + 2\lambda\frac{d\psi(x)}{dx} + \lambda^2\psi(x) + \frac{1}{vT_A}\left(\frac{d\psi(x)}{dx} + \lambda\psi(x)\right) + \frac{S(x)}{v^2}\psi(x)\right\} = 0 \tag{3.13}$$

The exponential function in Eq. (3.13) can be defined as follows:

$$e^{\lambda x} = \sum_{n=0}^{\infty}\frac{(\lambda x)^n}{n!} = F(x) \tag{3.14}$$

Equation (3.14) is extremely useful in obtaining approximations to complicated formulae, valid when x is small (Awojoyogbe, 2007). In particular, when

$$x = 4vT_A \tag{3.15}$$

Equation (3.14) becomes:

$$F(x) = 1 \tag{3.16}$$

Equation (3.13) can then be written as:

$$\frac{d^2\psi(x)}{dx^2} + 2\lambda\frac{d\psi(x)}{dx} + \lambda^2\psi(x) + \frac{1}{vT_A}\left(\frac{d\psi(x)}{dx} + \lambda\psi(x)\right) + \frac{S(x)}{v^2}\psi(x) = 0 \quad (3.17)$$

$$\frac{d^2\psi(x)}{dx^2} + \left(2\lambda + \frac{1}{vT_A}\right)\frac{d\psi(x)}{dx} + \left(\lambda^1 + \frac{\lambda}{vT_A}\right)\psi(x) + \frac{S(x)}{v^2}\psi(x) = 0 \quad (3.18)$$

provided that

$$\lambda = -\frac{1}{2vT_A} \quad (3.19)$$

$$\frac{d^2\psi(x)}{dx^2} - \frac{1}{v^2}\left(\frac{1}{4T_A^2}\right)\psi(x) + \frac{S(x)}{v^2}\psi(x) = 0 \quad (3.20)$$

$$\frac{d^2\psi(x)}{dx^2} - \frac{1}{v^2}\left(\frac{1}{4T_A^2}\right)\psi(x) + \frac{1}{v^2}\left(\gamma^2 B_1^2(x) + T_g\right)\psi(x) = 0$$

$$\frac{d^2\psi(x)}{dx^2} + \frac{1}{v^2}\left(\gamma^2 B_1^2(x) + T_g - T_R\right)\psi(x) = 0 \quad (3.21)$$

where $T_g = \frac{1}{T_1 T_2}$ and $T_R = \frac{1}{4T_A^2}$.

If we then design the radio frequency field such that:

$$\gamma B_1(x) = i\Omega\gamma Gx \quad (3.22)$$

where Ω is a dimensionless weighting parameter.

We have the following equation:

$$\frac{d^2\psi(x)}{dx^2} + \frac{1}{v^2}\left(T_g - T_R - \Omega^2\gamma^2 G^2 x^2\right)\psi(x) = 0 \quad (3.23)$$

where G is the strength of the gradient field.

We shall make the following assumption to further simplify Eq. (3.23)

$$\alpha = \frac{\Omega\gamma G}{v} \quad (3.24)$$

$$\beta = \frac{T_g - T_R}{v^2} \quad (3.25)$$

Equation (3.23) becomes:

$$\frac{d^2\psi}{dx^2} + (\beta - \alpha^2 x^2)\psi = 0 \quad (3.26)$$

Setting:

$$\xi = \sqrt{\alpha}x \tag{3.27}$$

The following equations are obtained from this assumption (Eisberg & Resnick, 1985).

$$\frac{d\psi}{dx} = \frac{d\xi}{dx}\frac{d\psi}{d\xi} = \sqrt{\alpha}\frac{d\psi}{d\xi} \tag{3.28}$$

$$\frac{d^2\psi}{dx^2} = \frac{d\psi}{dx} = \frac{d\xi}{dx}\frac{d}{d\xi}\left(\frac{d\psi}{dx}\right) = \alpha\frac{d^2\psi}{d\xi^2} \tag{3.29}$$

Hence, Eq. (3.26) becomes:

$$\frac{d^2\psi}{d\xi^2} + \left(\frac{\beta}{\alpha} - \xi^2\right)\psi = 0 \tag{3.30}$$

In order to ensure that the transverse magnetization is measurable at all points, $\psi(x)$ has to be finite as $|\xi| \to \infty$. Consequently, we need to first consider the case for very large $|\xi|$ (noticeable at very small fluid velocity and/or when x is just few centimetres). Therefore, for very large $|\xi|$, $\frac{\beta}{\alpha} - \xi^2 \approx -\xi^2$

$$\frac{d^2\psi}{d\xi^2} = \xi^2\psi \tag{3.31}$$

The general solution to Eq. (3.31) is:

$$\psi = A_1 e^{-\frac{\xi^2}{2}} + A_2 e^{\frac{\xi^2}{2}} \tag{3.32}$$

where A_1 and A_2 are arbitrary constants. For ψ to be measurable at large values of $|\xi|$, $A_2 = 0$ and Eq. (3.32) reduces to:

$$\psi = A_1 e^{-\frac{\xi^2}{2}}; \quad |\xi| \to \infty \tag{3.33}$$

This expression suggests that we seek for a solution to the generalized equation in the form:

$$\psi(\xi) = A_1 e^{-\frac{\xi^2}{2}}H(\xi) \tag{3.34}$$

These solutions are expected to be valid for all ξ such that $H(\xi)$ must be functions slowly varying in comparison to $e^{-\frac{\xi^2}{2}}$ as $|\xi| \to \infty$ so as to ensure Eq. (3.33) agree with Eq. (3.34). From Eq. (3.34),

$$\frac{d\psi}{d\xi} = -A_1\xi e^{-\frac{\xi^2}{2}}H + A_1 e^{-\frac{\xi^2}{2}}\frac{dH}{d\xi} \tag{3.35}$$

$$\frac{d^2\psi}{d\xi^2} = -A_1\xi e^{-\frac{\xi^2}{2}}H + A_1\xi^2 e^{-\frac{\xi^2}{2}}H - 2A_1\xi e^{-\frac{\xi^2}{2}}\frac{dH}{d\xi} + A_1 e^{-\frac{\xi^2}{2}}\frac{d^2H}{d\xi^2}$$
$$= A_1\xi e^{-\frac{\xi^2}{2}}\left(-H + \xi^2 H - 2\xi\frac{H}{d\xi} + \frac{d^2H}{d\xi^2}\right) \tag{3.36}$$

Then Eq. (3.30) becomes:

$$A_1\xi e^{-\frac{\xi^2}{2}}\left(-H + \xi^2 H - 2\xi\frac{dH}{d\xi} + \frac{d^2H}{d\xi^2}\right) + \left(\frac{\beta}{\alpha} - \xi^2\right)A_1\xi e^{-\frac{\xi^2}{2}}H = 0 \tag{3.37}$$

$$\left(-H + \xi^2 H - 2\xi\frac{dH}{d\xi} + \frac{d^2H}{d\xi^2}\right) + \left(\frac{\beta}{\alpha} - \xi^2\right)H = 0 \tag{3.38}$$

$$\frac{d^2H}{d\xi^2} - 2\xi\frac{dH}{d\xi} + \left(\frac{\beta}{\alpha} - 1\right)H = 0 \tag{3.39}$$

Setting:

$$\frac{\beta}{\alpha} = 2n + 1 \tag{3.40}$$

Equation (3.39) becomes:

$$\frac{d^2H}{d\xi^2} - 2\xi\frac{dH}{d\xi} + 2nH = 0 \tag{3.41}$$

Equation (3.41) is the Hermite differential equation whose solution (Abramowitz & Stegun, 1964) is known as the Hermite polynomial, H_n. Finally, the solution to Eq. (3.30) is

$$\psi(\xi) = A_1 e^{-\frac{\xi^2}{2}}H_n(\xi) \tag{3.42}$$

or

$$\psi(x) = A_1 e^{-\frac{\gamma\Omega G}{2v}x^2}H_n\left(\sqrt{\frac{\gamma\Omega_n G}{v}}x\right) \tag{3.43}$$

If we then use Eqs. (3.43) and (3.19) in (3.12),

$$M_y(x) = A_1 e^{-\frac{\gamma\Omega_n G}{2v}x^2}e^{-\frac{1}{2vT_A}x}H_n\left(\sqrt{\frac{\gamma\Omega_n G}{v}}x\right) \tag{3.44}$$

where

$$n = \frac{T_g - T_R - \Omega_n \gamma G v}{2\Omega_n \gamma G v} \tag{3.45}$$

3.3 Development of Non-gradient MRI

In order to understand the polynomial function $H_n(\xi) = H_n\left(\sqrt{\frac{\gamma \Omega_n G}{v}}x\right)$, we note that (Prudnikov et al., 1986; Zwillinger, 1997)

$$H_n(\xi) = (-j)^2 \frac{2^n e^{\xi^2}}{\sqrt{\pi}} \int\limits_{\infty}^{-\infty} e^{-z^2 + j2\xi z} z^n dz \tag{3.46}$$

where n = 0, 1, 2, 3, 4, ... and j = ±1, ±2, ±3, ±4, ±5, ...

It is convenient to define j as j = −1 and n = 2; so that Eq. (3.46) becomes:

$$H_2(\xi) = \frac{4e^{\xi^2}}{\sqrt{\pi}} \int\limits_{\infty}^{-\infty} e^{-z^2 + j2\xi z} z^2 dz \tag{3.47}$$

where

$$\int\limits_{\infty}^{-\infty} e^{-z^2 + j2\xi z} z^2 dz = \frac{\sqrt{\pi}}{2}\left(1 + 2\xi^2\right)e^{\xi^2} \tag{3.48}$$

From Eqs. (3.47) and (3.48), we write:

$$H_2(\xi) = 2e^{\xi^2}\left(1 + 2\xi^2\right)e^{\xi^2} = 2e^{2\xi^2} + 4\xi^2 e^{2\xi^2} \tag{3.49}$$

Hence,

$$M_y(x)\big|_{n=2} = 4A_1 e^{-\frac{\gamma \Omega_2 G}{2v}x^2} e^{-\frac{1}{2vT_A}x}\left(\frac{1}{2}e^{2\xi^2} + \xi^2 e^{2\xi^2}\right) \tag{3.50}$$

$$M_y(x)\big|_{n=2} = 4A_1 e^{-\frac{\gamma \Omega_2 G}{2v}x^2} e^{-\frac{1}{2vT_A}x}\left(\frac{1}{2}e^{2\frac{\gamma \Omega_2 G}{v}x^2} + \frac{\gamma \Omega_2 G}{v}x^2 e^{2\frac{\gamma \Omega_2 G}{v}x^2}\right) \tag{3.51}$$

$$M_y(x)\big|_{n=2} = A_1 e^{-\frac{1}{2vT_A}x}\left(2e^{\frac{3\gamma \Omega_2 G}{2v}x^2} + 4\frac{\gamma \Omega_2 G}{v}x^2 e^{\frac{3\gamma \Omega_2 G}{2v}x^2}\right) \tag{3.52}$$

Introducing equation (3.15) in Eq. (3.52) leads to:

$$M_y(x)\big|_{n=2} = 0.135335A_1 \left(2e^{24\gamma\Omega_2 GvT_A^2} + 64\gamma\Omega_2 GvT_A^2 e^{24\gamma\Omega_2 GvT_A^2} \right) \tag{3.53}$$

For the scenario being described here ($n = 2$),

$$\Omega_2\gamma Gv = \frac{T_g - T_R}{5} \tag{3.54}$$

The NMR transverse magnetization can easily be determined as a function of Hermite polynomial from Eq. (3.53) as:

$$M_y(x)\big|_{n=2} = 0.135335A_1 \left(2e^{\frac{24}{5}(T_g-T_R)T_A^2} + \frac{64}{5}\left(T_g - T_R\right)T_A^2 e^{\frac{24}{5}(T_g-T_R)T_A^2} \right) \tag{3.55}$$

From Eq. (3.15), the fluid velocity takes the form:

$$v = \frac{x}{4T_A} \tag{3.56}$$

Hence, according to Eq. (3.54), the excitation radiofrequency takes the form:

$$\omega_{1,2} = \frac{4}{5\Omega_2}\left(T_g - T_R\right)T_A \tag{3.57}$$

This expression forms the basis of our RFID system for tissues with identification features T_A, T_g and T_R (which are all functions of T_1 and T_2 relaxation times). For general consideration, Eq. (3.45) may be written in the form:

$$\Omega_n\gamma Gv = \frac{T_g - T_R}{(2n + 1)} \tag{3.58}$$

$$\Omega_n\gamma Gx = \frac{4}{(2n + 1)}\left(T_g - T_R\right)T_A \tag{3.59}$$

Setting:

$$\omega_1 = \gamma Gx \tag{3.60}$$

The RFID system becomes:

$$\omega_{1,n} = \frac{4}{(2n + 1)\Omega_n}\left(T_g - T_R\right)T_A \tag{3.61}$$

3.4   A New Method for Diffusion MRI

The Rodrigues Formula is written as (Abramowitz & Stegun, 1964; Zwillinger, 1997):

$$H_n(\xi) = (-1)^n \exp\left(-\xi^2\right) \frac{d^n}{dx^n} \exp\left(-\xi^2\right) \tag{3.62}$$

The first few Hermite polynomials are:

$$
\begin{aligned}
H_0(\xi) &= 1 \\
H_1(\xi) &= 2\xi \\
H_2(\xi) &= 4\xi^2 - 2 \\
H_3(\xi) &= 8\xi^3 - 12\xi \\
H_2(\xi) &= 16\xi^4 - 48\xi + 12
\end{aligned}
\tag{3.63}
$$

In terms of x, these Hermite polynomials are expressed as:

$$
\begin{aligned}
H_0(x) &= 1 \\
H_1(x) &= 2\sqrt{\frac{\gamma\Omega G}{v}}x \\
H_2(x) &= 4\frac{\gamma\Omega G}{v}x^2 - 2 \\
H_3(x) &= 8\left(\frac{\gamma\Omega G}{v}\right)^{\frac{3}{2}} - 12\sqrt{\frac{\gamma\Omega G}{v}}x \\
H_2(x) &= 16\left(\frac{\gamma\Omega G}{v}\right)^2 x^4 - 48\sqrt{\frac{\gamma\Omega G}{v}}x + 12
\end{aligned}
\tag{3.64}
$$

The corresponding transverse magnetizations are given as:

$$
\begin{aligned}
M_y(x)\big|_{n=0} &= A_1 e^{-\frac{\gamma\Omega G}{2v}x^2} e^{-\frac{1}{2vT_A}x} \\
M_y(x)\big|_{n=1} &= 2A_1 x\sqrt{\frac{\gamma\Omega G}{v}}\, e^{-\frac{\gamma\Omega G}{2v}x^2} e^{-\frac{1}{2vT_A}x} \\
M_y(x)\big|_{n=2} &= \left(4\frac{\gamma\Omega G}{v}x^2 - 2\right) A_1 e^{-\frac{\gamma\Omega G}{2v}x^2} e^{-\frac{1}{2vT_A}x} \\
M_y(x)\big|_{n=3} &= \left(8\left(\frac{\gamma\Omega G}{v}\right)^{\frac{3}{2}} - 12\sqrt{\frac{\gamma\Omega G}{v}}x\right) A_1 e^{-\frac{\gamma\Omega G}{2v}x^2} e^{-\frac{1}{2vT_A}x} \\
M_y(x)\big|_{n=4} &= \left(16\left(\frac{\gamma\Omega G}{v}\right)^2 x^4 - 48\sqrt{\frac{\gamma\Omega G}{v}}x + 12\right) A_1 e^{-\frac{\gamma\Omega G}{2v}x^2} e^{-\frac{1}{2vT_A}x}
\end{aligned}
\tag{3.65}
$$

Traditionally, the average distance travelled by a molecule in time τ, depends on the diffusion coefficient D, for a specific tissue (identified by relaxation parameters T_1 and T_2) compartment within a voxel is (Price, 1997):

$$x^2 = 2D\tau = \frac{D\tau}{\eta} \tag{3.66}$$

where η is a dimensionless constant. From Eqs. (3.15) and (3.66), the diffusion coefficient D can be defined as:

$$D = \frac{8v^2 T_A^2}{\tau} \tag{3.67}$$

Specifically, for n = 2, the NMR signal can be defined from Eqs. (3.65) as:

$$M_y(\tau)\big|_{n=2} = 2A_1 \left(2\frac{\gamma\Omega G}{v}\frac{D\tau}{\eta} - 1 \right) e^{-\frac{\gamma\Omega G D\tau}{2v\,\eta}} e^{-\frac{1}{2vT_A}\sqrt{\frac{D\tau}{\eta}}} \tag{3.68}$$

Equation (3.66) can be useful to accurately describe diffusion MRI especially in highly restricted biological systems in terms of the Hermite polynomials. Specifically, in Eq. (3.64) if $\eta < 1$ (the short time limit), the particle does not diffuse far enough during τ to feel the effects of restriction.

From Eqs. (3.59), we have:

$$x^2 = \frac{16(T_g - T_R)^2 T_A^2}{(2n+1)^2 \Omega_n^2 \gamma^2 G^2} \tag{3.69}$$

According to Eq. (3.66), Eq. (3.69) becomes:

$$D = \frac{8(T_g - T_R)^2 T_A^2}{(2n+1)^2 \Omega_n^2 \gamma^2 G^2 \tau} \tag{3.70}$$

$$D = \frac{8(T_g - T_R)^2 T_A^2 \delta^2}{(2n+1)^2 \Omega_n^2 \gamma^2 G^2 \delta^2 \tau} \tag{3.71}$$

Using Eq. (3.4) in Eq. (3.71) leads to:

$$D = \frac{8(T_g - T_R) T_A^2 \delta^2}{(2n+1)^2 \Omega_n^2 b} \tag{3.72}$$

From Eq. (3.2), Eq. (3.72) may be modified as follows:

$$-\frac{1}{b}\ln\left(\frac{S}{S_0}\right) = \frac{8\left(T_g - T_R\right)^2 T_A^2 \delta^2}{(2n+1)^2 \Omega_n^2 b} \tag{3.73}$$

or

$$\left(T_g - T_R\right)^2 T_A^2 = -\frac{(2n+1)^2 \Omega_n^2}{8\delta^2}\ln\left(\frac{S}{S_0}\right) \tag{3.74}$$

Equation (3.74) is given as:

$$\frac{(T_1 - T_2)^2}{4(T_1 T_2)(T_1 + T_2)} = \frac{(2n+1)^2 \Omega_n^2}{8\delta^2}\ln\left(\frac{S}{S_0}\right) \tag{3.75}$$

It is interesting to note that Eq. (3.75) can be used to extract relaxometric information from diffusion weighted images using the following expression:

$$T_1 = \frac{-T_2 - 1T_2^2 S_T - 2\sqrt{2T_2^3 S_T + T_2^4 S_T^2}}{-1 + 4T_2 S_T} \quad\text{or}\quad \frac{-T_2 - 2T_2^2 S_T + 2\sqrt{2T_2^3 S_T + T_2^4 S_T^2}}{-1 + 4T_2 S_T} \tag{3.76}$$

whichever is positive $S_T = \frac{(2n+1)^2 \Omega_n^2}{8\delta^2}\ln\left(\frac{S}{S_0}\right)$ and one of the relaxation times must be known in order to determine the other.

For a given diffusion imaging sections for which S_0 is the image signal without gradient ($b = 0$) and S is image signal with gradient, we can determine the radiofrequency identifying every part of the diffusion image. From Eq. (3.61), we write:

$$\left(T_g - T_R\right) T_A = \frac{(2n+1)\Omega_n \omega_{1,n}}{4} \tag{3.77}$$

Substituting for Eq. (3.77) in Eq. (3.74), we have:

$$\frac{(2n+1)^2 \Omega_n^2 \omega_{1,n}^2}{16} = -\frac{(2n+1)^2 \Omega_n^2}{8\delta^2}\ln\left(\frac{S}{S_0}\right) \tag{3.78}$$

Therefore,

$$\omega_1^2 = -\frac{2}{\delta^2}\ln\left(\frac{S}{S_0}\right) \tag{3.79}$$

The parameter n has been removed from the radiofrequency of Eq. (3.78) because the terms on the right-hand side of this equation are general terms for a given diffusion image.

3.5   Computational Analysis

In order to analyze the results obtained in earlier sections, we shall make use of relaxometry (T_1 and T_2 relaxation times) data and diffusion image data. The relaxometry data are grouped into those for normal tissues and those from neuro-oncology. The cancer-based relaxometry are examined in three different types of tissues; that is, solid tumor, peritumoral white matter and contralateral white matter (Badve et al., 2017). These NMR relaxometry data are presented in Tables 3.1 and 3.2.

Nuclear magnetic resonance signals detected during or after the applied RF pulse are obtained from the voltage induced in detection coils as a result of the precession of nuclear spins around the uniform magnetic field (B_0). Precession normally takes place as the nuclei's intrinsic Larmor frequency after the application of RF pulse. These intrinsic frequencies are dependent on the molecular signatures of the tissue under investigation via T_1 and T_2 relaxation times. Hence, we may assume that the parameter $\omega_{1,n}$ gives us a means of quantifying this intrinsic frequency.

Setting $n = 2$, we can write that the RF identification at this level is given from Eq. (3.57) as:

$$\omega_1^{RFID} = \frac{4}{5}\left(T_g - T_R\right)T_A \tag{3.80}$$

where

$$\Omega_2 = -1 \tag{3.81}$$

Table 3.1 T_1 and T_2 relaxation times of brain tissues obtained at 1.5 T and 3.0 T (Lu et al., 2005)

	$B_0 = 1.5$ T		$B_0 = 3.0$ T		Number
Brain region	T_1 (ms)	T_2 (ms)	T_1 (ms)	T_2 (ms)	of patients
Frontal white matter	556 ± 20	79 ± 2	699 ± 38	69 ± 2	10
Corpus callosum genu	556 ± 38	76 ± 3	721 ± 68	65 ± 2	10
Corpus callosum splenium	608 ± 55	86 ± 5	748 ± 64	75 ± 5	10
Occipital white matter	616 ± 32	92 ± 4	758 ± 49	81 ± 3	10
Frontal gray matter	1048 ± 61	99 ± 4	1209 ± 109	88 ± 3	10
Occipital gray matter	989 ± 44	90 ± 4	1122 ± 117	79 ± 5	10
Caudate	1009 ± 48	85 ± 2	1258 ± 55	69 ± 2	10
Putamen	875 ± 43	81 ± 3	1102 ± 40	66 ± 3	10
Thalamus	821 ± 34	80 ± 2	986 ± 33	67 ± 2	10

Table 3.2 T_1 and T_2 relaxation times determined from various tumor regions and contralateral white matter in three tumor types (Badve et al., 2017)

Solid tumor			
	T_1 (ms)	T_2 (ms)	Number of patients
Glioblastoma multiforme	1630 ± 247	138 ± 22	17
Metastasis	1324 ± 273	105 ± 27	8
Low grade glioma	1600 ± 197	172 ± 53	6
Peritumoral white matter			
Glioblastoma multiforme	1578 ± 331	140 ± 27	17
Metastasis	1382 ± 188	119 ± 27	8
Low grade glioma	1066 ± 218	102 ± 43	6
Contralateral white matter			
Glioblastoma multiforme	927 ± 133	69 ± 9	17
Metastasis	911 ± 39	72 ± 6	8
Low grade glioma	873 ± 61	72 ± 17	6

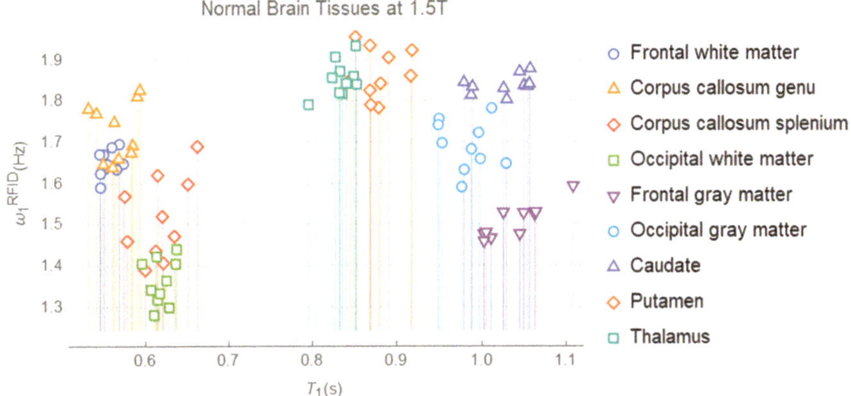

Fig. 3.1 Profile of ω_1^{RFID} across the range of T_1 in normal brain tissues

A wolfram Mathematica program was developed to randomly distribute the measurements in Table 3.1 across the ten patients such that each data point represents one patient. The generated data was then used to compute ω_1^{RFID}. Profiles of this angular frequency are given in Figs. 3.1 and 3.2.

In a similar manner, the measurements in Table 3.2 were distributed across the number of patients and the data generated was used to compute ω_1^{RFID} using Eq. (3.80). The resulting profiles are given in Figs. 3.3, 3.4, 3.5, 3.6, 3.7 and 3.8.

The MR transverse magnetization associated with this angular frequency is given in Eq. (3.55). Distributed relaxation time measurements across the patients were too few for the computation of transverse magnetization. Hence, we distributed these measurements over 200 data points and the generated data was used to calculate the

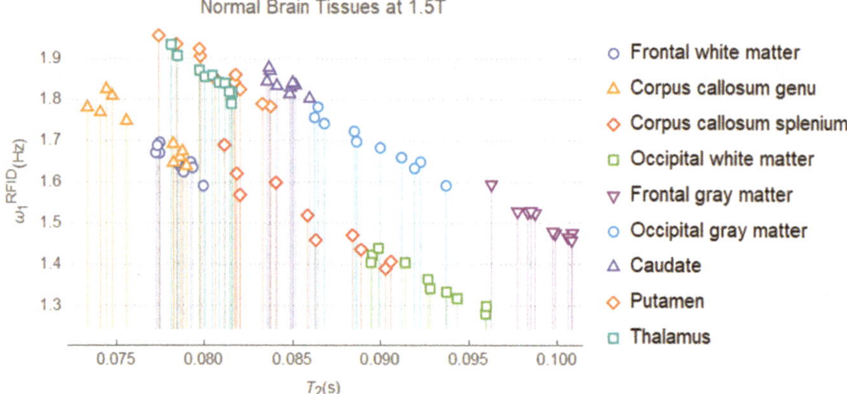

Fig. 3.2 Profile of ω_1^{RFID} across the range of T_2 in normal brain tissues

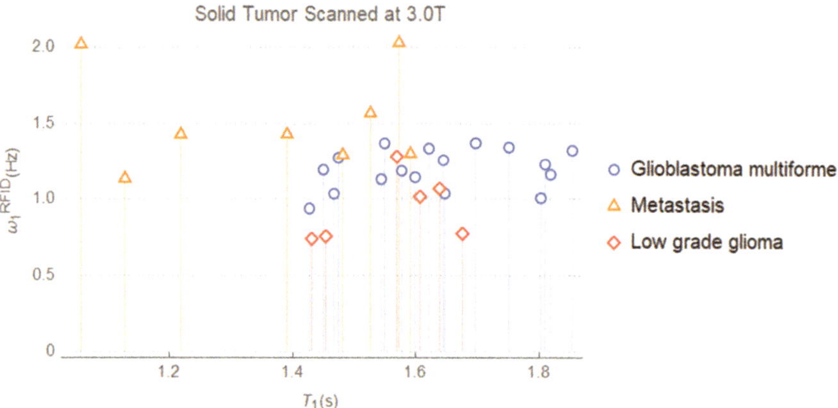

Fig. 3.3 Profile of ω_1^{RFID} across the range of T_1 for three tumor types found in a solid tumor region

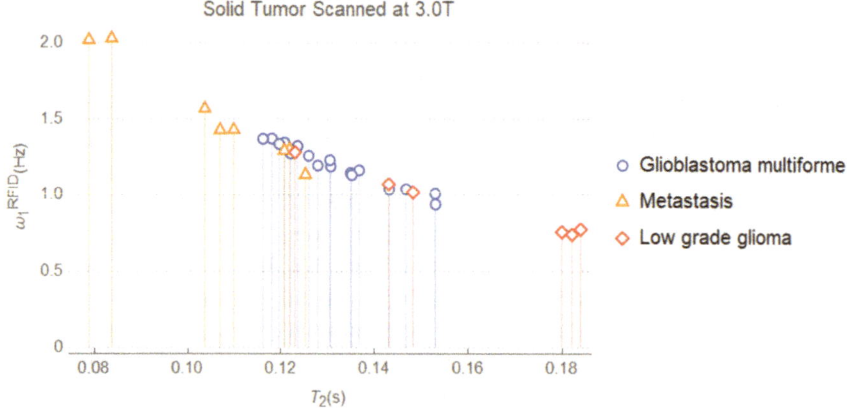

Fig. 3.4 Profile of ω_1^{RFID} across the range of T_2 for three tumor types found in a solid tumor region

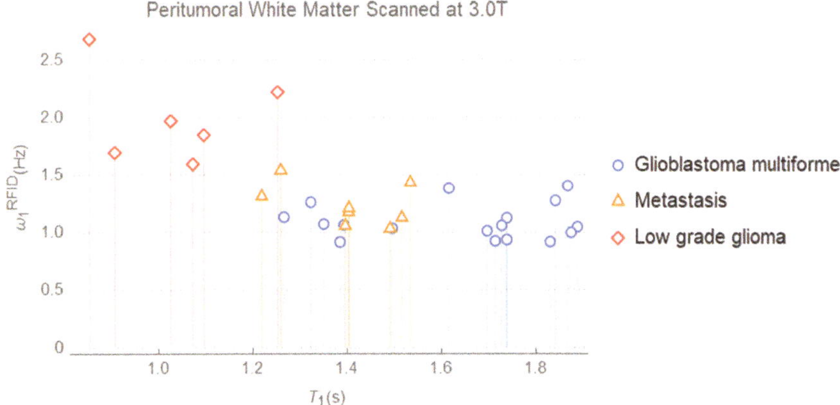

Fig. 3.5 Profile of ω_1^{RFID} across the range of T_1 for three tumor types found in peritumoral white matter region

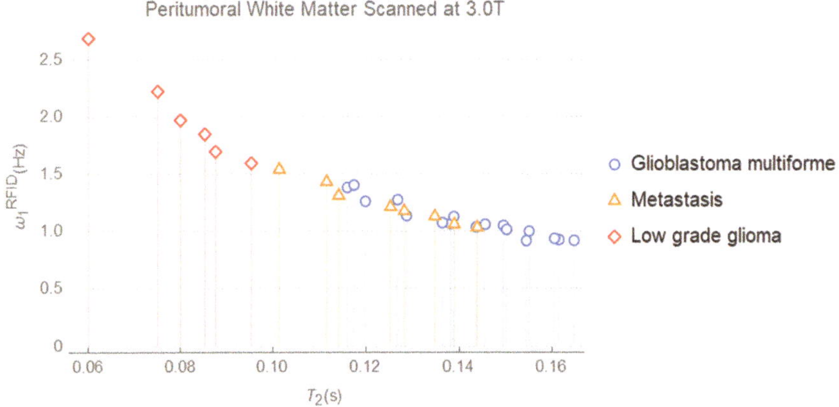

Fig. 3.6 Profile of ω_1^{RFID} across the range of T_2 for three tumor types found in peritumoral white matter region

MR signal at $n = 2$. The signal is presented as bubble charts in Figs. 3.9, 3.10, 3.11 and 3.12.

The applications of models developed in this study to clinical diffusion magnetic resonance imaging were demonstrated by making use of diffusion weighted images with and without diffusion gradients. These images were scanned on a 3.0-Tesla Discovery MR750 (GE Healthcare Systems, Milwaukee, WI) with a gradient strength of 0.050 T/m and an 8-channel high resolution receiver head coil (Zeng et al., 2018) using a single-shot echo planer imaging where: number of sections $= 24$, section thickness $= 4$ mm, field of view (FOV) $= 240 \times 240$ mm, matrix $= 256 \times 256$, in-plane resolution $= 0.94 \times 0.94$ mm, repetition time (TR) $= 3$ s, echo time (TE) $= 107.7$ ms, phase FOV $= 1.00$, flip angle $= 90$ and pixel

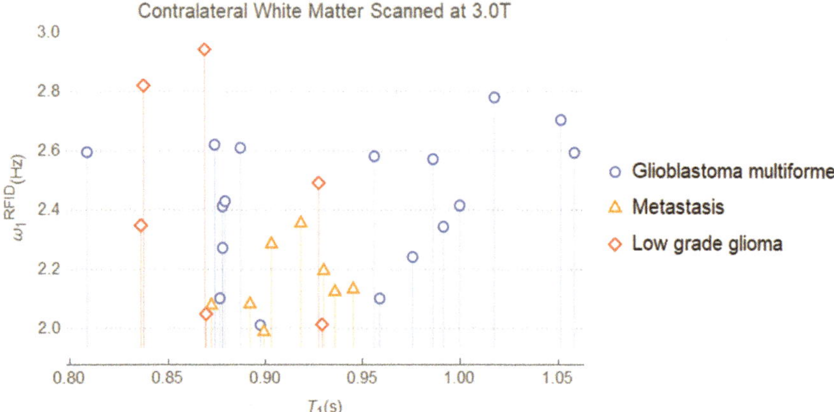

Fig. 3.7 Profile of ω_1^{RFID} across the range of T_1 for three tumor types found in contralateral white matter region

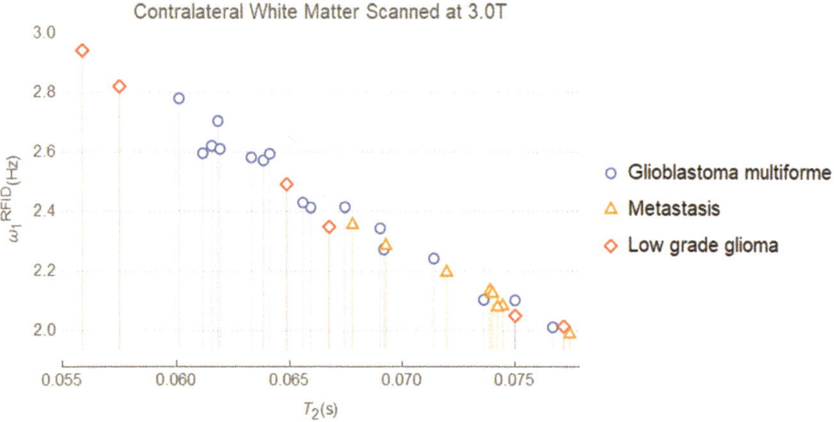

Fig. 3.8 Profile of ω_1^{RFID} across the range of T_2 for three tumor types found in contralateral white matter region

bandwidth = 1953.1 Hz/pixel. The signal intensities for these images were then extracted with the aid of ImageJ software. The extracted data were used to reconstruct original images (Figs. 3.13a, b) and also computed the resonant angular frequencies according to Eq. (3.79). The computed data were used to produce Fig. 3.13c. In this image, we made use of $\delta = 3$ ms and utilized the absolute values of the calculated angular frequency because some of the extracted data points produced angular frequency as complex numbers.

Fig. 3.9 MR transverse magnetization distribution across T_1 and T_2 relaxation times for (**a**) Frontal white matter (**b**) Corpus callosum genu (**c**) Corpus callosum splenium (**d**) Occipital white matter (**e**) Frontal gray matter (**f**) Occipital gray matter (**g**) Caudate (**h**) Putamen (**i**) Thalamus

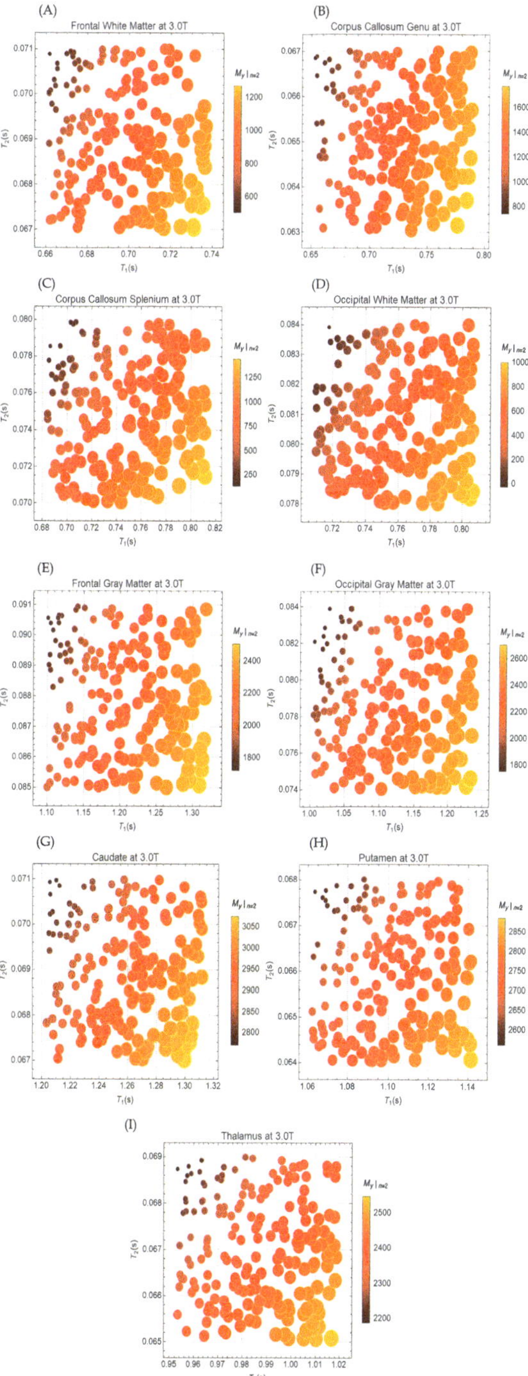

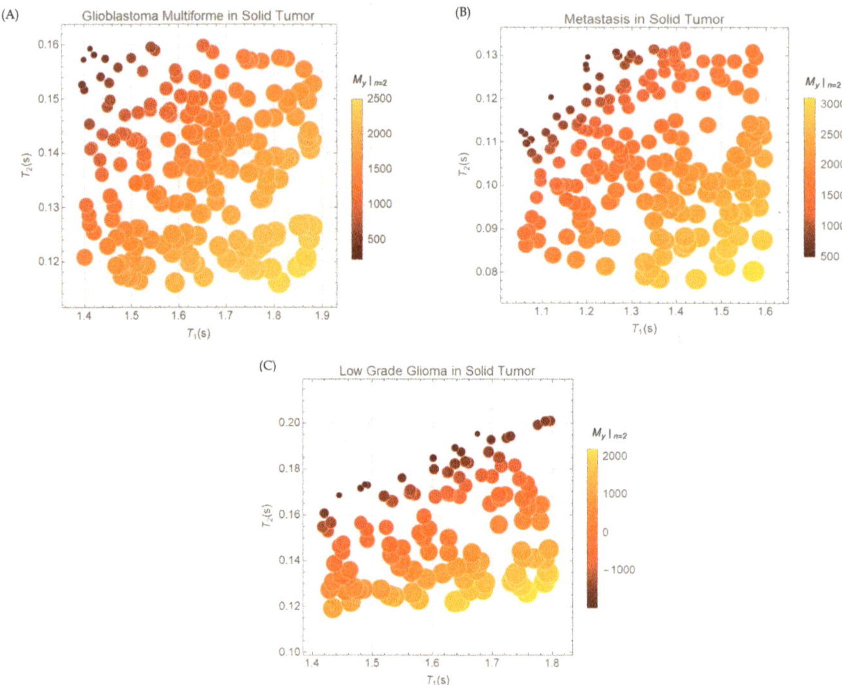

Fig. 3.10 MR transverse magnetization distribution across T_1 and T_2 relaxation times in solid tumor for (**a**) glioblastoma multiforme (**b**) metastasis (**c**) low grade glioma

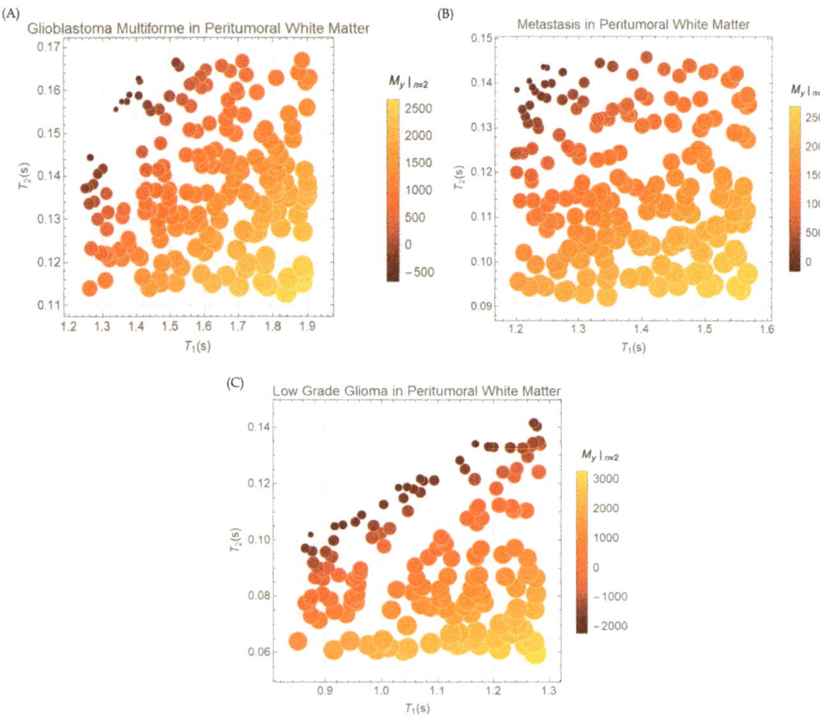

Fig. 3.11 MR transverse magnetization distribution across T_1 and T_2 relaxation times in peritumoral white matter for (**a**) glioblastoma multiforme (**b**) metastasis (**c**) low grade glioma

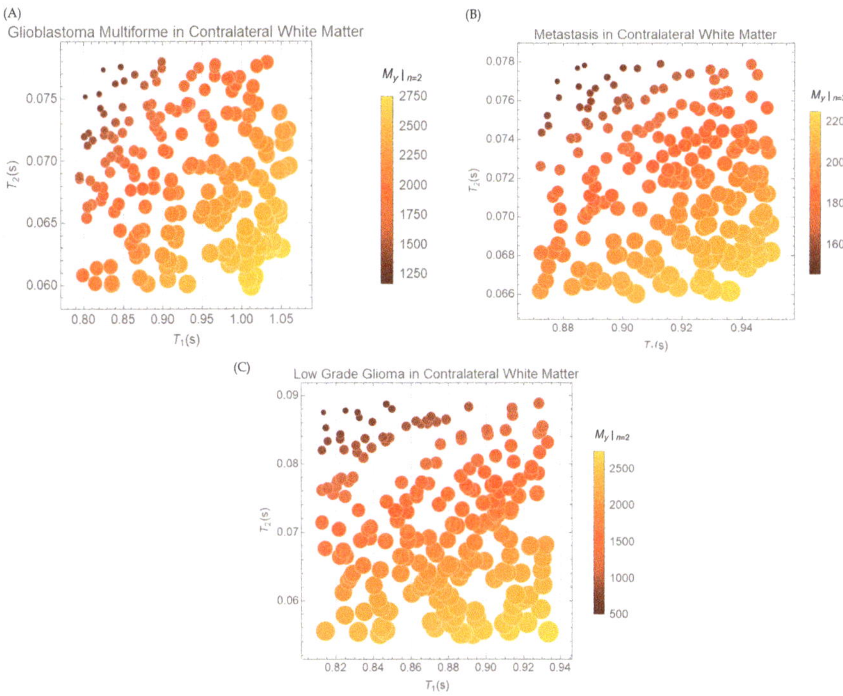

Fig. 3.12 MR transverse magnetization distribution across T_1 and T_2 relaxation times in contralateral white matter for (**a**) glioblastoma multiforme (**b**) metastasis (**c**) low grade glioma

3.6 Discussion

In this study, a method for tissue identification via their molecular signatures or MR fingerprints (T_1 and T_2 relaxation times) has been developed by tuning the applied radiofrequency to the intrinsic resonant frequency under the influence of B_0 magnetic field. Based on Eq. (3.9), analytical solutions in terms of Hermite polynomials were then obtained such that the order of this polynomial is given in Eq. (3.45). Fortunately, this polynomial order provides a means of relating the radiofrequency to appropriate T_1 and T_2 values of any tissue. This leads to the derivation of Eq. (3.61) which is now the basis of RFID of tissues as presented in this study. For example, by setting n = 2, Eqs. (3.57) and (3.53) give the expression for quantifying the identification frequency and NMR transverse magnetization respectively.

We have employed the relaxometry data given in Tables 3.1, 3.2, Eq. (3.80) and Eq. (3.81) to demonstrate RFID for normal brain tissues at 1.5 T. This is shown in Figs. 3.1 and 3.2. Despite the obvious overlap of the relaxation times across the brain tissues as shown in Table 3.1, we observe unique ω_1 clusters with which molecular imaging of the tissues can be easily done. In addition, T_1 provided a better performance in tissue identification (Fig. 3.1) when compared to those given by T_2. Points of overlap can easily be earmarked for further studies when this method is applied for clinical tissue imaging.

Figures 3.3, 3.4, 3.5, 3.6, 3.7 and 3.8 show the angular frequency profiles for different types of brain tumors in various tissues as measured at a uniform field of 3.0 T. In a solid tumor (Fig. 3.3), T_1 has better performance at inter-tumor detection as compared to T_2. In peritumoral white matter (Figs. 3.5 and 3.6), low grade glioma has impressive detection because its angular frequency did not overlap with those of other tumors. In contralateral white matter (Figs. 3.7 and 3.8), glioblastoma multiforme has the best detection performance despite some few overlaps. These profiles show that T_1 should be the first point of call for tissue identification using the method developed in this study.

The NMR transverse magnetization at this level ($n = 2$) as given in Eq. (3.55) has been illustrated as bubble charts in Fig. 3.9. Each tissue has unique distribution of NMR signal across the relaxation times especially the magnitudes of the transverse magnetization. Brain tissues with small volumes appear to have large transverse magnetization in which caudate has the largest followed by the putamen. Figures 3.10, 3.11 and 3.12 have been obtained with the data of Table 3.2. These charts showed interesting MR signal distribution also with comparable values of transverse magnetization. As shown in Figs. 3.10c and 3.11c, the transverse magnetizations of low-grade glioma found in solid tumor and peritumoral white matter are compressed towards lower values of T_2. This behaviour was not found in the case of contralateral white matter and the charts show that the model is not just able to detect a particular type of brain tumour but also can give information on the specific tissues in which these tumours are found. The transverse magnetization values of glioblastomas multiforme are higher in contralateral white matter (Fig. 3.12a) when compared to solid tumor and peritumoral white matter. As shown in Figs. 3.10c and 3.11a, c, low grade glioma solid tumor and peritumoral white matter as well as glioblastomas multiforme in peritumoral white matter have negative values of transverse magnetization at higher values of T_2. All these results show that MR Relaxometry can do more than just measurement of relaxation features; it can be used for real-time imaging in conjunction with appropriate computational models.

As an illustration of this study, we have extracted MR signal values from diffusion weighted images with and without diffusion gradients using ImageJ software. A wolfram Mathematica computer program was developed to reconstruct the extracted data. The data from the original images were reconstructed and presented in Figs. 3.13a, b. The extracted data from these images were then used to compute the resonant angular frequency according to Eq. (3.79). The reconstructed ω_1 map is given in Fig. 3.13c. It can be observed that some brain features that were somehow concealed in Fig. 3.13b were shown in Fig. 3.13c. It is important to note that Fig. 3.13c is not a replacement for Fig. 3.13b but rather a compliment. In Fig. 3.13c, the image background features were brought to the fore and this may indicate suppression of higher contrast features of Fig. 3.13b so as to delineate the features with low contrast. This could prove to be useful in whole body imaging without the need for the use of numerous MR image sequences to achieve fat/water suppression.

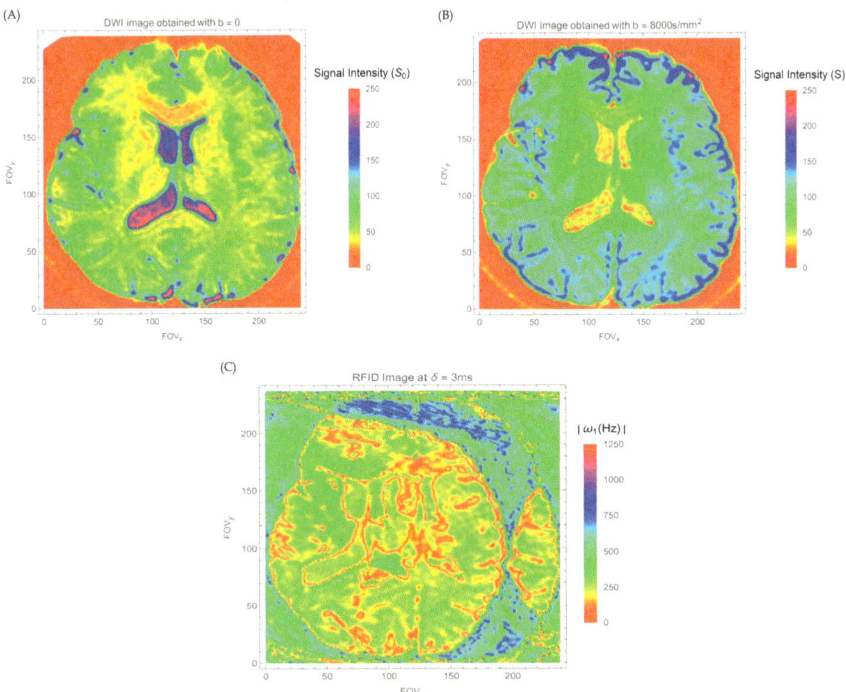

Fig. 3.13 Reconstructed (**a**) diffusion weighted image with b = 0 (**b**) diffusion weighted image with b = 8000 s/mm² (**c**) angular frequency image, using the extracted data

3.7   Conclusion

In this chapter, we have presented a new mathematical formulation for diffusion MRI developed by solving the Bloch NMR flow equation and obtaining the NMR transverse magnetization as a function of Hermite polynomials for detailed studies of processes taking place at molecular level in living tissues. The mathematical representation of the NMR signals is identical to those obtained using the experimentally tenable calculations of the Stejskal-Tanner procedures in Eq. (3.1). Although diffusion sequences based on Eq. (3.1) are very demanding on the scanner, the new generalised diffusion MRI based on Eqs. (3.44) and (3.45) can utilize the available large amounts of data for cloud computing, artificial intelligence, internet of things and machine learning. This is possible because the Stejskal-Tanner formulation for diffusion measurement has hidden MRI information that can be revealed by Eq. (3.68). The relaxation features of tissues can be used to compute diffusion coefficient maps using Eq. (3.71) while the resulting signals can be computed with Eq. (3.55) when n = 2. Interesting information can be revealed at higher values of Hermite polynomials of order, n. The valuable experimental data obtained from Eq. (3.1) over the last decades which is available in the literature can serve as initial experimental data for a comprehensive evaluation of Eq. (3.68) for smarter health care as demonstrated in Fig. 3.13.

The RFID model in Eq. (3.61) uniquely identifies a patient or a tissue through the T_1 and T_2 relaxation times and this can enable the tracking and tracing of patients at hospitals and by other appropriate authorities for patient body flow monitoring and intelligent medication monitoring. The results from this study can be used to develop a portable MR imaging system which can be implemented on portable NMR relaxometers. In situations where this may be difficult, tissue T_1 and T_2 identities can be transmitted wirelessly using radio waves from portable NMR relaxometers to MR imaging systems making use of the methods developed in this chapter for detailed analysis. However, it is important to explore the influence of the parameters n and Ω_n as developed in this chapter on RFID technique and the MR imaging method. This will be the focus of our next investigation.

Chapter 4
Radio-Frequency Identification System for Computational Magnetic Resonance Imaging of Blood Flow at Suction Points

Abstract In this chapter, the solutions of the Bloch NMR flow equations based on the perturbation theory have been used to analyze the asymptotic suction profile and described the physical quantities that affect the boundary layer thickness of blood flow system. We study the flow properties of the time-independent Bloch NMR flow equations which describe the dynamics of blood flow under the influence of radio frequency identification (RFID) system subject to the resonance condition at Larmor frequency. The NMR transverse magnetization and signal are obtained in terms of Hermite polynomials that display flow features at suction points accordingly in term of NMR parameters in an exciting way.

Keywords Bloch NMR flow equation · Perturbation theory · Radio frequency identification (RFID) system · Hermite polynomials · Blood flow at suction points

4.1 Introduction to Blood Flow and MRI

The blood is the transport system by which oxygen and nutrients reach the body's cells, and waste materials are expelled. In addition, blood carries hormones, which control body processes, and antibodies, which fight invading germs. In general, the brain uses more oxygen when it is active, and local blood flow increases to supply even more oxygen than is required. This over-supply leaves its carrier, hemoglobin, more oxygenated and the magnetic distortion by deoxyhemoglobin decreases. This is the basis of the blood oxygen level dependent (BOLD) effect and of most functional magnetic resonance imaging or fMRI, a noninvasive way to assess brain function and neurological diseases (Alsop & Detre, 1998; Wong et al., 1997; Detre & Wang, 2002; Calamante et al., 2002; Jacobs et al., 2005; Dickerson & Sperling, 2005; Thulborn et al., 1982; Turner et al., 1991; Kwong et al., 1992). It has been shown that the BOLD effect correlates directly with electrical communication among nerve cells, that is, synaptic activity.

The heart is the pump that keeps this transport system moving. It is the center of suction in the circulatory system and the blood is kept in a swirl by the action of the

M. O. Dada, B. O. Awojoyogbe, *Computational Molecular Magnetic Resonance Imaging for Neuro-oncology*, Biological and Medical Physics, Biomedical Engineering, https://doi.org/10.1007/978-3-030-76728-0_4

heart for various cardiovascular and myocardial functions (Awojoyogbe et al., 2016a; Moore et al., 1999; Dawson & Lauterbur, 2008; Edvardsen et al., 2006; Baer et al., 1987; Osman et al., 1999, 2000, Parthasarathy, 2006; Atalar & McVeigh, 1994; Young & Axel, 1992; Guttman et al., 1994; Denney, 1997; Acosta-Martínez et al., 2013; Nishimura et al., 2012; Ma et al., 2016; Hoit, 2011). The blood cannot be pushed through the body system until it has first been drawn to the heart by suction from all parts of the body. Similarly, the heart cannot push the blood back through the body again after the blood has been drawn into it by suction unless at the terminals of every current where there is a suction point to attract it. The heart would become stagnant if it were not for the innumerable suction points in the body flow system. There must be suction points at the end of each blood current that permits the blood to be drawn into a space of lesser density. The space as well as the suction is created by the action of the part moved which in turn draws the blood toward and into it. Drawing blood into a part by increased suction not only feeds the cells but tends to increase their number, by causing the growth of new tissue, for cells will continue to extend them as long as space, suction and nourishment are provided. The internal swirl of blood flow goes on as long as the suction points are maintained and the heart itself is not injured by sudden or excessive strain.

In this chapter, the solutions of the Bloch NMR flow equations using the perturbation theory have been used to analyze the asymptotic suction profile and described the physical quantities that affect the boundary layer thickness. We study the flow properties of the time-independent Bloch NMR flow equations which describe the dynamics of blood flow under the influence of radio frequency identification (RFID) system subject to the resonance condition at Larmor frequency. Equation (2.269) can predict analytically and precisely the blood flow behaviour at suction points in the circulatory system by applying the existing knowledge of fluid dynamics. The solutions of Bloch NMR flow equations are based on the perturbation theory as a model; to analyze the asymptotic suction profile of a blood flow system with uniform suction across the flow (Tilton & Cortelezzi, 2015) in terms of NMR parameters and then, to describe the NMR parameter that affects the boundary layer thickness of the flow model. The NMR parameter (a constant) is obtained in term of the RF B_1 magnetic field, T_1 and T_2 relaxation times. The suction points display flow features which are analyzed in terms of the NMR parameter. It is shown that the perturbation solutions of the Bloch flow model equation are equivalent to the exact solutions. We obtained the NMR transverse magnetization and signal in terms of Hermite polynomials that display flow features at suction points accordingly in term of NMR parameters in an exciting way.

4.2 Analytical Method

In this study, we have solved the time-independent Bloch NMR flow equation by transforming the transverse magnetization M_y into the radio frequency (RF) and wave vector domains with a transformation constant 'a^2' which must be positive. We study the flow properties of the differential equations obtained from Bloch NMR flow equations, which describe the dynamics of fluid flow under the influence of RF magnetic field (Osman et al., 1999; Parthasarathy, 2006; Atalar & McVeigh, 1994; Young & Axel, 1992; Guttman et al., 1994; Denney, 1997; Awojoyogbe & Dada, 2011a, b; Awojoyogbe & Dada, 2018; Awojoyogbe et al., 2010, 2011a, b; Dada et al., 2009, 2010; Awojoyogbe & Boubaker, 2009; Awojoyogbe, 2002, 2003, 2004, 2007):

$$\frac{d^2 M_y}{dx^2} + \frac{T_0}{v}\frac{dM_y}{dx} + \frac{1}{v^2}\left(\gamma^2 B_1^2(x) + \frac{1}{T_1 T_2}\right)M_y = \frac{M_0 \gamma B_1(x)}{v^2 T_1} \tag{4.1}$$

where $T_0 = \frac{1}{T_1} + \frac{1}{T_2}$, subject to the following conditions:

1. $M_o \neq M_z$, a situation which holds well in general and in particular when the RF $B_1(x)$ field is strong say of the order of 1.0 gauss or more.
2. Before entering signal detector coil, fluid particles have magnetization $M_x = 0$, $M_y = 0$.
3. If $B_1(x)$ is large; $B_1(x) >> 1$ gauss or more so that M_y of the fluid bolus changes appreciably from M_0.
4. for steady flow, $\frac{\partial M_y}{\partial t} = 0$

γ denotes the gyromagnetic ratio of fluid spins; $\omega/2\pi$ is the RF excitation frequency; f_0/γ is the off-resonance field in the rotating frame of reference. T_1 and T_2 are the spin-lattice and spin-spin relaxation times respectively, the reciprocals of T_1 and T_2 are defined as relaxation rates. RF B_1 is the spatially varying magnetic field and v is the fluid flow velocity. However, when the constant RF $B_1(x) = B_1$ field is applied, M_y has a maximum value when RF B_1 is maximum and $M_o \approx 0$. At the point when maximum NMR signal is received (maximum values of M_y and B_1 respectively), Eq. (4.1) becomes

$$\frac{d^2 M_y}{dx^2} + \frac{T_0}{v}\frac{dM_y}{dx} + \frac{1}{v^2}\left(\frac{1}{T_1 T_2} + \gamma^2 B_1^2\right)M_y = 0 \tag{4.2}$$

where

$$\frac{\gamma^2 B_1^2}{v^2} >> \frac{1}{T_1 T_2} \tag{4.3}$$

It may be very important to determine the parameter

$$F = \frac{\gamma^2 B_1^2}{v^2} M_y \tag{4.4}$$

When the magnitudes of B_1 and M_y have maximum values, the constant fluid velocity v, can be measured. We can write Eq. (4.2) as

$$F(\omega_1) = \alpha(\omega_1)M_y = \vec{k}^2 M_y \tag{4.5}$$

where

$$\gamma B_1(x) = \omega_1(x) \tag{4.6}$$

Using Eq. (4.5), the following transformations holds

$$\frac{dM_y}{dx} = \frac{dM_y}{d\omega_1}\frac{d\omega_1}{dx} \tag{4.7}$$

$$\frac{d^2 M_y}{dx^2} = \left(\frac{d\omega_1}{dx}\right)^2 \frac{d^2 M_y}{d\omega_1{}^2} + \frac{dM_y}{d\omega_1}\frac{d^2\omega_1}{dx^2} \tag{4.8}$$

Equation (4.1) therefore becomes

$$\frac{d^2 M_y}{d\omega_1{}^2} + A\frac{dM_y}{d\omega_1} + a^2(1 + \alpha(\omega_1))M_y = 0 \tag{4.9}$$

or

$$\frac{d^2 M_y}{d\omega_1{}^2} + A\frac{dM_y}{d\omega_1} + a^2\left(1 + \frac{\omega^2}{v^2}\right)M_y = 0 \tag{4.10}$$

where

$$A = \mu\left[\frac{\frac{d^2\omega_1}{dx^2} + \frac{T_0}{v}\frac{d\omega_1}{dx}}{\left(\frac{d\omega_1}{dx}\right)^2}\right] \tag{4.11}$$

and

$$a^2 = \frac{\frac{1}{v^2}\left(\frac{1}{T_1 T_2}\right)}{\left(\frac{d\omega_1}{dx}\right)^2} \tag{4.12}$$

$$\frac{d\omega_1}{dx} = \sqrt{\frac{\frac{1}{v^2}\left(\frac{1}{T_1 T_2}\right)}{a^2}} \tag{4.13}$$

$$\frac{d^2\omega_1}{dx^2} = 0 \tag{4.14}$$

$$\omega_1 = \sqrt{\frac{\frac{1}{v^2}\left(\frac{1}{T_1 T_2}\right)}{a^2}} x + c \tag{4.15}$$

From Eq. (4.15), the parameter A in Eq. (4.11) becomes:

$$A = \mu \frac{T_0 \sqrt{T_g}}{a}$$

Based on fundamental mathematical procedures, parameter A must be a constant and a^2 a positive constant respectively. Parameter μ is less or equal to unity ($\mu \leq 1$). In order to obtain detail information in terms of the Hermite polynomials from Eq. (4.13), we need to make $\mu = 0$ so that Eq. (4.13) becomes:

$$\frac{d^2 M_y}{d\omega_1^2} + a^2 (1 + \alpha(\omega_1)) M_y = 0 \tag{4.16}$$

$$\frac{d^2 M_y}{d\omega_1^2} + \left(a^2 + a^2 \frac{\gamma^2 B_1^2}{v^2} \right) M_y = 0$$

$$\frac{d^2 M_y}{d\omega_1^2} + \left(\varepsilon - \beta^2 \omega_1^2 \right) M_y = 0 \tag{4.17}$$

where

$$\begin{pmatrix} \varepsilon = a^2 \\ \beta^2 = \dfrac{a^2}{v_m^2} \\ \beta = \dfrac{a}{v_m} \end{pmatrix} \tag{4.18}$$

We consider the solution of Eq. (4.16) for an arbitrary complex initial velocity (such that $v = iv_m$) under the following transformation variable in Eq. (4.17):

$$k = \sqrt{\beta} \omega_1 \tag{4.19}$$

$$dk = \sqrt{\beta} d\omega_1 \tag{4.20}$$

The inverse of Eq. (4.20) gives:

$$\frac{d}{dk} = \frac{d}{\sqrt{\beta} d\omega_1}$$

$$\frac{d^2}{dk^2} = \frac{1}{\beta}\frac{d^2}{d\omega_1^2} \tag{4.21}$$

Equation (4.18) becomes

$$\frac{d^2 M_y}{dk^2} + \left(\frac{\varepsilon}{\beta} + k^2\right) M_y = 0 \tag{4.22}$$

An asymptotic solution of Eq. (4.22) can be obtained for a very large value of k. Thus, when $k >> \frac{\varepsilon}{\beta}$, Eq. (4.22), reduces to,

$$\frac{d^2 M_y}{dk^2} = +k^2 M_y = 0 \tag{4.23}$$

A general solution to Eq. (4.23) is given in the form:

$$M_y = Ae^{-\frac{k^2}{2}} + Be^{\frac{k^2}{2}} \tag{4.24}$$

$$\frac{dM_y}{dk^2} = A(-k)^2 e^{-\frac{k^2}{2}} - Ae^{-\frac{k^2}{2}} + Bk^2 e^{\frac{k^2}{2}} + Be^{\frac{k^2}{2}}$$

$$\frac{dM_y}{dk^2} = A\left(k^2 - 1\right)e^{-\frac{k^2}{2}} + B\left(k^2 + 1\right)e^{\frac{k^2}{2}} \tag{4.25}$$

In Eqs. (4.24) and (4.25), as $k \rightarrow \infty$, $B \rightarrow 0$. Thus, Eq. (4.25) becomes 0, while Eq. (4.24) becomes

$$M_y = \exp\left(-\frac{k^2}{2}\right) \tag{4.26}$$

This suggests that the general solution for Eq. (4.22) contains the factor $\exp^{-\frac{k^2}{2}}$ and the form of solution may be written as:

$$M_y = F(k)\exp^{-\frac{k^2}{2}} \tag{4.27}$$

The function F(k) satisfies the Hermite's differential equation:

$$\frac{d^2 F}{dk^2} - 2k\frac{dF}{dk} + \left(\frac{\varepsilon}{\beta} - 1\right)F = 0 \tag{4.28}$$

The solution to Eq. (4.28) is obtained by assuming that the function F can be expanded as power series in k, thus:

$$F = a_0 + a_1 k + a_2 k^2 + a_3 k^3 + a_4 k^4 + \cdots \tag{4.29}$$

Equations (4.29) and (4.28) lead to a formula for determination of the coefficients,

$$a_{l+2} = \frac{-\left(\frac{\alpha}{\beta} - 1 - 2k\right)}{(l+1)(l+2)} a_l \tag{4.30}$$

where l is an integer. This is known as recursion formula for Eq. (4.28). It permits the calculation of coefficient a_{l+2} of the term k^{l+2} in term of the coefficient a_l of k^l; that is, we can obtain a_2, a_4, a_6, in terms of a_0 (even numbers) and a_3, a_5, a_7, etc., in terms of a_1 (odd numbers). It can be shown that if the power series $F(k)$ is infinitely long, the higher order terms will dominate and the function $F(k)$ will behave like the Taylor series expansion. It is observed that the series in Eq. (4.29) breaks off after a finite number of terms for a certain value of l (say n) when,

$$\frac{\varepsilon}{\beta} = \frac{2n+1}{l} = a v_m \tag{4.31}$$

$$v_m = (2n+1)\frac{1}{al} \tag{4.32}$$

so that the numerator in Eq. (4.30) vanishes; n may be 1, 2, 3, $\ldots$ The resulting series with a finite number of terms is a polynomial, called the Hermite polynomial, $H(k)$. The complete solution of Eq. (4.22) is then,

$$M_y(k) = N \cdot H(k) \cdot \exp\left(-\frac{k^2}{2}\right) \tag{4.33}$$

where $H(k)$ is usually called Hermite polynomial; N is the normalization factor. The function in Eq. (4.33) is finite and single valued for any value of k, being controlled by the factor $\left(-\frac{k^2}{2}\right)$ and decreases continuously as k, increases; hence, it is an acceptable function representing the NMR/MRI transverse magnetization. In general, the Hermite polynomials are defined as:

$$H_n(k) = (-1)^n \exp\left(k^2\right) \frac{d^n \exp\left(-k^2\right)}{dk^n} \tag{4.34}$$

where n is the degree of the polynomial that corresponds to the vibrational number of harmonic oscillators, that is, the highest power of k in it.

4.3   Analysis of Blood Flow at Suction Points

In this section, we have analysed the asymptotic suction profile and described the physical quantities that affect the boundary layer thickness based on the solutions of the Bloch NMR flow equations using the perturbation theory. We study the flow properties of Eq. (4.5) which describes the dynamics of fluid flow under the influence of RF magnetic field subject to the resonance condition at Larmor frequency. From Eq. (4.8) when $\mu = 0$, we can write,

$$\frac{d^2\omega_1}{dx^2} + \frac{T_0}{v}\frac{d\omega_1}{dx} \tag{4.35}$$

In Eq. (4.35), ω_1 is defined as the asymptotic suction profile or the suction RF $B_1(x)$ field. The purpose here is to solve Eq. (4.35) by perturbation technique and apply the solution for qualitative analysis of flow at suction points. We also describe the properties of the asymptotic suction profile, and find the NMR physical quantities that affect the boundary layer thickness. A mathematical flow model with uniform suction across the flow is considered as shown in Fig. 4.1. The upper plate is moving parallel to itself at speed v_o and the lower plate is stationary. The distance between the plates is h, and there is a uniform downward suction velocity v_i, with the fluid coming in through the upper plate and going out through the bottom. For all variables,

$$\frac{\partial}{\partial x} = 0 \tag{4.36}$$

and the non-dimensional equation is

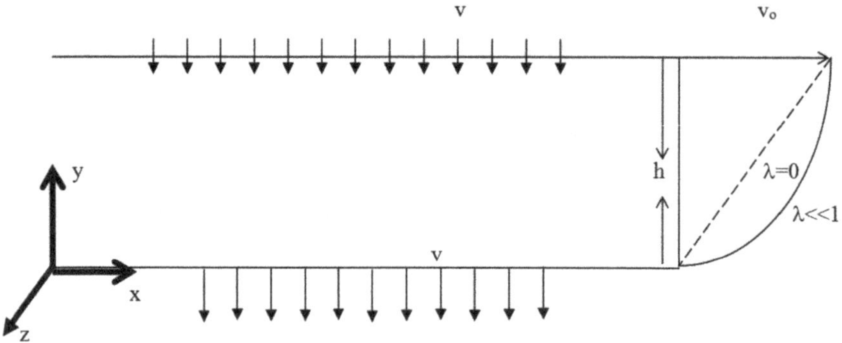

Fig. 4.1   The asymptotic profile in uniform suction of a fluid flow model for $\lambda = 0$ and $\lambda \ll 1$

$$\frac{\partial v_i}{\partial y} = 0 \tag{4.37}$$

$$\frac{d^2\omega_1}{\partial y^2} + \frac{1}{v_1}\frac{d\omega_1}{dy} = 0 \tag{4.38}$$

subject to the conditions:

$$v_{i1}(0) = v_1(1) = -u \tag{4.39}$$

$$\omega_1(0) = 0 \tag{4.40}$$

$$\omega_1(1) = 1 \tag{4.41}$$

where

$$\sigma = \frac{v_0 h}{v_1}, u = \frac{u'}{v_0} \tag{4.42}$$

We denote dimensional variables by primed and non-dimensional variables to obtain the following:

$$y = \frac{y'}{h}, \quad M_y = \frac{M'_y}{v_0}, \quad v_1 = \frac{v'_1}{v_0} \tag{4.43}$$

The velocity component normal to the wall $v_1(y)$ is spatially constant and equal to the suction velocity $-u$, that is

$$v_1(y) = -u \tag{4.43}$$

The velocity component that is parallel to the wall varies with the distance from the wall as shown in Fig. 4.1. Equation (4.39) shows that the lateral flow is independent of y. Based on Eq. (4.40) and Fig. 4.1, we can write Eq. (4.37) as

$$\frac{d^2\omega_1}{dy^2} + \lambda\frac{d\omega_1}{dy} = 0 \tag{4.44}$$

Equation (4.44) is the equation of motion for the flow system such that:

$$\lambda = u\sigma = \frac{u'h}{v_1} = \frac{T_0}{v} = \frac{alT_0}{(2n+1)} \tag{4.45}$$

In Eq. (4.45), λ is the suction parameter and σ is the Reynolds number. We assume that the suction velocity is small, so that $\lambda \ll 1$. Although an exact solution can be obtained for Eq. (4.44), an exact solution may not exist in problems that are more complicated. We have therefore chosen to apply the perturbation solution in integral powers of λ, of the form:

$$\omega_1(y) = \omega_{10}(y) + \lambda\omega_{11}(y) + \lambda^2\omega_{12}(y) + O(\lambda^3) \tag{4.46}$$

By the substitution of Eq. (4.46) into Eqs. (4.44), (4.39) and (4.40), we obtain

$$\frac{d^2\omega_{10}}{dy^2} + \lambda\left(\frac{d\omega_{10}}{dy} + \frac{d^2\omega_{11}}{dy^2}\right) + \lambda^2\left(\frac{d\omega_{11}}{dy} + \frac{d^2\omega_{12}}{dy^2}\right) + O(\lambda^3) = 0 \tag{4.47}$$

subject to the conditions:

$$\omega_{10}(0) + \lambda\omega_{11}(0) + \lambda^2\omega_{12}(0) + O(\lambda^3) = 0 \tag{4.48}$$

$$\omega_{10}(1) + \lambda\omega_{11}(1) + \lambda^2\omega_{12}(1) + O(\lambda^3) = 0 \tag{4.49}$$

By equating terms with like powers of λ in Eqs. (4.47)–(4.49), the complete solution of Eq. (4.44) up to order λ^2 is obtained as:

$$\omega_1(y) = y + \frac{\lambda}{2}(y(1-y)) + O(\lambda^2) \tag{4.50}$$

Since the suction parameter in Eq. (4.45) is very small ($\lambda \ll 1$), Eq. (4.50) becomes:

$$\omega_1(y) = y + \frac{\lambda}{2}(y(1-y)) \tag{4.51}$$

The NMR transverse magnetization can be completely derived analytically in terms of the Hermite polynomials and suction parameter from Eqs. (4.33), (4.34) and (4.51) as:

$$M_y(y) = N(-1)^n \exp\left(-\frac{\beta(\omega_1(y))^2}{2}\right) \exp\beta(\omega_1(y))^2 \frac{d^n \exp\left(-\beta(\omega_1(y))^2\right)}{dy^n} \tag{4.52}$$

$$M_y(y) = N(-1)^n \exp\frac{a}{v_i}(\omega_1(y))^2 \frac{d^n \exp\left(-\frac{a}{v_i}(\omega_1(y))^2\right)}{dy^n} \exp\left(-\frac{\frac{a}{v}(\omega_1(y))^2}{2}\right) \tag{4.53}$$

$$M_y(y) = N(-1)^n \exp \frac{a}{2n+1} (\omega_1(y))^2 \frac{d^n \exp\left(-\frac{a}{2n+1}(\omega_1(y))^2\right)}{dy^n} \exp\left(-\frac{\frac{a^2}{2n+1}(\omega_1(y))^2}{2}\right)$$

$$(4.54)$$

4.4   Computational Analysis of Blood Flow at Suction Points

Using the expression of Eq. (4.54), we have the profiles displayed in Figs. 4.2, 4.3, 4.4, 4.5, 4.6, 4.7, 4.8, 4.9, 4.10, 4.11 and 4.12. It is observed from Eq. (4.9) that a^2 is always a positive constant as long as the RF gradient $\frac{d\omega_1}{dx}$ is applied in the complex domain so that a^2 remains positive and real. From Figs. 4.3, 4.4, 4.5, 4.6, 4.7, 4.8, 4.9, 4.10 and 4.11, the value of a is very important to the entire flow process. Parameter a is given in seconds and controls the system. Parameter n is a discrete integer of the flow velocity. The NMR signal decreases asymptotically as a function of the Hermite polynomials along the distance h, between the plates when n $= 0$ and increases when n > 0. The suction parameter λ is inversely proportional to the square root of the positive parameter a and both are functions of T_1 and T_2 relaxation rates. Equation (4.32) shows that for a flow velocity, only discrete values (integer-plus-half

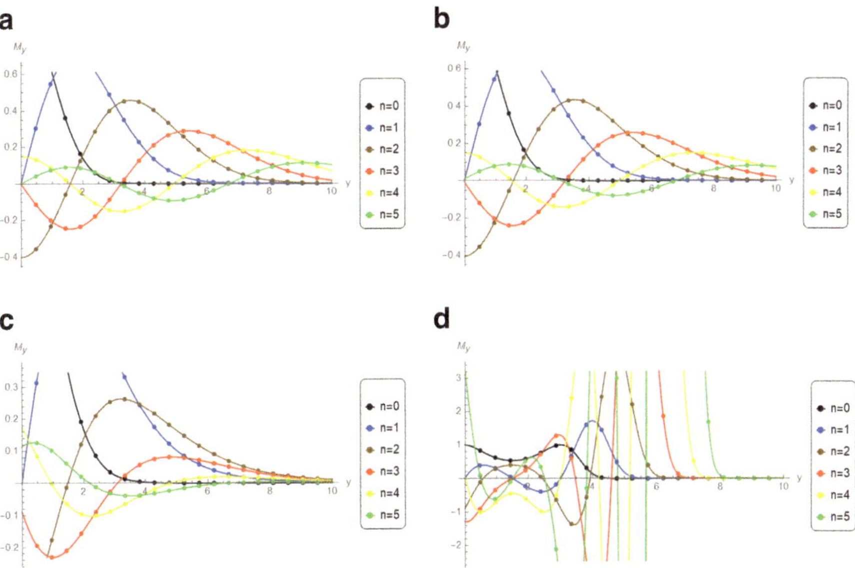

Fig. 4.2 Plots of NMR transverse magnetization within the plates as shown in Fig. 4.1 for different capture velocities (depicted by n) and parameter $a = 1$ at suction parameter (**a**) $\lambda = 0.001$ (**b**) $\lambda = 0.01$ (**c**) $\lambda = 0.1$ (**d**) $\lambda = 1$

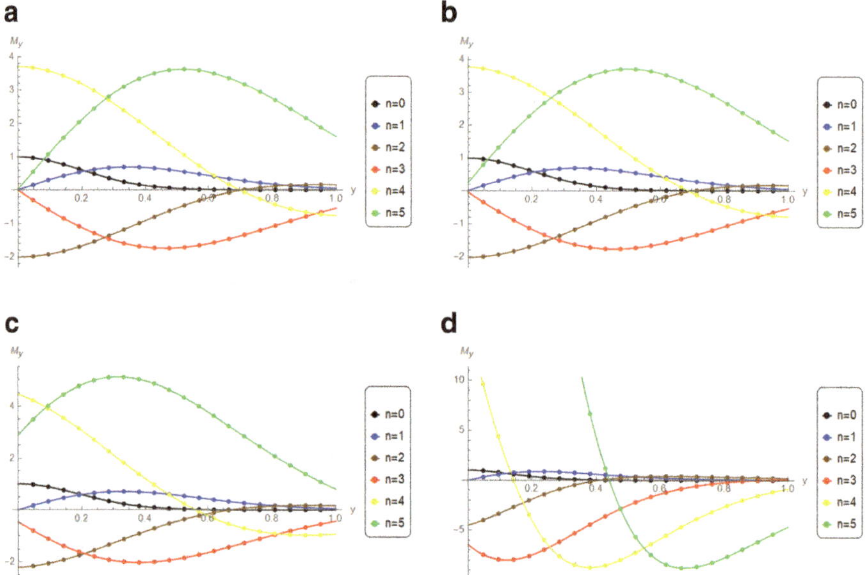

Fig. 4.3 Plots of NMR transverse magnetization within the plates as shown in Fig. 4.1 for different capture velocities (depicted by n) and parameter $a = 5$ at suction parameter (**a**) $\lambda = 0.001$ (**b**) $\lambda = 0.01$ (**c**) $\lambda = 0.1$ (**d**) $\lambda = 1$

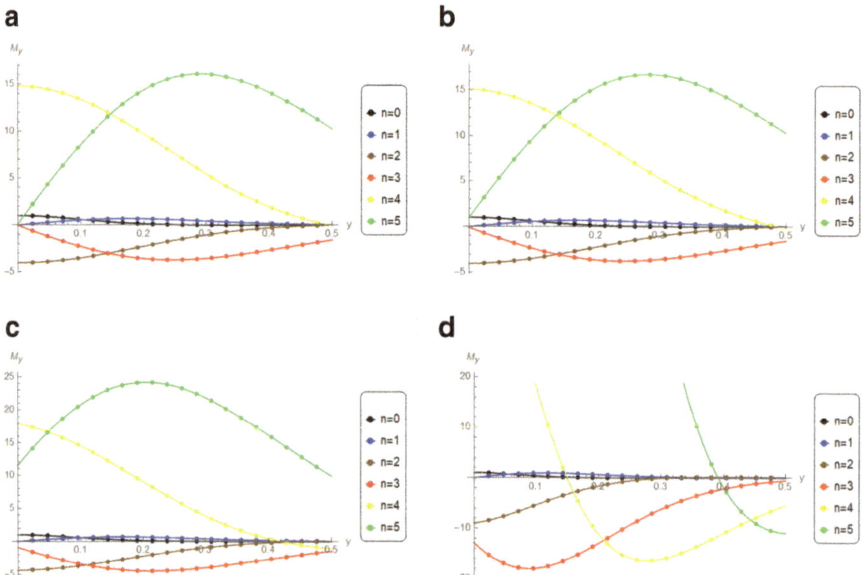

Fig. 4.4 Plots of NMR transverse magnetization within the plates as shown in Fig. 4.1 for different capture velocities (depicted by n) and parameter $a = 10$ at suction parameter (**a**) $\lambda = 0.001$ (**b**) $\lambda = 0.01$ (**c**) $\lambda = 0.1$ (**d**) $\lambda = 1$

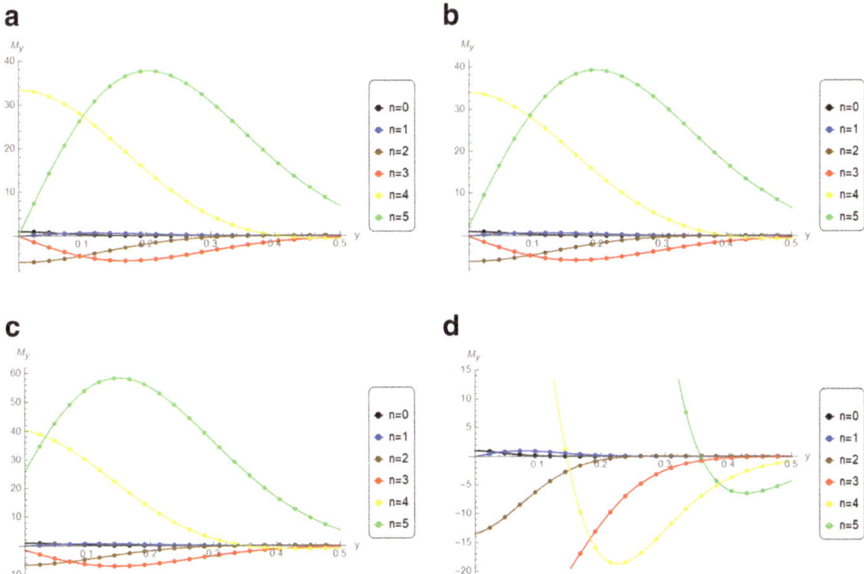

Fig. 4.5 Plots of NMR transverse magnetization within the plates as shown in Fig. 4.1 for different capture velocities (depicted by n) and parameter $a = 15$ at suction parameter (**a**) $\lambda = 0.001$ (**b**) $\lambda = 0.01$ (**c**) $\lambda = 0.1$ (**d**) $\lambda = 1$

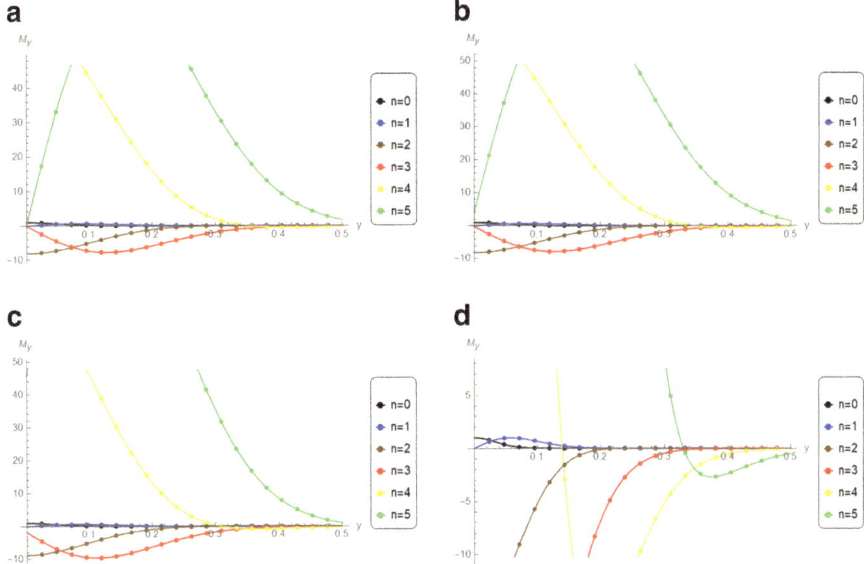

Fig. 4.6 Plots of NMR transverse magnetization within the plates as shown in Fig. 4.1 for different capture velocities (depicted by n) and parameter $a = 20$ at suction parameter (**a**) $\lambda = 0.001$ (**b**) $\lambda = 0.01$ (**c**) $\lambda = 0.1$ (**d**) $\lambda = 1$

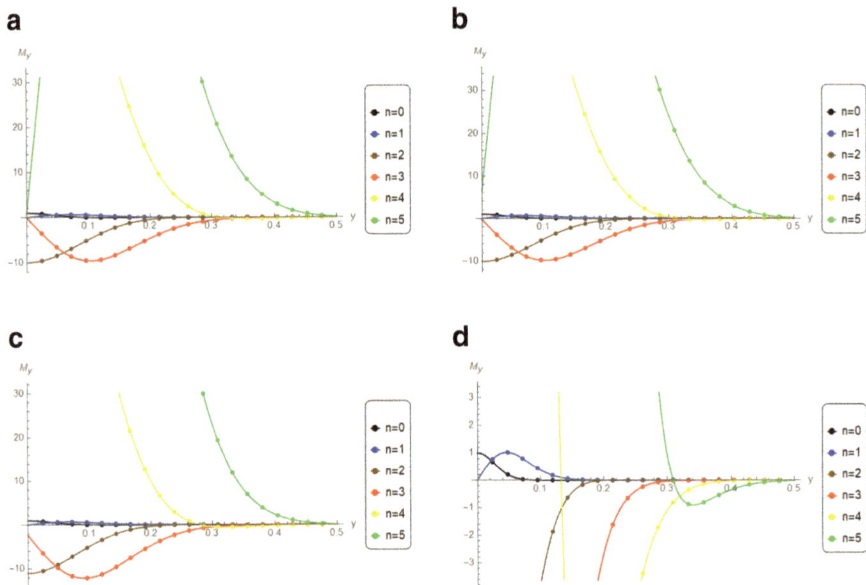

Fig. 4.7 Plots of NMR transverse magnetization within the plates as shown in Fig. 4.1 for different capture velocities (depicted by n) and parameter $a = 25$ at suction parameter (**a**) $\lambda = 0.001$ (**b**) $\lambda = 0.01$ (**c**) $\lambda = 0.1$ (**d**) $\lambda = 1$

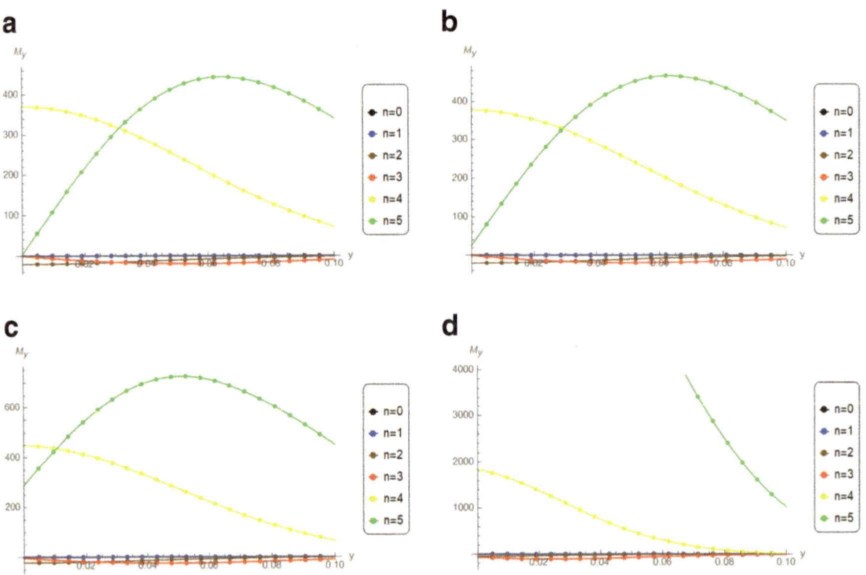

Fig. 4.8 Plots of NMR transverse magnetization within the plates as shown in Fig. 4.1 for different capture velocities (depicted as n) and parameter $a = 50$ at suction parameter (**a**) $\lambda = 0.001$ (**b**) $\lambda = 0.01$ (**c**) $\lambda = 0.1$ (**d**) $\lambda = 1$

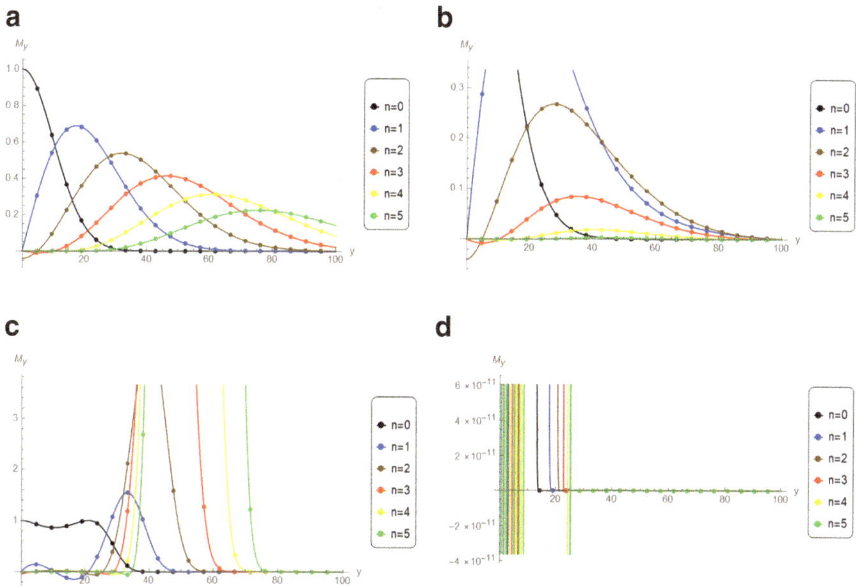

Fig. 4.9 Plots of NMR transverse magnetization within the plates as shown in Fig. 4.1 for different capture velocities (depicted as n) and parameter $a = 0.1$ at suction parameter (**a**) $\lambda = 0.001$ (**b**) $\lambda = 0.01$ (**c**) $\lambda = 0.1$ (**d**) $\lambda = 1$

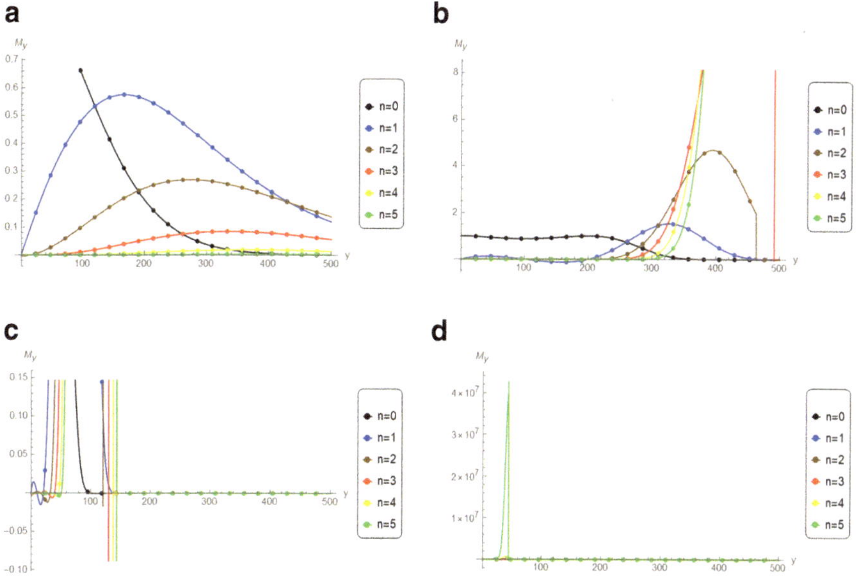

Fig. 4.10 Plots of NMR transverse magnetization within the plates as shown in Fig. 4.1 for difference capture velocities (depicted as n) and parameter $a = 0.01$ at suction parameter (**a**) $\lambda = 0.001$ (**b**) $\lambda = 0.01$ (**c**) $\lambda = 0.1$ (**d**) $\lambda = 1$

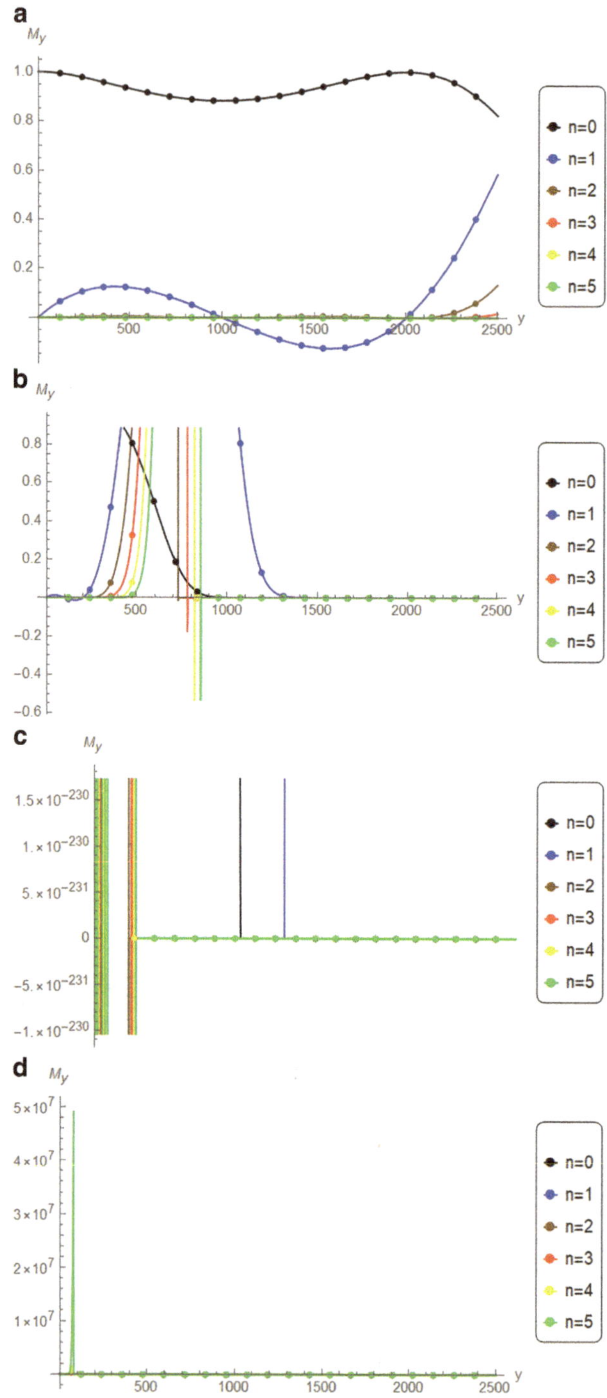

Fig. 4.11 Plots of NMR transverse magnetization within the plates as shown in Fig. 4.1 for different capture velocities (depicted as n) and parameter $a = 0.001$ at suction parameter (**a**) $\lambda = 0.001$ (**b**) $\lambda = 0.01$ (**c**) $\lambda = 0.1$ (**d**) $\lambda = 1$

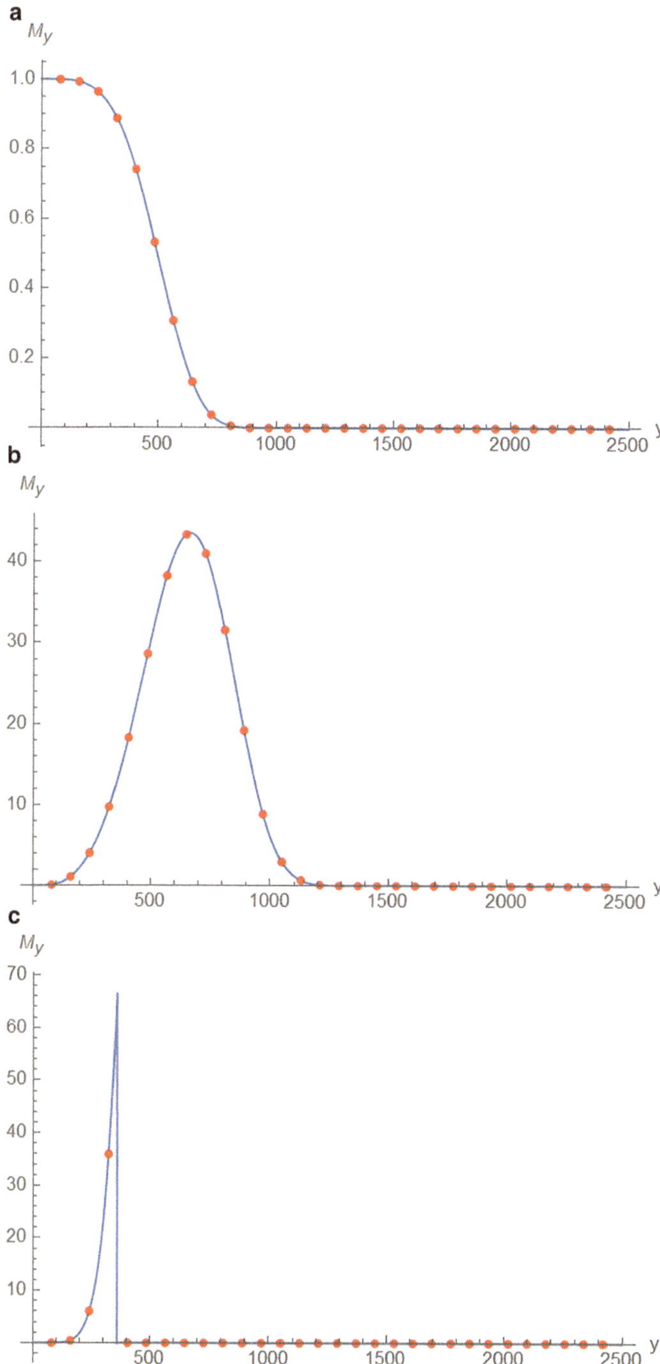

Fig. 4.12 Plots of NMR transverse magnetization within the plates as shown in Fig. 4.1 at suction parameter $\lambda = 0.1$ with parameter $a = 0.0001$ for difference capture velocities (**a**) n = 0 (**b**) n = 1 (**c**) n = 2 (**d**) n = 3 (**e**) n = 4 (**f**) n = 5

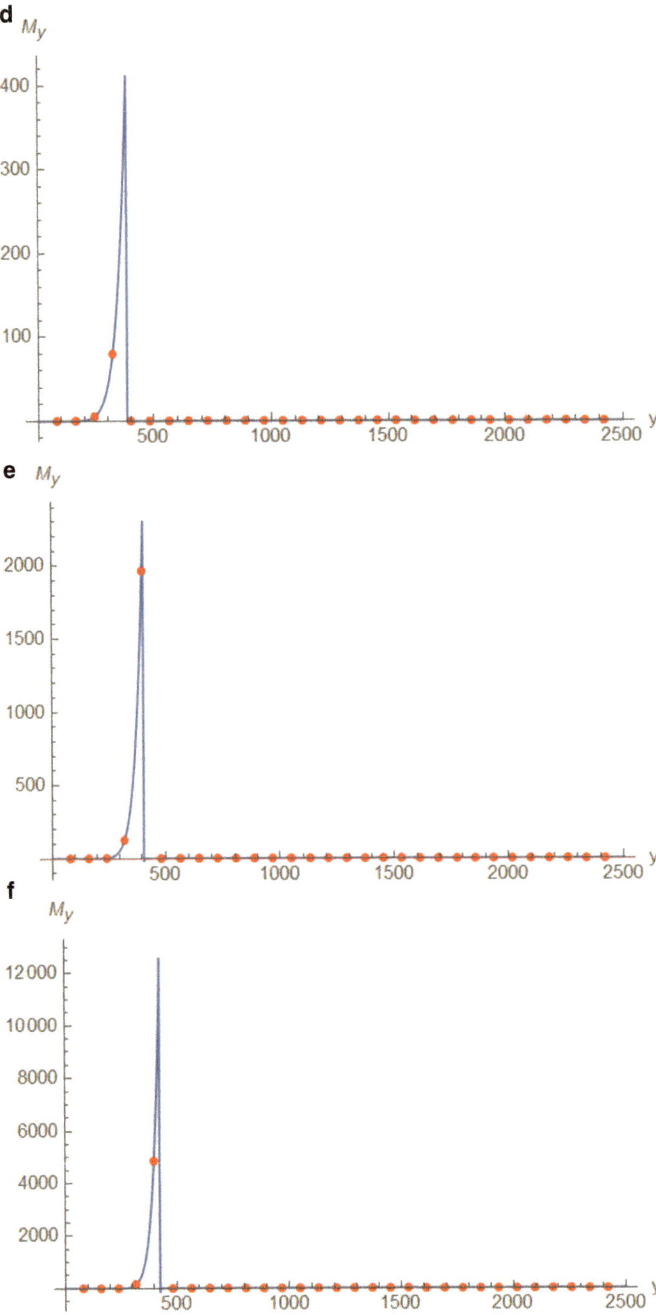

Fig. 4.12 (continued)

multiples of 1/a) are possible. This is a general feature of systems with confined particles. These discrete flow velocities are equally spaced. The lowest achievable value (the flow velocity when n = 0 is called the ground state) is not equal to the minimum, but $\frac{1}{2a}$. This value is not arbitrary though the initial flow velocity is arbitrary. The ground state flow velocity is 1/2a above that of the initial value. The computer analyses of Eq. (4.54) are presented in Figs. 4.4, 4.5, 4.6, 4.7, 4.8, 4.9, 4.10 and 4.11. It is noticed that the suction parameter λ has significant influence on the NMR signal. Fluid flow can thus be analysed at discrete velocities at specified suction points and suction parameter a.

Figures 4.2, 4.3, 4.4, 4.5, 4.6, 4.7, 4.8, 4.9, 4.10 and 4.11 show the behaviour of NMR transverse magnetization within the plates for different values of flow velocity and suction parameter a, for $a > 1$ and $a < 1$ respectively. The NMR signal for each value of a, λ and n is unique and displays features of specific mathematical functions. For example, the NMR transverse magnetization for different flow velocities with parameter $a = 0.1$ at suction parameter $\lambda = 0.001$ shown in Fig. 4.8a looks to be more of Bessel functions and can be analysed accordingly. This makes the analysis in this study unique, exciting and easily adaptable to solving specific real-life problems.

In Fig. 4.12, the NMR signal is normally distributed when n is an integer greater than zero with different peaks. The real power of Figs. 4.2, 4.3, 4.4, 4.5, 4.6, 4.7, 4.8, 4.9, 4.10, 4.11 and 4.12 is derived from the fact that they can be used to formulate the same basic mathematics to (approximately) describe many flow systems. This is the signature of this study, and any time there is a flow system the velocity of which has a minimum at some position (position or suction point), the results of this study can be used. The vibration of atoms in a solid, the motion of two atoms bound together in a molecule, the motion of a cloud of ultra-cold atoms held in a magnetic trap, or the flexing of a microscopic membrane will all show the same characteristic behaviour: a characteristic resonance frequency, a total velocity that can increase or decrease only by integer multiples of that frequency, and a lowest state showing this very non-classical zero-point motion. The density plots of NMR transverse magnetization against h (within the plates) and capture parameter a at different suction parameters for discrete values n = 0, 1, 2, 3, 4, 5 are shown in Figs. 4.13, 4.14, 4.15, 4.16, 4.17, 4.18, 4.19 and 4.20.

Based on Eq. (4.32), the minimum capture velocity is $v = \frac{1}{al}$ as captured in Figs. 4.2a, 4.3a, 4.4a, 4.5a, 4.6a, 4.7a, 4.8a, 4.9a, 4.10a, 4.11a, 4.12a, 4.13a, 4.14a, 4.15a, 4.16a, 4.17a, 4.18a, 4.19a and 4.20a). From the density plots shown in Figs. 4.13, 4.14, 4.15, 4.16, 4.17, 4.18, 4.19 and 4.20, it can be observed that the contrast is better at higher flow velocity values. This is because the particles with higher capture velocity are more classical than those with lower capture velocity values. The desired contrast for any experimental set up can be achieved by careful design of the flow system based on the values of parameters a, λ and n displayed in Figs. 4.13, 4.14, 4.15, 4.16, 4.17, 4.18, 4.19 and 4.20.

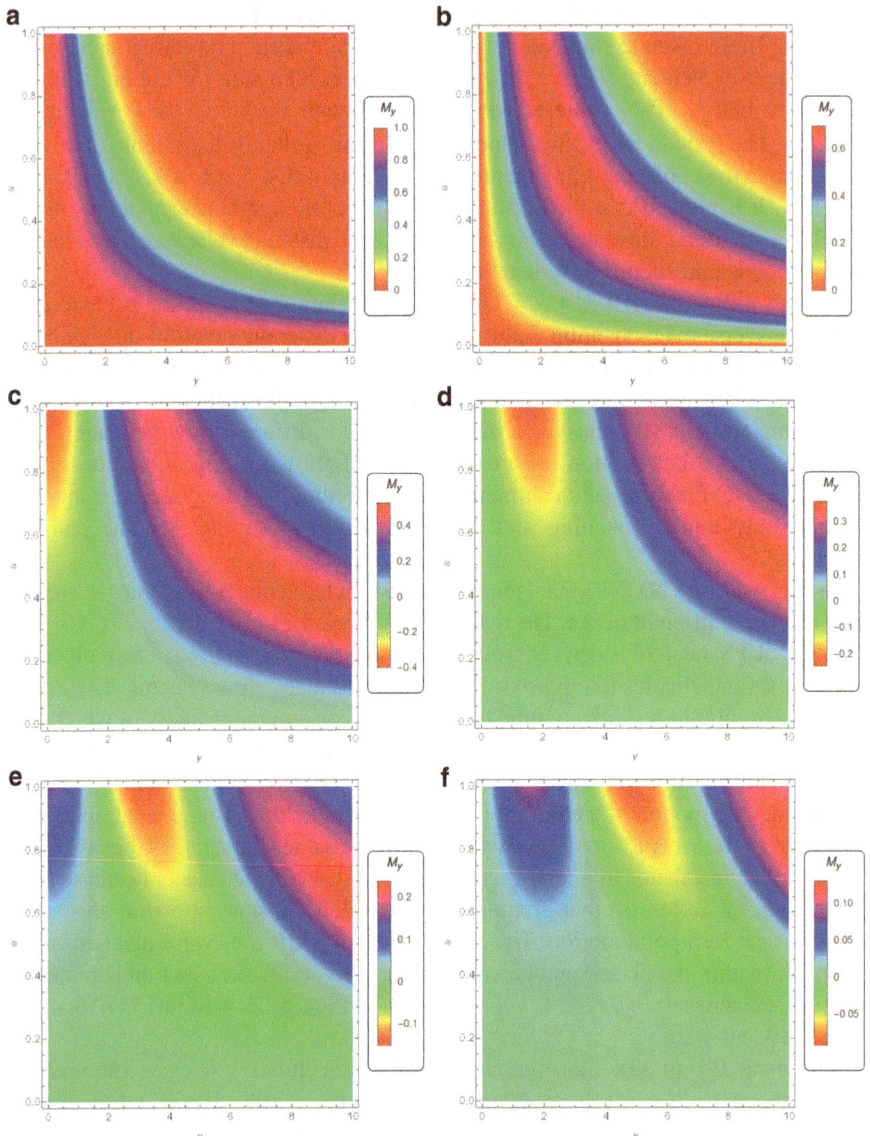

Fig. 4.13 Density plots of NMR transverse magnetization against h = y (within the plates) and capture parameter a at suction parameter $\lambda = 0.001$ for velocity parameters (**a**) n = 0 (**b**) n = 1 (**c**) n = 2 (**d**) n = 3 (**e**) n = 4 (**f**) n = 5

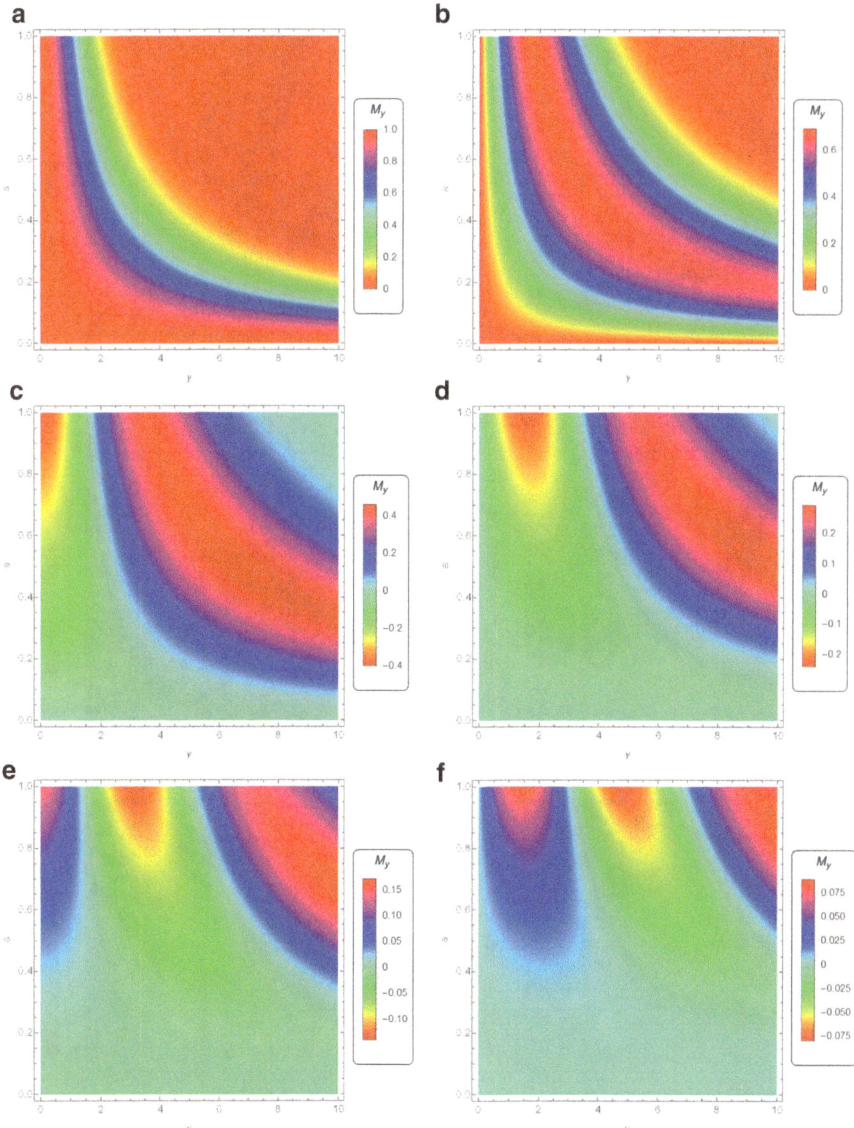

Fig. 4.14 Density plots of NMR transverse magnetization against h = y (within the plates) and capture parameter a at suction parameter $\lambda = 0.01$ for velocity parameters (**a**) n = 0 (**b**) n = 1 (**c**) n = 2 (**d**) n = 3 (**e**) n = 4 (**f**) n = 5

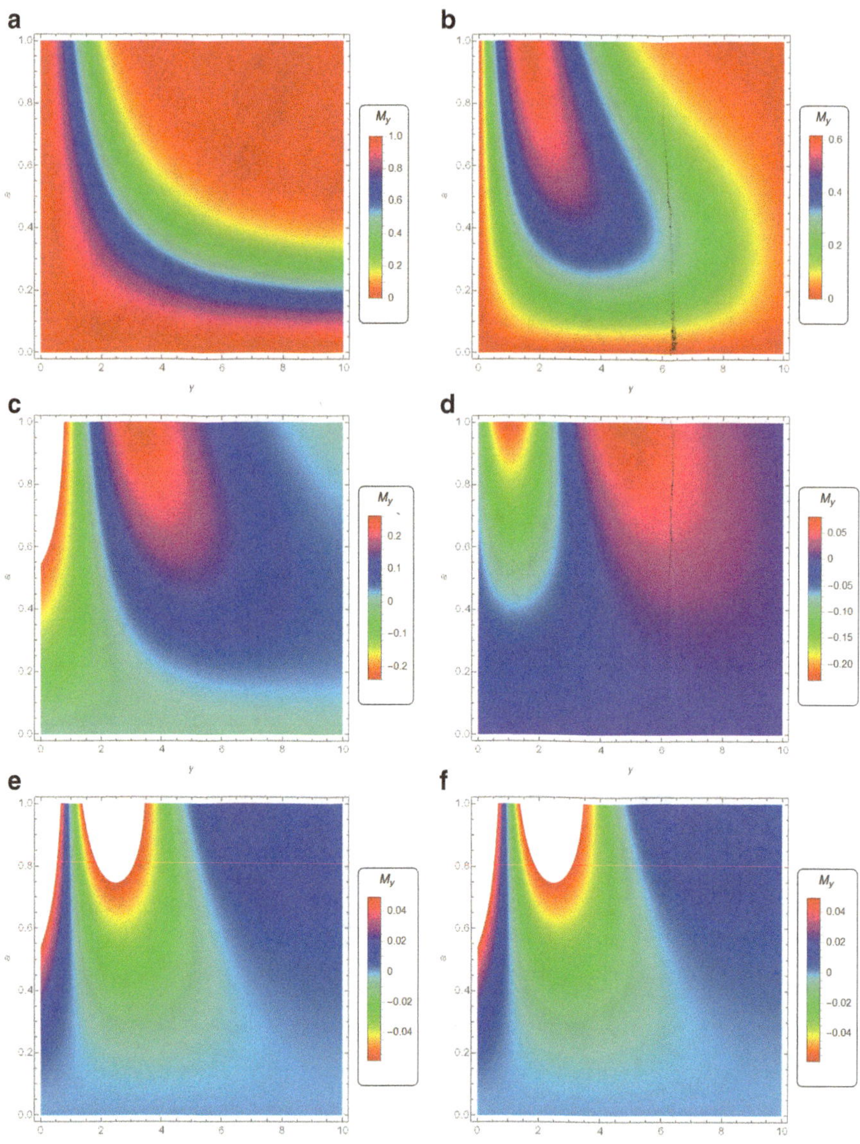

Fig. 4.15 Density plots of NMR transverse magnetization against h = y (within the plates) and capture parameter *a* at suction parameter $\lambda = 0.1$ for velocity parameters (**a**) n = 0 (**b**) n = 1 (**c**) n = 2 (**d**) n = 3 (**e**) n = 4 (**f**) n = 5

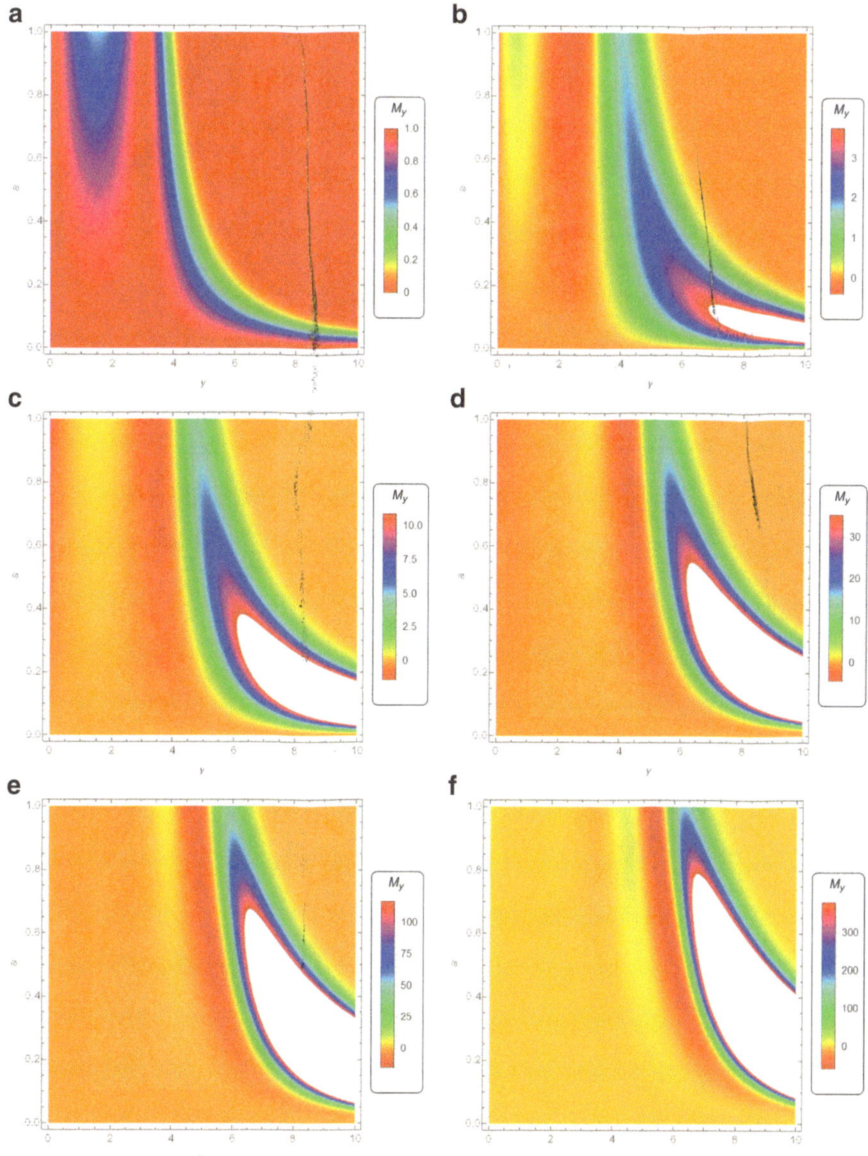

Fig. 4.16 Density plots of NMR transverse magnetization against h = y (within the plates) and capture parameter *a* at suction parameter λ = 1 for velocity parameters (**a**) n = 0 (**b**) n = 1 (**c**) n = 2 (**d**) n = 3 (**e**) n = 4 (**f**) n = 5

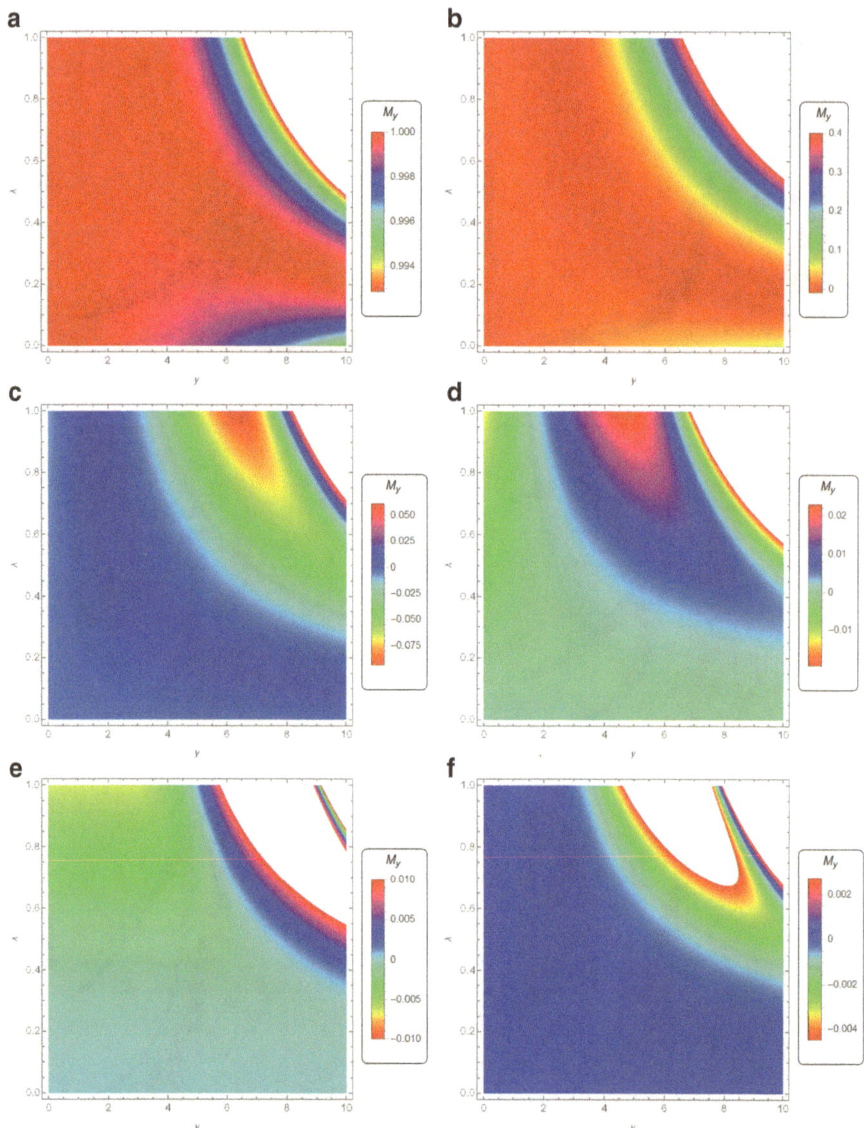

Fig. 4.17 Density plots of NMR transverse magnetization against h = y (within the plates) and λ at capture parameter a = 0.01 for velocity parameters (**a**) n = 0 (**b**) n = 1 (**c**) n = 2 (**d**) n = 3 (**e**) n = 4 (**f**) n = 5

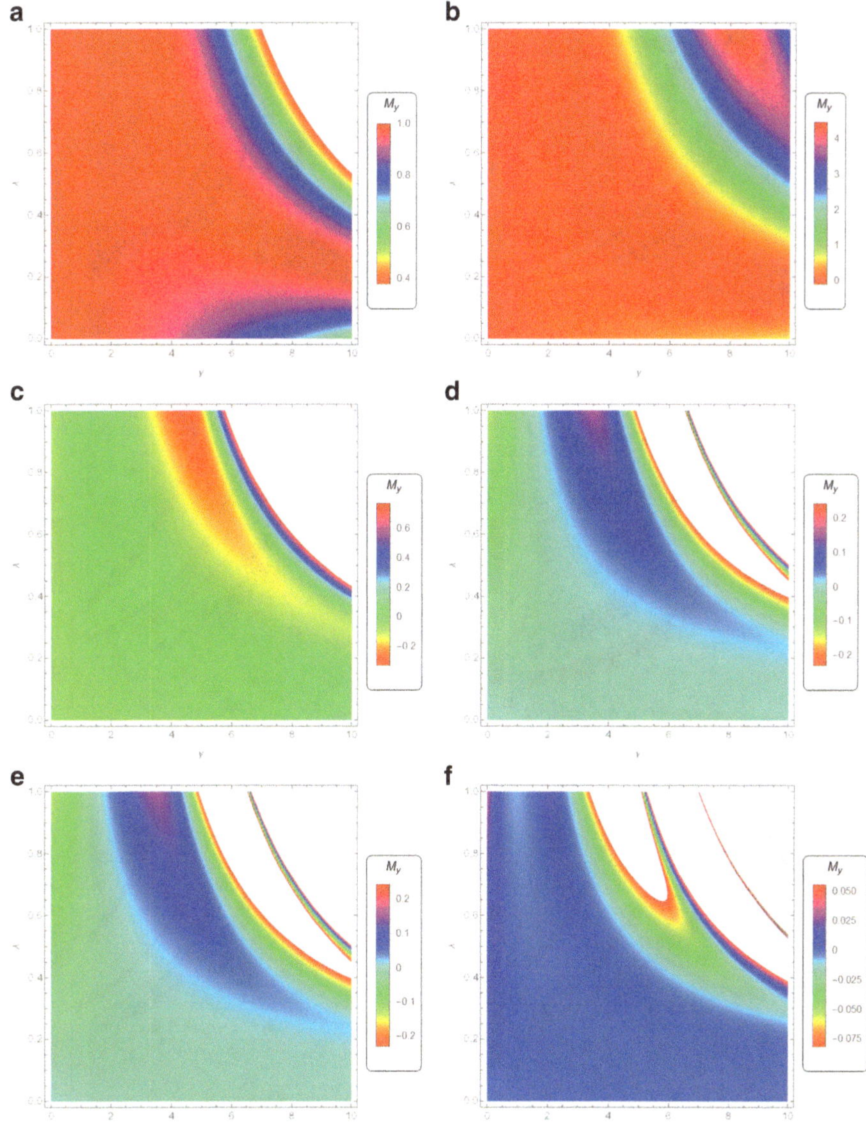

Fig. 4.18 Density plots of NMR transverse magnetization against h = y (within the plates) and λ at capture parameter a = 0.1 for velocity parameters (**a**) n = 0 (**b**) n = 1 (**c**) n = 2 (**d**) n = 3 (**e**) n = 4 (**f**) n = 5

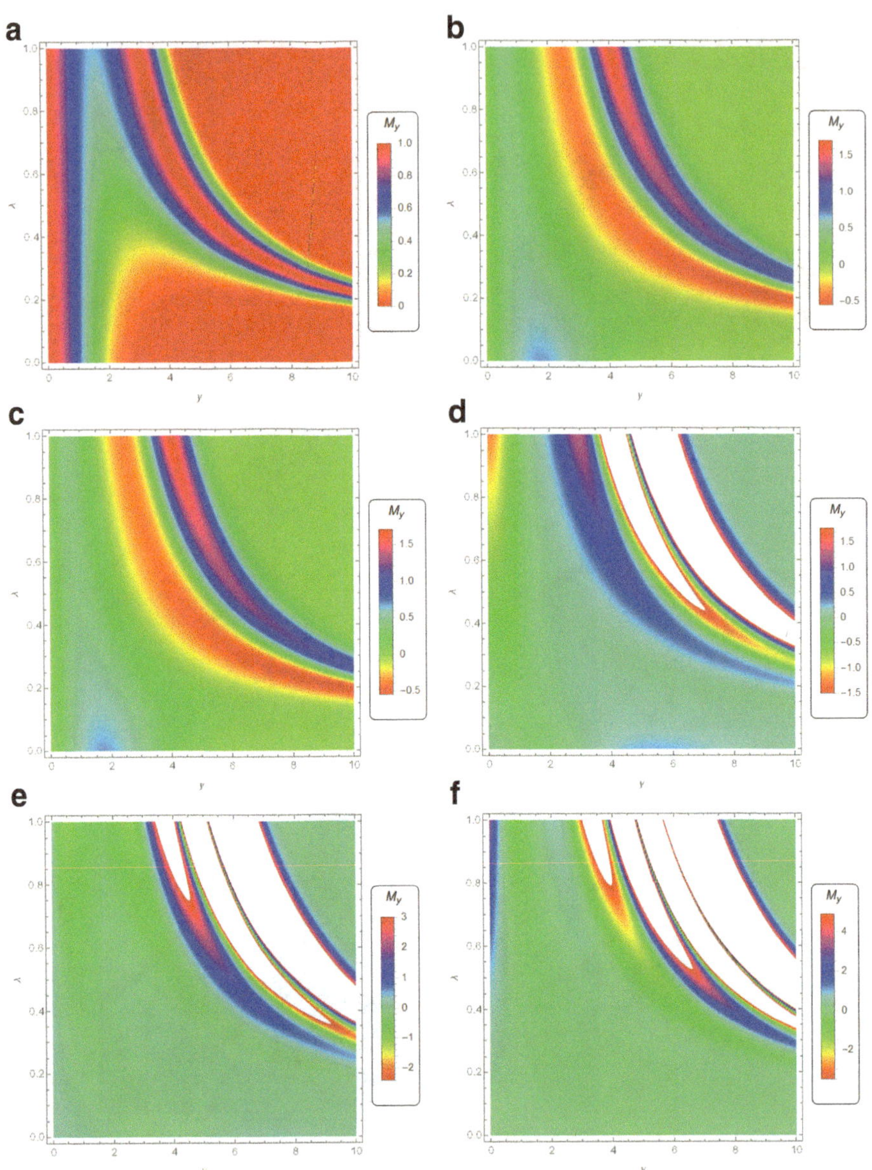

Fig. 4.19 Density plots of NMR transverse magnetization against h = y (within the plates) and λ at capture parameter a = 1 for velocity parameters (**a**) n = 0 (**b**) n = 1 (**c**) n = 2 (**d**) n = 3 (**e**) n = 4 (**f**) n = 5

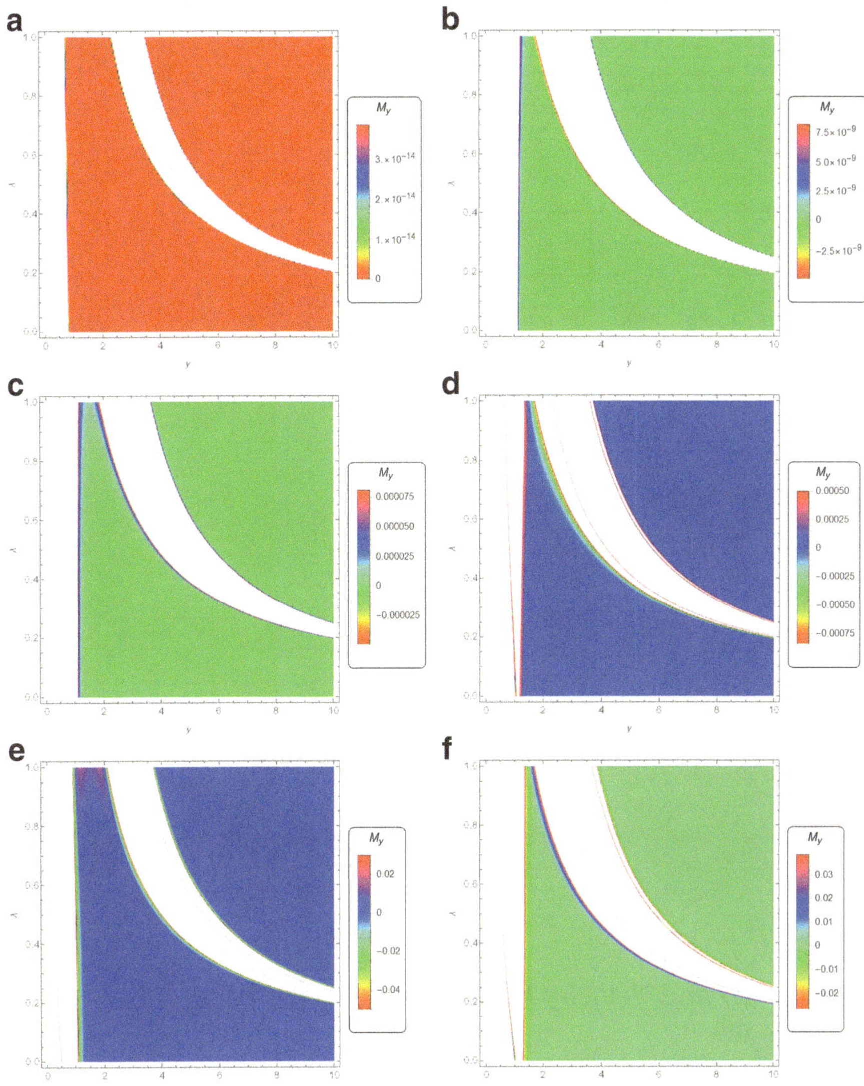

Fig. 4.20 Density plots of NMR transverse magnetization against h = y (within the plates) and λ at capture parameter a = 10 for velocity parameters (**a**) n = 0 (**b**) n = 1 (**c**) n = 2 (**d**) n = 3 (**e**) n = 4 (**f**) n = 5

Table 4.1 NMR transverse magnetized RFIDs and velocities for different tissues and parameter $a = 1$, x $= 0.001$ at n $= 1, 2, 3, 4, 5$

Tissue	T_2 (s)	T_1 (s)	T_o (s^{-1})	T_g (s^{-2})	a	n	x(m)	v (ms^{-1})	RF ID
Gray brain matter	0.07	0.66	15.80087	9.428571	1	0	0.001	1	0.010299
						1		3	0.030896
						2		5	0.051493
						3		7	0.07209
						4		9	0.092687
						5		11	0.113284
White brain matter	0.088	0.76	12.67943	8.636364	1	0	0.001	1	0.010761
						1		3	0.032282
						2		5	0.053803
						3		7	0.075324
						4		9	0.096845
						5		11	0.118366
Liver	0.029	0.39	37.04686	13.44828	1	0	0.001	1	0.008623
						1		3	0.025869
						2		5	0.043116
						3		7	0.060362
						4		9	0.077608
						5		11	0.094855
Kidney	0.034	0.47	31.53942	13.82353	1	0	0.001	1	0.008505
						1		3	0.025516
						2		5	0.042527
						3		7	0.059537
						4		9	0.076548
						5		11	0.093558

4.5 RF ID and Velocity Data Generation for Different Tissues

The flow and NMR relaxation-dependent radiofrequencies can easily be calculated. For few selected human tissues, the computations are shown in Table 4.1.

4.6 Conclusion

We have developed the Bloch NMR flow equation into two second-order differential equations which can predict fluid flow at suction points analytically and precisely. Equation (4.35) is solved by the method of perturbation theory: (1) to analyse the asymptotic suction profile for flow with a uniform suction across the flow in terms of NMR parameters; and (2) to describe the NMR parameter that affects the boundary layer thickness of the flow model. The suction points display flow features which are analysed accordingly in term of NMR parameters as shown in Figs. 4.2, 4.3, 4.4, 4.5, 4.6, 4.7, 4.8, 4.9, 4.10, 4.11 and 4.12 where the suction parameter λ, the parameter a and values of n significantly influence the NMR transverse magnetization. It is observed that when $a \ll$ n, the flow velocity is quite large especially when $a \ll 1$. The flow velocity is very small for $a \gg$ n.

The NMR transverse magnetization satisfies the nth-order Hermite polynomial and the even/odd parity as shown in Eqs. (4.33) and (4.34). It is interesting to note that the NMR signal follows a normal distribution within the plates when n > 0, but asymptotically decreases from the peak at the origin when n = 0. These are interesting features which should be explored further for fluid flow analysis.

Equation (4.32) is valid in the regime for the generation of NMR signal, and it can provide a comprehensive description of MRI experimental results by the definition of scaling parameter $l = 1.0$ m^{-1}. It should be noted that the mathematical formulation for flow at suction points based on Bloch's NMR flow equation considers the parameter A ($\mu = 0$) in Eq. (4.7) to be equal to zero. This makes Eqs. (4.17–4.32) to be appropriate for flow system where β is a positive constant which can be written as $\beta = \frac{1}{\alpha}$. The parameter α is called the acceleration of a particle when its flow velocity increases to a higher value (due to suction). This increase in flow velocity is expected so that parameter β remains finite.

In addition to the analysis of blood flow at suction points for various medical purposes, the analytical model presented in this study may be applicable to functional Magnetic Resonance Imaging (fMRI), a technique for determining which parts of the brain are activated by different types of physical sensation or activity, such as sight, sound or the movement of a subject's fingers. This "brain mapping"may be achieved by setting up a special wearable MRI scanner so that the increased blood flow to the activated areas of the brain by suction may be accurately analysed. The behaviour of the asymptotic suction profile can then be used to relate information from blood flow at suction points directly to information about brain function and brain structure and to diagnose blood-related diseases in order to formulate for patient care plans. Such wearable devices for diagnosis of brain tumors and heart disease can be developed, using newly available technologies in artificial intelligence, deep learning and IoT. In time, wearable scanners could replace bulky MRI machines. They would enable constant monitoring of patients, and improve the treatment of cancer, cardiovascular diseases, mental disease, and neurodegenerative diseases.

Parameter a as defined in Eq. (4.9) and analysed in Tables 4.1, 4.2, 4.3, 4.4, 4.5, 4.6, 4.7, 4.8, 4.9, 4.10 and 4.11 exerts great influence on the RFID system. When $a \gg 1$, the RFID system can be very small (Tables 4.1, 4.2, 4.3, 4.4, 4.5, 4.6 and 4.7)

Table 4.2 NMR transverse magnetized RFIDs and velocities for different tissues and parameter $a = 5$, x $= 0.001$ at n $= 1, 2, 3, 4, 5$

Tissue	T_2 (s)	T_1 (s)	T_0 (s^{-1})	T_g (s^{-2})	a	n	x(m)	v (ms^{-1})	RF ID
Gray brain matter	0.07	0.66	15.80087	9.428571	5	0	0.001	0.2	0.000412
						1		0.6	0.001236
						2		1	0.00206
						3		1.4	0.002884
						4		1.8	0.003707
						5		2.2	0.004531
White brain matter	0.088	0.76	12.67943	8.636364	5	0	0.001	0.2	0.00043
						1		0.6	0.001291
						2		1	0.002152
						3		1.4	0.003013
						4		1.8	0.003874
						5		2.2	0.004735
Liver	0.029	0.39	37.04686	13.44828	5	0	0.001	0.2	0.000345
						1		0.6	0.001035
						2		1	0.001725
						3		1.4	0.002414
						4		1.8	0.003104
						5		2.2	0.003794
Kidney	0.034	0.47	31.53942	13.82353	5	0	0.001	0.2	0.00034
						1		0.6	0.001021
						2		1	0.001701
						3		1.4	0.002381
						4		1.8	0.003062
						5		2.2	0.003742

Table 4.3 NMR transverse magnetized RF IDs and velocities for different tissues and parameter $a = 10$, x $= 0.001$ at n $= 1, 2, 3, 4, 5$

Tissue	T_2 (s)	T_1 (s)	T_0 (s^{-1})	T_g (s^{-2})	a	n	x(m)	v (ms^{-1})	RF ID
Gray brain matter	0.07	0.66	15.80087	9.428571	10	0	0.001	0.1	0.000103
						1		0.3	0.000309
						2		0.5	0.000515
						3		0.7	0.000721
						4		0.9	0.000927
						5		1.1	0.001133
White brain matter	0.088	0.76	12.67943	8.636364	10	0	0.001	0.1	0.000108
						1		0.3	0.000323
						2		0.5	0.000538
						3		0.7	0.000753
						4		0.9	0.000968
						5		1.1	0.001184
Liver	0.029	0.39	37.04686	13.44828	10	0	0.001	0.1	8.62E-05
						1		0.3	0.000259
						2		0.5	0.000431
						3		0.7	0.000604
						4		0.9	0.000776
						5		1.1	0.000949
Kidney	0.034	0.47	31.53942	13.82353	10	0	0.001	0.1	8.62E-05
						1		0.3	0.000259
						2		0.5	0.000431
						3		0.7	0.000604
						4		0.9	0.000765
						5		1.1	0.000936

Table 4.4 Table of NMR transverse magnetized RFIDs and velocities for different tissues and parameter $a = 15$, x $= 0.001$ at n $= 1, 2, 3, 4, 5$

Tissue	T_2 (s)	T_1 (s)	T_o (s^{-1})	T_g (s^{-2})	a	n	x(m)	v (ms^{-1})	RF ID
Gray brain matter	0.07	0.66	15.80087	9.428571	15	0	0.001	0.066667	4.58E-05
						1		0.2	0.000137
						2		0.333333	0.000229
						3		0.466667	0.00032
						4		0.6	0.000412
						5		0.733333	0.000503
White brain matter	0.088	0.76	12.67943	8.636364	15	0	0.001	0.066667	4.78E-05
						1		0.2	0.000137
						2		0.333333	0.000239
						3		0.466667	0.000335
						4		0.6	0.00043
						5		0.733333	0.000526
Liver	0.029	0.39	37.04686	13.44828	15	0	0.001	0.066667	3.83E-05
						1		0.2	0.000115
						2		0.333333	0.000192
						3		0.466667	0.000268
						4		0.6	0.000345
						5		0.733333	0.000422
Kidney	0.034	0.47	31.53942	13.82353	15	0	0.001	0.066667	3.78E-05
						1		0.2	0.000115
						2		0.333333	0.000192
						3		0.466667	0.000265
						4		0.6	0.000345
						5		0.000416	0.000416

Table 4.5 NMR transverse magnetized RFIDs and velocities for different tissues and parameter $a = 20$, x $= 0.001$ at n $= 1, 2, 3, 4, 5$

Tissue	T_2 (s)	T_1 (s)	T_o (s^{-1})	T_g (s^{-2})	a	n	x(m)	v (ms^{-1})	RF ID
Gray brain matter	0.07	0.66	15.80087	9.428571	20	0	0.001	0.05	2.57E-05
						1		0.15	7.72E-05
						2		0.25	0.000129
						3		0.35	0.00018
						4		0.45	0.000232
						5		0.55	0.000283
White brain matter	0.088	0.76	12.67943	8.636364	20	0	0.001	0.05	2.69E-05
						1		0.15	8.07E-05
						2		0.25	0.000129
						3		0.35	0.000188
						4		0.45	0.000242
						5		0.55	0.000296
Liver	0.029	0.39	37.04686	13.44828	20	0	0.001	0.05	2.16E-05
						1		0.15	6.47E-05
						2		0.25	0.000108
						3		0.35	0.000151
						4		0.45	0.000194
						5		0.55	0.000237
Kidney	0.034	0.47	31.53942	13.82353	20	0	0.001	0.05	2.13E-05
						1		0.15	6.38E-05
						2		0.25	0.000108
						3		0.35	0.000151
						4		0.45	0.000191
						5		0.55	0.000234

Table 4.6 NMR transverse magnetized RF IDs and velocities for different tissues and parameter $a = 25$, x $= 0.001$ at n $= 1, 2, 3, 4, 5$

Tissue	T_2 (s)	T_1 (s)	T_o (s^{-1})	T_g (s^{-2})	a	n	x(m)	v (ms^{-1})	RF ID
Gray brain matter	0.07	0.66	15.80087	9.428571	25	0	0.001	0.04	1.65E-05
						1		0.12	4.94E-05
						2		0.2	8.24E-05
						3		0.28	0.000115
						4		0.36	0.000148
						5		0.44	0.000181
White brain matter	0.088	0.76	12.67943	8.636364	25	0	0.001	0.04	1.72E-05
						1		0.12	5.17E-05
						2		0.2	8.61E-05
						3		0.28	0.000115
						4		0.36	0.000148
						5		0.44	0.000189
Liver	0.029	0.39	37.04686	13.44828	25	0	0.001	0.04	1.38E-05
						1		0.12	4.14E-05
						2		0.2	6.9E-05
						3		0.28	9.66E-05
						4		0.36	0.000124
						5		0.44	0.00015
Kidney	0.034	0.47	31.53942	13.82353	25	0	0.001	0.04	1.36E-05
						1		0.12	4.14E-05
						2		0.2	6.8E-05
						3		0.28	9.53E-05
						4		0.36	0.000122
						5		0.44	0.00015

Table 4.7 Table of NMR transverse magnetized RF IDs and velocities for different tissues and parameter $a = 50$, x $= 0.001$ at n $= 1, 2, 3, 4, 5$

Tissue	T_2 (s)	T_1 (s)	T_o (s^{-1})	T_g (s^{-2})	a	n	x(m)	v (ms^{-1})	RF ID
Gray brain matter	0.07	0.66	15.80087	9.428571	50	0	0.001	0.02	4.12E-06
						1		0.06	1.24E-05
						2		0.1	2.06E-05
						3		0.14	2.88E-05
						4		0.18	3.71E-05
						5		0.22	4.53E-05
White brain matter	0.088	0.76	12.67943	8.636364	50	0	0.001	0.02	4.3E-06
						1		0.06	1.29E-05
						2		0.1	2.15E-05
						3		0.14	3.01E-05
						4		0.18	3.87E-05
						5		0.22	4.73E-05
Liver	0.029	0.39	37.04686	13.44828	50	0	0.001	0.02	3.45E-06
						1		0.06	1.03E-05
						2		0.1	1.72E-05
						3		0.14	2.41E-05
						4		0.18	3.1E-05
						5		0.22	3.79E-05
Kidney	0.034	0.47	31.53942	13.82353	50	0	0.001	0.02	3.4E-06
						1		0.06	1.03E-05
						2		0.1	1.72E-05
						3		0.14	2.41E-05
						4		0.18	3.1E-05
						5		0.22	3.74E-05

Table 4.8 NMR transverse magnetized RF IDs and velocities for different tissues and parameter $a = 0.1$, $x = 0.001$ at $n = 1, 2, 3, 4, 5$

Tissue	T_2 (s)	T_1 (s)	T_o (s^{-1})	T_g (s^{-2})	a	n	x(m)	v (ms^{-1})	RF ID
Gray brain matter	0.07	0.66	15.80087	9.428571	0.1	0	0.001	10	1.029857
						1		30	3.089572
						2		50	5.149287
						3		70	7.209001
						4		90	9.268716
						5		110	11.32843
White brain matter	0.088	0.76	12.67943	8.636364	0.1	0	0.001	10	1.076055
						1		30	3.228166
						2		50	5.380276
						3		70	7.532386
						4		90	9.684497
						5		110	11.83661
Liver	0.029	0.39	37.04686	13.44828	0.1	0	0.001	10	0.862316
						1		30	2.586949
						2		50	4.311582
						3		70	6.036215
						4		90	7.760848
						5		110	9.485481
Kidney	0.034	0.47	31.53942	13.82353	0.1	0	0.001	10	0.850532
						1		30	2.551595
						2		50	4.252659
						3		70	5.953722
						4		90	7.654786
						5		110	9.355849

Table 4.9 NMR transverse magnetized RF IDs and velocities for different tissues and parameter $a = 0.01$, x $= 0.001$ at n $= 1, 2, 3, 4, 5$

Tissue	T_2 (s)	T_1 (s)	T_o (s^{-1})	T_g (s^{-2})	a	n	x(m)	v (ms^{-1})	RF ID
Gray brain matter	0.07	0.66	15.80087	9.428571	0.01	0	0.001	100	102.9857
						1		300	308.9572
						2		500	514.9287
						3		700	720.9001
						4		900	926.8716
						5		1100	1132.843
White brain matter	0.088	0.76	12.67943	8.636364	0.01	0	0.001	100	107.6055
						1		300	322.8166
						2		500	538.0276
						3		700	753.2386
						4		900	968.4497
						5		1100	1183.661
Liver	0.029	0.39	37.04686	13.44828	0.01	0	0.001	100	86.23165
						1		300	258.6949
						2		500	431.1582
						3		700	603.6215
						4		900	776.0848
						5		1100	948.5481
Kidney	0.034	0.47	31.53942	13.82353	0.01	0	0.001	100	85.05317
						1		300	255.1595
						2		500	425.2659
						3		700	595.3722
						4		900	765.4786
						5		1100	935.5849

Table 4.10 NMR transverse magnetized RF IDs and velocities for different tissues and parameter $a = 0.001$, $x = 0.001$ at $n = 1, 2, 3, 4, 5$

Tissue	T_2 (s)	T_1 (s)	T_o (s^{-1})	T_g (s^{-2})	a	n	x(m)	v (ms^{-1})	RF ID
Gray brain matter	0.07	0.66	15.80087	9.428571	0.001	0	0.001	1000	10298.57
						1		3000	30895.72
						2		5000	51492.87
						3		7000	72090.01
						4		9000	92687.16
						5		11,000	113284.3
White brain matter	0.088	0.76	12.67943	8.636364	0.001	0	0.001	1000	10760.55
						1		3000	32281.66
						2		5000	53802.76
						3		7000	75323.86
						4		9000	96844.97
						5		11,000	118366.1
Liver	0.029	0.39	37.04686	13.44828	0.001	0	0.001	1000	8623.165
						1		3000	25869.49
						2		5000	43115.82
						3		7000	60362.15
						4		9000	77608.48
						5		11,000	94854.81
Kidney	0.034	0.47	31.53942	13.82353	0.001	0	0.001	1000	8505.317
						1		3000	25515.95
						2		5000	42526.59
						3		7000	59537.22
						4		9000	76547.86
						5		11,000	93558.49

Table 4.11 NMR transverse magnetized RF IDs and velocities for different tissues and parameter $a = 0.001$, $x = 0.001$ at $n = 1, 2, 3, 4, 5$

Tissue	T_2 (s)	T_1 (s)	T_0 (s^{-1})	T_g (s^{-2})	a	n	x(m)	v (ms^{-1})	RF ID
Gray brain matter	0.07	0.66	15.80087	9.428571	0.0001	0	0.001	10,000	1029857
						1		30,000	3089572
						2		50,000	5149287
						3		70,000	7209001
						4		90,000	9268716
						5		110,000	11328430
White brain matter	0.088	0.76	12.67943	8.636364	0.0001	0	0.001	10,000	1076055
						1		30,000	3228166
						2		50,000	5380276
						3		70,000	7532386
						4		90,000	9684497
						5		110,000	11836607
Liver	0.029	0.39	37.04686	13.44828	0.0001	0	0.001	10,000	862316.5
						1		30,000	2586949
						2		50,000	4311582
						3		70,000	6036215
						4		90,000	7760848
						5		110,000	9485481
Kidney	0.034	0.47	31.53942	13.82353	0.0001	0	0.001	10,000	850531.7
						1		30,000	2551595
						2		50,000	4252659
						3		70,000	5953722
						4		90,000	7654786
						5		110,000	9355849

for low flow velocities and when $a \ll 1$, RFID systems can be extremely large for large flow velocities (Tables 4.9, 4.10 and 4.11). Therefore, our next investigation will focus on how parameter a can be utilized effectively within an IoT platform for smarter healthcare in general and neurological computing in particular.

Chapter 5
A Computational MRI Based on Bloch's NMR Flow Equation, MRI Fingerprinting and Python Deep Learning for Classifying Adult Brain Tumors

Abstract No single neuroimaging technique currently lends itself to easy, reliable and consistent use in day-to-day settings to differentiate intra-axial brain tumors based on origin and histopathologic grading. Tissue discrimination is a more significant challenge after treatment, because pseudo-progression, pseudo-response and development of radiation necrosis are real possibilities. In this chapter we have demonstrated practical ways in which computational MRI can be used to provide innovative data-driven algorithms for the prediction, simplification and characterization of brain tumors as well as to delineate their intrinsic mechanisms based on Bloch's Nuclear Magnetic Resonance (NMR) flow equation, Bessel functions and Bessel properties. The results have motivated data-driven predictions required for statistical learning or machine learning and have led to the application of deep learning for the classification of brain tumors. These can help significantly to accurately delineate tumor margins in order to ensure appropriate diagnosis, treatment and prognostic information for early differentiation between primary and metastatic malignant brain tumors.

Keywords Bloch NMR flow equation · Bessel functions · MRI fingerprinting · Machine learning · Python deep learning · Brain tumors

5.1 Introduction

Glioblastoma multiforme (GBM) is the most common primary age-adjusted malignant brain tumor while brain metastases form about half of all intracranial neoplasm cases (Badve et al., 2017). In order to provide medical researchers and practitioners with appropriate diagnostic and treatment methods, prognostic information provided by early differentiation between primary and metastatic malignant brain tumors is necessary and important. Furthermore, identifying primarily vasogenic edema found around metastatic lesions and another kind of edema, that is, those with neoplastic cellular infiltration (found around glioblastomas) can help significantly to accurately delineate tumor margins. This information is extremely important for treatment planning so as to ensure better outcomes for patients (Badve et al., 2017; Giese et al., 2003).

It is of utmost importance that healthcare givers have information on the different biological features associated with various tumor types. This would definitely lead to a better understanding of complex aggressive tumor behaviour and lack of response to treatment, normally encountered in glioblastomas (Smith & Ironside, 2007). Advanced magnetic resonance (MR) imaging techniques such as diffusion tensor imaging (DTI), perfusion imaging and MR spectroscopy have shown promise in showing contrasts between brain metastases, glioblastomas and regions of peritumoral infiltration (Badve et al., 2017; Cha et al., 2007).

However, a simple, rapid, quantitative and non-invasive method for probing tissue features in patients with brain tumors is currently not available. Such a method should be able to extract information from minute changes in the tissue microenvironment that cannot be observed by the human eye on standard qualitative clinical images.

Magnetic Resonance fingerprinting (MRF) is a new MR imaging technique in which pseudo-randomized acquisition parameters are used to simultaneously quantify multiple tissue properties, including T_1 and T_2 relaxation times. The MR sequence design leads to signal evolution in each image voxel as a function of T_1 and T_2 of tissue in specific voxels. With the use of Bloch's NMR flow equations, a dictionary of all possible signal evolutions is simulated with acquisition parameters from sequences employed and all possible T_1 and T_2 combinations. The selection of the dictionary entry that best correlates with the voxel signal time course identifies the best dictionary match for each voxel. The T_1 and T_2 times used to construct the dictionary entry are identified as the relaxation time measurements for that voxel (Ma et al., 2013).

Imaging techniques such as perfusion imaging have demonstrated usefulness in showing contrasts between metastases and GBMs among various glioma grades (Cha et al., 2007). DTI has shown impressive promise in delineating tumor margins as well as identifying tumor infiltration. The combination of positron emission tomography (PET) imaging and fluorodeoxyglucose (FDG) (termed FDG-PET) has been modestly successful in differentiating glioma grades; however, a major shortcoming is the lack of specificity and significant background uptake (Wong et al., 2002; Pauleit et al., 2009). Molecular imaging used together with newer PET agents holds some promise but large-scale evaluation studies are still needed in this area (Pauleit et al., 2009; Becherer et al., 2003).

In spite of these developments, no single neuroimaging technique currently exists that can be easily, reliably, and consistently used in a day-to-day setting to differentiate intra-axial brain tumors as a function of their origin and histopathologic grading. The problem of tissue discrimination is even more challenging in a post-therapy situation in which pseudo-progression, pseudo-response and development of radiation necrosis are possibilities (Clarke & Chang, 2009; Henson et al., 2008). A combination of these sophisticated neuroimaging techniques and conventional MR imaging can improve different aspects of brain tumor diagnostics; however, they involve significant financial and time limitations and technical challenges. In order to address this problem, this study aims to develop a computational MR model from Bloch's NMR flow equation, and employ MRF-derived relaxometry data for the computation of magnetic resonance transverse magnetization datasets to implement a Python deep learning algorithm for the classification of adult brain tumors.

5.2 Computational Modelling of the Bloch NMR Flow Equation

For this study, spin dynamics of arterial blood and surrounding tissues based on the time-independent Bloch NMR flow equation (Awojoyogbe, 2002, 2003, 2004, 2007) would be considered. Within a rotating frame of reference, the Larmor condition exists such that:

$$f_0 = \gamma B - \omega = 0 \tag{5.1}$$

When the bulk protons are moving at a variable velocity v(x), the Bloch NMR flow equations have been derived in Eq. (2.330) as follows:

$$v^2(x)\frac{d^2 M_y}{dx^2} + v(x)\left(T_0 + \frac{dv}{dx}\right)\frac{dM_y}{dx} + \left(\gamma^2 B_1^2(x) + T_g\right)M_y = \frac{M_0}{T_1}\gamma B_1(x) \tag{5.2}$$

where $T_0 = \frac{1}{T_1} + \frac{1}{T_2}$, $T_g = \frac{1}{T_1 T_2}$.

In Eq. (5.2), γ is the gyromagnetic ratio, v is the fluid velocity, T_1 and T_2 are spin-lattice and spin-spin relaxation times respectively, M_y is the NMR transverse magnetization. Equation (5.2) gives information about the system if appropriate boundary conditions are applied. The term in Eq. (5.2) is the forcing function. M_0 is the equilibrium magnetization.

If the RF field $B_1(x)$ is applied such that the transverse magnetization M_y is sampled at maximum magnitude, $M_0 \approx 0$. Equation (5.2) therefore becomes:

$$v^2(x)\frac{d^2 M_y}{dx^2} + v(x)\left(T_0 + \frac{dv}{dx}\right)\frac{dM_y}{dx} + \left(\gamma^2 B_1^2(x) + T_g\right)M_y = 0 \tag{5.3}$$

For simplicity, we may design the fluid velocity and the applied RF field in the form:

$$v(x) = \frac{x}{\delta} \quad \text{and} \quad B_1(x) = \frac{T_1}{T_2}Gx \tag{5.4}$$

$$\frac{x^2}{\delta^2}\frac{d^2M_y}{dx^2} + \frac{x}{\delta}\left(T_0 + \frac{1}{\delta}\right)\frac{dM_y}{dx} + \left(\frac{T_1^2}{T_2^2}\gamma^2G^2x^2 + T_g\right)M_y = 0 \tag{5.5}$$

$$x^2\frac{d^2M_y}{dx^2} + x(1 + \delta T_0)\frac{dM_y}{dx} + \left(\frac{T_1^2}{T_2^2}\gamma^2G^2\delta^2x^2 + \delta^2 T_g\right)M_y = 0 \tag{5.6}$$

If we set:

$$\xi = \frac{T_1}{T_2}\gamma G\delta \tag{5.7}$$

$$x^2\frac{d^2M_y}{dx^2} + x(1 + \delta T_0)\frac{dM_y}{dx} + \left(\xi^2 x^2 + \delta^2 T_g\right)M_y = 0 \tag{5.8}$$

Equation (5.8) is an equation transformable to the Bessel differential equation and the solution is:

$$M_y(x) = x^\alpha(C_1 J_n(\xi x) + C_2 Y_n(\xi x)) \tag{5.9}$$

where $\alpha = -\frac{\delta T_0}{2}$, $n = \frac{\delta}{2}\sqrt{T_0^2 - 4T_g}$; C_1 and C_2 are constants.

Since we require that the transverse magnetization be finite at all points x, it follows that $C_2 = 0$. Hence, we write:

$$M_y(x) = C_1 x^\alpha J_n(\xi x) \tag{5.10}$$

The NMR signal is defined as:

$$S = \int_0^\infty M_y(x)dx = \int_0^\infty C_1 x^\alpha J_n(\xi x)dx = 2^\alpha \xi^{-\alpha-1}\frac{\Gamma\left(\frac{1}{2} + \frac{1}{2}n + \frac{1}{2}\alpha\right)}{\Gamma\left(\frac{1}{2} + \frac{1}{2}n - \frac{1}{2}\alpha\right)} \tag{5.11}$$

provided that $|\operatorname{Re}(n)| - 1 < \alpha < \frac{1}{2}, \xi > 0$.

5.3   Relaxometry Data

For this study, we have concentrated on the solid tumor region defined as the enhancing region in tumors with post-contrast enhancement or an expansile fluid attenuated inversion recovery (FLAIR) hyperintensity region in non-enhancing or minimally enhancing tumors (Badve et al., 2017). MR fingerprinting-derived T_1 and T_2 relaxometry measurements for three common types of intra-axial brain tumors (glioblastomas, lower grade gliomas and metastases) have been given. The mean values of these relaxation measurements are presented in Table 5.1.

Table 5.1 Mean values of T_1 and T_2 from solid tumor regions in three tumor types scanned at $B_0 = 3.0$ T and radiofrequency pulse duration of 800 μs

Tissue type			
	Glioblastomas (GBM)	Metastases	Lower grade gliomas (LGG)
T_1(s)	1.639 ± 0.247	1.324 ± 0.273	1.600 ± 0.197
T_2(s)	0.138 ± 0.022	0.105 ± 0.027	0.172 ± 0.053

For our dataset, we shall distribute each relaxometry entry into 100 inputs using the Generating Data Online app (https://www.generatedata.com/). This app randomly distributes each entry between ranges of values as specified in Table 5.1.

5.4   Computation of Magnetic Resonance Signal Dataset

Based on the data generated in Table 5.1 and Eq. (5.10), we obtained the dataset presented in Table 5.2. The other parameters used are: gradient magnitude (G) $=40$ mT m^{-1}, RF pulse duration (δ) $= 800$ μs, x $= 5.2$ mm, gyromagnetic ratio (γ) $= 42666666.67$ T^{-1} m^{-1} and $C_1 = 1000$ Am^{-1}.

5.5   Data Visualization

In order to see the distribution of the data, it is important to visualize them from different points of view (Fig. 5.1).

5.6   Linear Regression

Linear regression is a linear approach to modelling the relationship between a scalar response (a dependent variable) and one or more explanatory variables (or independent variables). Figures 5.2 and 5.3 show how some of the data entries vary as a function of the other. For this analysis, it is noted that linear regression performed poorly for all the parameters in the datasets of Tables 5.2 and 5.3 except MR signal which appeared to show a linear relationship with relaxation rate T_0 (as illustrated in Fig. 5.3b). This implies that the relationship between the parameters in our dataset (especially Figs. 5.2a, b and 5.3a) have a complicated nature. Hence, we shall explore how machine learning could be used to explicate these complicated relationships.

Table 5.2 Magnetic resonance signal dataset calculated from MR fingerprinting measurements and Eq. (5.10)

T_1 relaxation time (s)	T_2 relaxation time (s)	T_0 relaxation rate (s^{-1})	Transverse magnetization (M_y)	Tumor type
1.414	0.133	8.226011	71.23105	Glioblastoma_Multiforme
1.867	0.156	6.945875	−70.4165	Glioblastoma_Multiforme
1.708	0.156	6.995737	2.116533	Glioblastoma_Multiforme
1.548	0.14	7.788852	−62.1793	Glioblastoma_Multiforme
1.439	0.116	9.315617	67.52559	Glioblastoma_Multiforme
1.713	0.153	7.119719	−89.6469	Glioblastoma_Multiforme
1.751	0.126	8.50761	−72.2422	Glioblastoma_Multiforme
1.801	0.146	7.404562	33.63957	Glioblastoma_Multiforme
1.636	0.149	7.322656	−18.061	Glioblastoma_Multiforme
1.462	0.132	8.259752	−70.5005	Glioblastoma_Multiforme
1.464	0.12	9.016393	−46.6987	Glioblastoma_Multiforme
1.732	0.152	7.156315	0.183453	Glioblastoma_Multiforme
1.751	0.149	7.282512	50.47213	Glioblastoma_Multiforme
1.632	0.128	8.425245	−16.9109	Glioblastoma_Multiforme
1.782	0.146	7.410482	−43.6337	Glioblastoma_Multiforme
1.417	0.131	8.339304	75.88854	Glioblastoma_Multiforme
1.438	0.152	7.274358	−90.7995	Glioblastoma_Multiforme
1.849	0.157	6.91026	36.50772	Glioblastoma_Multiforme
1.881	0.12	8.864965	−65.8513	Glioblastoma_Multiforme
1.839	0.128	8.356274	62.23343	Glioblastoma_Multiforme
1.67	0.149	7.310212	−88.0641	Glioblastoma_Multiforme
1.805	0.116	9.174706	−74.6287	Glioblastoma_Multiforme
1.722	0.143	7.573727	−87.0962	Glioblastoma_Multiforme
1.769	0.127	8.439307	−61.5561	Glioblastoma_Multiforme
1.764	0.123	8.696975	70.55777	Glioblastoma_Multiforme
1.674	0.16	6.847372	−30.7488	Glioblastoma_Multiforme
1.405	0.117	9.258752	−82.6966	Glioblastoma_Multiforme
1.599	0.124	8.689907	−80.0434	Glioblastoma_Multiforme
1.755	0.117	9.116809	35.41739	Glioblastoma_Multiforme
1.83	0.124	8.610964	−75.3636	Glioblastoma_Multiforme
1.505	0.157	7.033879	−26.5147	Glioblastoma_Multiforme
1.641	0.148	7.366141	−75.0627	Glioblastoma_Multiforme
1.712	0.148	7.340869	84.24418	Glioblastoma_Multiforme
1.815	0.12	8.884298	76.60474	Glioblastoma_Multiforme
1.537	0.12	8.983951	−49.0967	Glioblastoma_Multiforme
1.856	0.16	6.788793	88.63128	Glioblastoma_Multiforme
1.479	0.127	8.550148	87.32393	Glioblastoma_Multiforme
1.809	0.136	7.905733	69.01262	Glioblastoma_Multiforme
1.549	0.14	7.788435	−65.4837	Glioblastoma_Multiforme
1.462	0.123	8.814076	−30.3078	Glioblastoma_Multiforme

(continued)

Table 5.2 (continued)

T_1 relaxation time (s)	T_2 relaxation time (s)	T_0 relaxation rate (s^{-1})	Transverse magnetization (M_y)	Tumor type
1.43	0.117	9.246309	−34.5013	Glioblastoma_Multiforme
1.606	0.15	7.289332	91.57603	Glioblastoma_Multiforme
1.548	0.122	8.842716	20.11071	Glioblastoma_Multiforme
1.641	0.135	8.016792	−67.2913	Glioblastoma_Multiforme
1.517	0.131	8.292784	86.70794	Glioblastoma_Multiforme
1.559	0.124	8.705953	75.04193	Glioblastoma_Multiforme
1.542	0.141	7.740707	10.29882	Glioblastoma_Multiforme
1.401	0.156	7.124032	100.4106	Glioblastoma_Multiforme
1.434	0.148	7.454107	43.31237	Glioblastoma_Multiforme
1.527	0.131	8.288467	85.53145	Glioblastoma_Multiforme
1.653	0.122	8.801682	33.01069	Glioblastoma_Multiforme
1.757	0.128	8.381652	−61.7833	Glioblastoma_Multiforme
1.662	0.15	7.268351	−72.0633	Glioblastoma_Multiforme
1.739	0.156	6.9853	−89.4949	Glioblastoma_Multiforme
1.634	0.123	8.742076	63.03783	Glioblastoma_Multiforme
1.399	0.133	8.233593	6.21006	Glioblastoma_Multiforme
1.875	0.138	7.77971	11.64593	Glioblastoma_Multiforme
1.641	0.147	7.412106	−90.8046	Glioblastoma_Multiforme
1.476	0.116	9.298196	−1.35368	Glioblastoma_Multiforme
1.677	0.127	8.470319	23.69041	Glioblastoma_Multiforme
1.67	0.117	9.145811	80.78219	Glioblastoma_Multiforme
1.567	0.141	7.730361	−83.2086	Glioblastoma_Multiforme
1.661	0.123	8.732128	55.34912	Glioblastoma_Multiforme
1.547	0.137	7.945682	−60.2639	Glioblastoma_Multiforme
1.855	0.127	8.413099	−57.8975	Glioblastoma_Multiforme
1.752	0.138	7.817153	15.83538	Glioblastoma_Multiforme
1.834	0.153	7.081204	−76.7918	Glioblastoma_Multiforme
1.493	0.119	9.073154	81.70127	Glioblastoma_Multiforme
1.805	0.117	9.101025	−27.2731	Glioblastoma_Multiforme
1.842	0.146	7.392203	58.2924	Glioblastoma_Multiforme
1.829	0.143	7.539754	−39.6528	Glioblastoma_Multiforme
1.424	0.132	8.278005	85.23585	Glioblastoma_Multiforme
1.828	0.141	7.639245	−83.5395	Glioblastoma_Multiforme
1.797	0.122	8.753204	−78.9349	Glioblastoma_Multiforme
1.671	0.151	7.220961	−66.2726	Glioblastoma_Multiforme
1.441	0.132	8.26972	22.86995	Glioblastoma_Multiforme
1.821	0.131	8.182737	−71.084	Glioblastoma_Multiforme
1.475	0.122	8.874687	−85.4893	Glioblastoma_Multiforme
1.82	0.132	8.125208	−78.7062	Glioblastoma_Multiforme
1.701	0.119	8.991251	79.59399	Glioblastoma_Multiforme
1.824	0.133	8.067043	−56.8087	Glioblastoma_Multiforme

(continued)

Table 5.2 (continued)

T_1 relaxation time (s)	T_2 relaxation time (s)	T_0 relaxation rate (s^{-1})	Transverse magnetization (M_y)	Tumor type
1.713	0.156	6.994028	−18.6241	Glioblastoma_Multiforme
1.822	0.138	7.795224	22.61471	Glioblastoma_Multiforme
1.442	0.144	7.637926	35.40379	Glioblastoma_Multiforme
1.502	0.151	7.288296	72.54686	Glioblastoma_Multiforme
1.688	0.116	9.213107	−32.9992	Glioblastoma_Multiforme
1.59	0.134	8.091617	−17.9148	Glioblastoma_Multiforme
1.408	0.154	7.203734	27.60451	Glioblastoma_Multiforme
1.429	0.127	8.573806	−77.0489	Glioblastoma_Multiforme
1.667	0.117	9.146889	79.81872	Glioblastoma_Multiforme
1.67	0.132	8.17456	41.04611	Glioblastoma_Multiforme
1.47	0.133	8.199069	−60.0601	Glioblastoma_Multiforme
1.68	0.155	7.046851	66.48417	Glioblastoma_Multiforme
1.852	0.118	9.014533	−59.5357	Glioblastoma_Multiforme
1.446	0.135	8.09897	92.16649	Glioblastoma_Multiforme
1.662	0.122	8.798406	−9.40542	Glioblastoma_Multiforme
1.857	0.124	8.603019	22.8555	Glioblastoma_Multiforme
1.442	0.15	7.360148	−7.77663	Glioblastoma_Multiforme
1.879	0.131	8.165786	69.90272	Glioblastoma_Multiforme
1.611	0.159	6.910041	−42.5514	Glioblastoma_Multiforme
1.202	0.094	11.47024	−38.3744	Metastasis
1.552	0.082	12.83945	−5.04646	Metastasis
1.579	0.106	10.06727	−21.1907	Metastasis
1.238	0.119	9.211116	−64.7573	Metastasis
1.348	0.103	10.45058	−43.9971	Metastasis
1.308	0.107	10.11032	−33.3723	Metastasis
1.535	0.132	8.227223	88.9579	Metastasis
1.351	0.084	12.64495	73.17418	Metastasis
1.554	0.123	8.773582	50.07809	Metastasis
1.38	0.082	12.91976	58.14911	Metastasis
1.211	0.086	12.45367	18.46248	Metastasis
1.522	0.079	13.31526	−43.2037	Metastasis
1.363	0.111	9.742685	−0.08909	Metastasis
1.213	0.095	11.35072	−27.8288	Metastasis
1.325	0.131	8.388305	−31.7149	Metastasis
1.1	0.107	10.25489	−95.2799	Metastasis
1.176	0.108	10.1096	40.13794	Metastasis
1.32	0.132	8.333333	44.12875	Metastasis
1.192	0.083	12.88712	64.88468	Metastasis
1.375	0.106	10.16123	−83.1995	Metastasis
1.246	0.096	11.21923	−82.3359	Metastasis
1.456	0.107	10.03261	−0.34718	Metastasis

(continued)

Table 5.2 (continued)

T_1 relaxation time (s)	T_2 relaxation time (s)	T_0 relaxation rate (s^{-1})	Transverse magnetization (M_y)	Tumor type
1.323	0.116	9.376548	6.877921	Metastasis
1.086	0.107	10.2666	−53.1162	Metastasis
1.362	0.132	8.309972	−92.7287	Metastasis
1.596	0.13	8.318874	−1.54365	Metastasis
1.08	0.124	8.990442	−21.7662	Metastasis
1.264	0.117	9.338148	80.80712	Metastasis
1.592	0.091	11.61715	−45.7294	Metastasis
1.505	0.127	8.538468	−8.44893	Metastasis
1.218	0.087	12.31527	−28.2053	Metastasis
1.188	0.079	13.49998	53.31394	Metastasis
1.509	0.097	10.97197	−74.3629	Metastasis
1.409	0.08	13.20972	12.03159	Metastasis
1.475	0.104	10.29335	65.92807	Metastasis
1.167	0.103	10.56564	−40.1284	Metastasis
1.226	0.101	10.71665	−73.9578	Metastasis
1.239	0.122	9.003824	−56.4214	Metastasis
1.428	0.116	9.32097	18.95457	Metastasis
1.508	0.079	13.32136	−66.1934	Metastasis
1.335	0.094	11.38736	71.95031	Metastasis
1.124	0.088	12.25332	−30.3501	Metastasis
1.26	0.091	11.78266	−81.8184	Metastasis
1.274	0.126	8.721437	−29.5533	Metastasis
1.15	0.118	9.344141	74.19014	Metastasis
1.174	0.104	10.46717	−62.27	Metastasis
1.205	0.127	8.703891	−81.5522	Metastasis
1.452	0.088	12.05234	−75.0656	Metastasis
1.136	0.092	11.74985	40.77281	Metastasis
1.362	0.124	8.798731	−20.6722	Metastasis
1.476	0.105	10.20132	4.502151	Metastasis
1.218	0.112	9.749589	48.25949	Metastasis
1.585	0.079	13.28914	−66.1515	Metastasis
1.41	0.115	9.404872	−11.5027	Metastasis
1.13	0.099	10.98597	12.70309	Metastasis
1.316	0.106	10.19384	71.7167	Metastasis
1.319	0.115	9.453802	46.17437	Metastasis
1.2	0.117	9.380342	−93.6835	Metastasis
1.182	0.079	13.50425	15.68662	Metastasis
1.597	0.123	8.756255	−81.9695	Metastasis
1.528	0.1	10.65445	50.26699	Metastasis
1.361	0.084	12.63952	31.66882	Metastasis
1.241	0.104	10.42119	−56.4295	Metastasis

(continued)

Table 5.2 (continued)

T_1 relaxation time (s)	T_2 relaxation time (s)	T_0 relaxation rate (s^{-1})	Transverse magnetization (M_y)	Tumor type
1.541	0.091	11.63794	75.47696	Metastasis
1.288	0.108	10.03566	−53.089	Metastasis
1.077	0.12	9.261838	103.0235	Metastasis
1.122	0.093	11.64395	−89.7593	Metastasis
1.359	0.109	9.910147	85.77359	Metastasis
1.449	0.125	8.690131	89.98612	Metastasis
1.131	0.122	9.080895	−60.08	Metastasis
1.456	0.091	11.67582	75.10538	Metastasis
1.544	0.107	9.993463	35.03587	Metastasis
1.246	0.079	13.4608	−26.0863	Metastasis
1.413	0.082	12.90284	−43.0309	Metastasis
1.174	0.101	10.75278	91.934	Metastasis
1.199	0.12	9.167362	50.50385	Metastasis
1.422	0.121	8.967698	51.81068	Metastasis
1.519	0.102	10.46225	−24.0659	Metastasis
1.568	0.125	8.637755	84.4341	Metastasis
1.355	0.084	12.64277	63.17877	Metastasis
1.471	0.083	12.728	61.20454	Metastasis
1.419	0.086	12.33263	−77.5134	Metastasis
1.415	0.118	9.18129	−81.0596	Metastasis
1.34	0.096	11.16294	−51.1015	Metastasis
1.104	0.131	8.539385	−86.865	Metastasis
1.466	0.125	8.682128	64.44962	Metastasis
1.255	0.128	8.609313	96.71935	Metastasis
1.148	0.127	8.745096	89.33259	Metastasis
1.533	0.099	10.75333	−56.1132	Metastasis
1.47	0.13	8.37258	−54.4564	Metastasis
1.166	0.119	9.260994	95.83854	Metastasis
1.324	0.085	12.51999	−80.2145	Metastasis
1.566	0.127	8.512585	32.24685	Metastasis
1.561	0.131	8.274203	−49.072	Metastasis
1.368	0.094	11.36929	−35.4347	Metastasis
1.378	0.091	11.7147	82.16134	Metastasis
1.065	0.095	11.46528	−92.5756	Metastasis
1.07	0.098	11.13866	22.97191	Metastasis
1.128	0.088	12.25016	−56.8533	Metastasis
1.077	0.124	8.993021	−40.7171	Metastasis
1.547	0.151	7.268929	−90.5602	Lower_Grade_Glioma
1.701	0.169	6.505049	1.295143	Lower_Grade_Glioma
1.758	0.193	5.750175	49.88712	Lower_Grade_Glioma
1.759	0.157	6.937932	−88.6399	Lower_Grade_Glioma

(continued)

Table 5.2   (continued)

T_1 relaxation time (s)	T_2 relaxation time (s)	T_0 relaxation rate (s^{-1})	Transverse magnetization (M_y)	Tumor type
1.785	0.181	6.085086	95.94904	Lower_Grade_Glioma
1.535	0.139	7.84571	−55.259	Lower_Grade_Glioma
1.674	0.205	5.47542	85.07649	Lower_Grade_Glioma
1.786	0.154	7.053417	88.4925	Lower_Grade_Glioma
1.748	0.138	7.818459	32.69398	Lower_Grade_Glioma
1.796	0.162	6.729632	−74.4401	Lower_Grade_Glioma
1.649	0.198	5.656933	−23.0798	Lower_Grade_Glioma
1.458	0.219	5.252081	−93.7518	Lower_Grade_Glioma
1.579	0.212	5.350293	−28.0516	Lower_Grade_Glioma
1.73	0.147	7.380756	41.28489	Lower_Grade_Glioma
1.712	0.204	5.486073	−65.3519	Lower_Grade_Glioma
1.664	0.203	5.52707	68.98587	Lower_Grade_Glioma
1.449	0.194	5.84477	−43.6426	Lower_Grade_Glioma
1.689	0.164	6.689627	−94.3658	Lower_Grade_Glioma
1.58	0.145	7.529463	35.3302	Lower_Grade_Glioma
1.589	0.19	5.892485	−47.4061	Lower_Grade_Glioma
1.443	0.149	7.40441	40.41829	Lower_Grade_Glioma
1.569	0.202	5.587844	−63.1293	Lower_Grade_Glioma
1.672	0.217	5.206381	−95.228	Lower_Grade_Glioma
1.741	0.167	6.562406	−52.9951	Lower_Grade_Glioma
1.796	0.189	5.847798	−74.8622	Lower_Grade_Glioma
1.555	0.182	6.137592	−102.059	Lower_Grade_Glioma
1.782	0.205	5.439216	−33.5566	Lower_Grade_Glioma
1.736	0.135	7.983444	−70.2944	Lower_Grade_Glioma
1.483	0.191	5.909911	−65.0122	Lower_Grade_Glioma
1.572	0.219	5.202342	112.5284	Lower_Grade_Glioma
1.543	0.14	7.790945	−43.4086	Lower_Grade_Glioma
1.554	0.159	6.932809	84.70809	Lower_Grade_Glioma
1.74	0.152	7.15366	32.96776	Lower_Grade_Glioma
1.763	0.144	7.511659	−22.2004	Lower_Grade_Glioma
1.42	0.21	5.46613	−115.95	Lower_Grade_Glioma
1.728	0.217	5.186999	74.7291	Lower_Grade_Glioma
1.642	0.222	5.113518	12.39626	Lower_Grade_Glioma
1.6	0.167	6.613024	−29.9327	Lower_Grade_Glioma
1.548	0.172	6.459948	97.28402	Lower_Grade_Glioma
1.665	0.149	7.31201	−90.9626	Lower_Grade_Glioma
1.51	0.174	6.409378	−43.4968	Lower_Grade_Glioma
1.576	0.149	7.345927	43.23529	Lower_Grade_Glioma
1.784	0.225	5.004983	54.09704	Lower_Grade_Glioma
1.486	0.178	6.290925	−37.3716	Lower_Grade_Glioma
1.551	0.132	8.220503	51.44249	Lower_Grade_Glioma

(continued)

Table 5.2 (continued)

T_1 relaxation time (s)	T_2 relaxation time (s)	T_0 relaxation rate (s^{-1})	Transverse magnetization (M_y)	Tumor type
1.707	0.21	5.347728	98.85576	Lower_Grade_Glioma
1.438	0.148	7.452167	59.14055	Lower_Grade_Glioma
1.455	0.119	9.090647	−31.807	Lower_Grade_Glioma
1.526	0.217	5.263603	49.103	Lower_Grade_Glioma
1.499	0.157	7.036538	−50.9764	Lower_Grade_Glioma
1.567	0.137	7.937432	27.36947	Lower_Grade_Glioma
1.717	0.135	7.989819	2.056211	Lower_Grade_Glioma
1.762	0.129	8.319475	−29.7695	Lower_Grade_Glioma
1.604	0.141	7.71564	−11.8583	Lower_Grade_Glioma
1.663	0.13	8.293631	−40.8307	Lower_Grade_Glioma
1.79	0.214	5.231556	−48.2148	Lower_Grade_Glioma
1.601	0.208	5.432302	−98.1742	Lower_Grade_Glioma
1.49	0.12	9.004474	71.65378	Lower_Grade_Glioma
1.546	0.14	7.789688	−55.0932	Lower_Grade_Glioma
1.443	0.145	7.589552	70.41394	Lower_Grade_Glioma
1.588	0.217	5.238018	69.36157	Lower_Grade_Glioma
1.714	0.154	7.076937	−86.825	Lower_Grade_Glioma
1.653	0.213	5.299796	−67.2934	Lower_Grade_Glioma
1.431	0.222	5.203317	65.01856	Lower_Grade_Glioma
1.59	0.153	7.164879	−69.7753	Lower_Grade_Glioma
1.695	0.19	5.853128	97.48393	Lower_Grade_Glioma
1.41	0.195	5.837425	108.2415	Lower_Grade_Glioma
1.538	0.214	5.323092	112.909	Lower_Grade_Glioma
1.495	0.206	5.523265	100.1845	Lower_Grade_Glioma
1.775	0.161	6.77456	−45.2505	Lower_Grade_Glioma
1.617	0.187	5.966023	−62.9516	Lower_Grade_Glioma
1.754	0.167	6.558149	−4.31199	Lower_Grade_Glioma
1.466	0.199	5.707254	35.28313	Lower_Grade_Glioma
1.656	0.205	5.481914	106.5644	Lower_Grade_Glioma
1.625	0.179	6.201977	67.67693	Lower_Grade_Glioma
1.657	0.178	6.221478	−77.9776	Lower_Grade_Glioma
1.694	0.204	5.49228	−5.11174	Lower_Grade_Glioma
1.734	0.143	7.569708	−77.4738	Lower_Grade_Glioma
1.625	0.2	5.615385	99.88717	Lower_Grade_Glioma
1.477	0.188	5.996197	1.375445	Lower_Grade_Glioma
1.674	0.127	8.471387	9.934051	Lower_Grade_Glioma
1.407	0.22	5.256187	96.34361	Lower_Grade_Glioma
1.547	0.154	7.139919	14.61066	Lower_Grade_Glioma
1.472	0.14	7.822205	3.203005	Lower_Grade_Glioma
1.499	0.155	7.118724	31.65709	Lower_Grade_Glioma
1.409	0.172	6.523677	72.06508	Lower_Grade_Glioma

(continued)

Table 5.2 (continued)

T_1 relaxation time (s)	T_2 relaxation time (s)	T_0 relaxation rate (s^{-1})	Transverse magnetization (M_y)	Tumor type
1.585	0.211	5.370251	−71.865	Lower_Grade_Glioma
1.552	0.12	8.977663	−84.6459	Lower_Grade_Glioma
1.521	0.181	6.182324	−71.6642	Lower_Grade_Glioma
1.719	0.154	7.07524	−90.6999	Lower_Grade_Glioma
1.594	0.199	5.652478	95.60084	Lower_Grade_Glioma
1.728	0.191	5.814306	82.43275	Lower_Grade_Glioma
1.406	0.122	8.907959	71.72102	Lower_Grade_Glioma
1.545	0.122	8.843971	34.29852	Lower_Grade_Glioma
1.689	0.175	6.306352	18.55257	Lower_Grade_Glioma
1.436	0.192	5.904712	−50.8051	Lower_Grade_Glioma
1.62	0.135	8.024691	−80.5195	Lower_Grade_Glioma
1.772	0.17	6.446687	−53.8762	Lower_Grade_Glioma
1.783	0.122	8.757574	−61.1101	Lower_Grade_Glioma
1.604	0.127	8.497457	52.12411	Lower_Grade_Glioma

5.7 Machine Learning

Machine learning enables computers to learn automatically without human intervention or assistance in such a way that its actions are adjusted accordingly. For this study, the dataset presented in Tables 5.2 and 5.3 need to be tested. The data will be split into three parts: train, test and validation sets, using the scikit-learn Python library. Specifically, we wish to predict the given T_1/T_2 relaxation times, T_0 relaxation rate/transverse magnetization and T0 relaxation rate/NMR signal according to the type of brain tumor based on Eqs. (5.10) and (5.11). The tumour types would be converted to categorical values using label encoding.

5.7.1 Logistic Regression

Logistic regression is a statistical method for analyzing a dataset in which there are one or more independent variables that determine an outcome (Hosmer et al., 2013). The outcome is measured with a dichotomous variable (in which there are only two possible outcomes). With the dataset, the logistic regression model produced tumor classification accuracy of 80% for NMR transverse magnetization M_y and 78.89% for NMR signal S as given in the dataset of Tables 5.2 and 5.3. A summary of classification accuracies for logistic regression and other machine learning models are given in Table. 5.4.

(A)

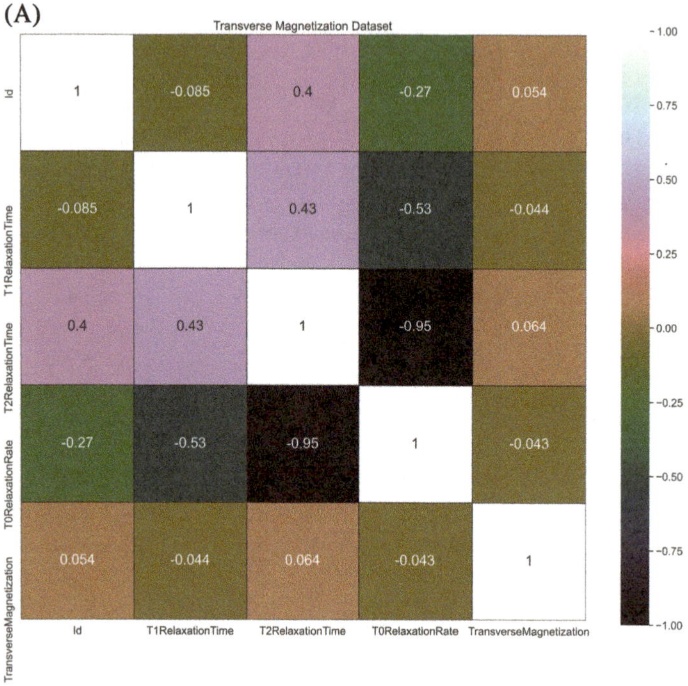

(B)

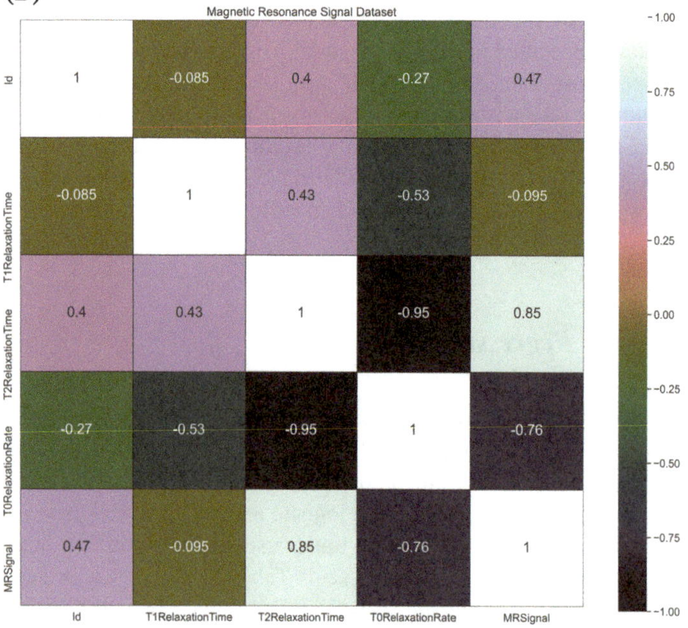

Fig. 5.1 Correlation matrices of features in the dataset of (**a**) Table 5.2, (**b**) Table 5.3

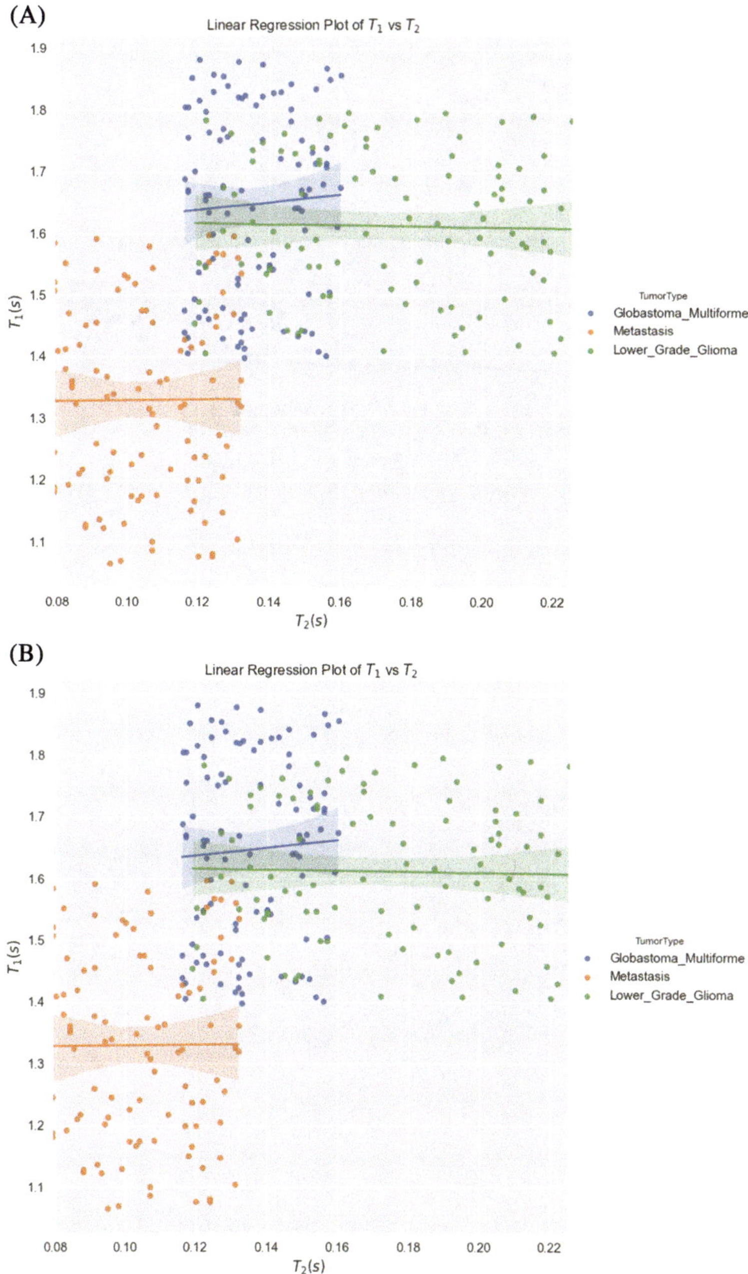

Fig. 5.2 Linear regression plots of T_1 against T_2 using the datasets of (**a**) Table 5.2, (**b**) Table 5.3

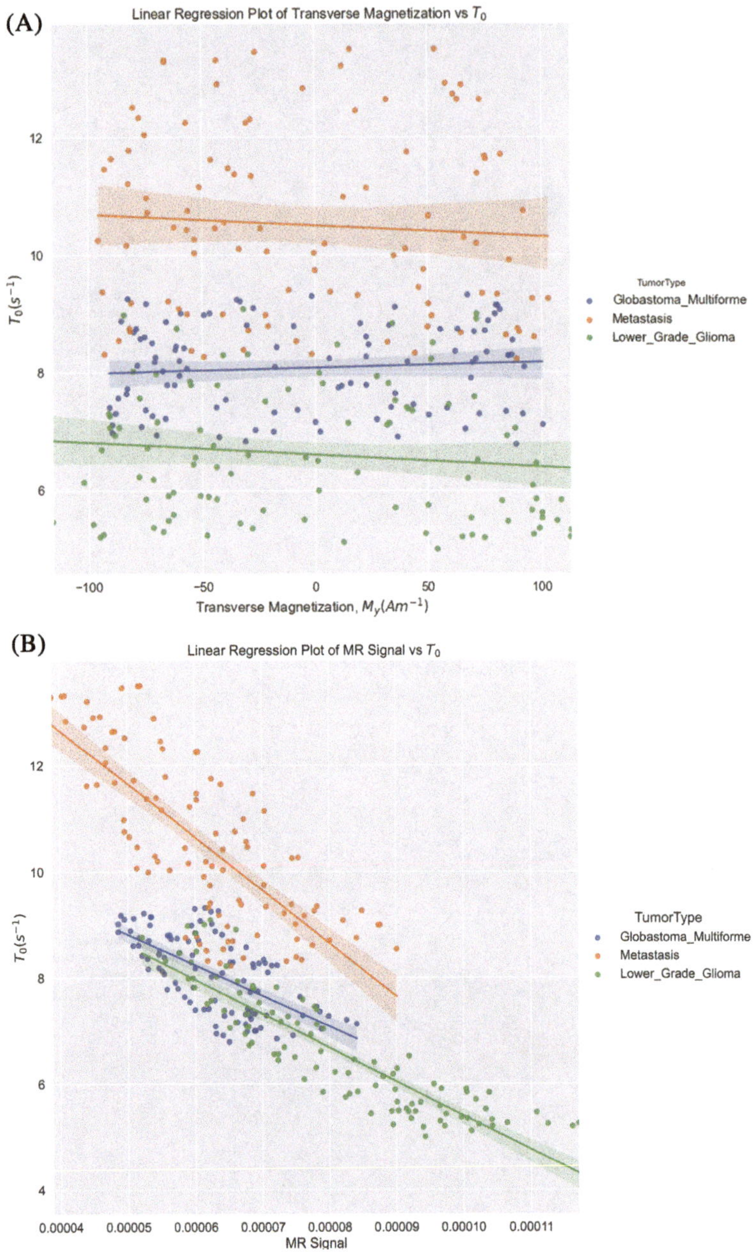

Fig. 5.3 Linear regression plots of T_0 against (**a**) transverse magnetization, (**b**) MR signal using the datasets of Tables 5.2 and 5.3

Table 5.3 Magnetic resonance signal dataset calculated from MR fingerprinting measurements and Eq. (5.11)

T_1 relaxation time (s)	T_2 relaxation time (s)	T_0 relaxation rate (s^{-1})	MR signal (S)	T_1 relaxation time (s)
1.414	0.133	8.226011	7.14E−05	Glioblastoma_Multiforme
1.867	0.156	6.945875	6.31E−05	Glioblastoma_Multiforme
1.708	0.156	6.995737	6.90E−05	Glioblastoma_Multiforme
1.548	0.14	7.788852	6.85E−05	Glioblastoma_Multiforme
1.439	0.116	9.315617	6.15E−05	Glioblastoma_Multiforme
1.713	0.153	7.119719	6.75E−05	Glioblastoma_Multiforme
1.751	0.126	8.50761	5.47E−05	Glioblastoma_Multiforme
1.801	0.146	7.404562	6.13E−05	Glioblastoma_Multiforme
1.636	0.149	7.322656	6.89E−05	Glioblastoma_Multiforme
1.462	0.132	8.259752	6.86E−05	Glioblastoma_Multiforme
1.464	0.12	9.016393	6.25E−05	Glioblastoma_Multiforme
1.732	0.152	7.156315	6.63E−05	Glioblastoma_Multiforme
1.751	0.149	7.282512	6.43E−05	Glioblastoma_Multiforme
1.632	0.128	8.425245	5.96E−05	Glioblastoma_Multiforme
1.782	0.146	7.410482	6.20E−05	Glioblastoma_Multiforme
1.417	0.131	8.339304	7.02E−05	Glioblastoma_Multiforme
1.438	0.152	7.274358	7.99E−05	Glioblastoma_Multiforme
1.849	0.157	6.91026	6.41E−05	Glioblastoma_Multiforme
1.881	0.12	8.864965	4.86E−05	Glioblastoma_Multiforme
1.839	0.128	8.356274	5.29E−05	Glioblastoma_Multiforme
1.67	0.149	7.310212	6.75E−05	Glioblastoma_Multiforme
1.805	0.116	9.174706	4.90E−05	Glioblastoma_Multiforme
1.722	0.143	7.573727	6.29E−05	Glioblastoma_Multiforme
1.769	0.127	8.439307	5.46E−05	Glioblastoma_Multiforme
1.764	0.123	8.696975	5.31E−05	Glioblastoma_Multiforme
1.674	0.16	6.847372	7.21E−05	Glioblastoma_Multiforme
1.405	0.117	9.258752	6.35E−05	Glioblastoma_Multiforme
1.599	0.124	8.689907	5.90E−05	Glioblastoma_Multiforme
1.755	0.117	9.116809	5.09E−05	Glioblastoma_Multiforme
1.83	0.124	8.610964	5.16E−05	Glioblastoma_Multiforme
1.505	0.157	7.033879	7.88E−05	Glioblastoma_Multiforme
1.641	0.148	7.366141	6.82E−05	Glioblastoma_Multiforme
1.712	0.148	7.340869	6.54E−05	Glioblastoma_Multiforme
1.815	0.12	8.884298	5.04E−05	Glioblastoma_Multiforme
1.537	0.12	8.983951	5.95E−05	Glioblastoma_Multiforme
1.856	0.16	6.788793	6.50E−05	Glioblastoma_Multiforme
1.479	0.127	8.550148	6.53E−05	Glioblastoma_Multiforme
1.809	0.136	7.905733	5.70E−05	Glioblastoma_Multiforme
1.549	0.14	7.788435	6.85E−05	Glioblastoma_Multiforme
1.462	0.123	8.814076	6.40E−05	Glioblastoma_Multiforme
1.43	0.117	9.246309	6.24E−05	Glioblastoma_Multiforme

(continued)

Table 5.3 (continued)

T_1 relaxation time (s)	T_2 relaxation time (s)	T_0 relaxation rate (s^{-1})	MR signal (S)	T_1 relaxation time (s)
1.606	0.15	7.289332	7.06E−05	Glioblastoma_Multiforme
1.548	0.122	8.842716	6.00E−05	Glioblastoma_Multiforme
1.641	0.135	8.016792	6.24E−05	Glioblastoma_Multiforme
1.517	0.131	8.292784	6.56E−05	Glioblastoma_Multiforme
1.559	0.124	8.705953	6.05E−05	Glioblastoma_Multiforme
1.542	0.141	7.740707	6.93E−05	Glioblastoma_Multiforme
1.401	0.156	7.124032	8.41E−05	Glioblastoma_Multiforme
1.434	0.148	7.454107	7.81E−05	Glioblastoma_Multiforme
1.527	0.131	8.288467	6.52E−05	Glioblastoma_Multiforme
1.653	0.122	8.801682	5.62E−05	Glioblastoma_Multiforme
1.757	0.128	8.381652	5.54E−05	Glioblastoma_Multiforme
1.662	0.15	7.268351	6.82E−05	Glioblastoma_Multiforme
1.739	0.156	6.9853	6.77E−05	Glioblastoma_Multiforme
1.634	0.123	8.742076	5.73E−05	Glioblastoma_Multiforme
1.399	0.133	8.233593	7.22E−05	Glioblastoma_Multiforme
1.875	0.138	7.77971	5.58E−05	Glioblastoma_Multiforme
1.641	0.147	7.412106	6.78E−05	Glioblastoma_Multiforme
1.476	0.116	9.298196	6.00E−05	Glioblastoma_Multiforme
1.677	0.127	8.470319	5.76E−05	Glioblastoma_Multiforme
1.67	0.117	9.145811	5.34E−05	Glioblastoma_Multiforme
1.567	0.141	7.730361	6.82E−05	Glioblastoma_Multiforme
1.661	0.123	8.732128	5.64E−05	Glioblastoma_Multiforme
1.547	0.137	7.945682	6.72E−05	Glioblastoma_Multiforme
1.855	0.127	8.413099	5.21E−05	Glioblastoma_Multiforme
1.752	0.138	7.817153	5.97E−05	Glioblastoma_Multiforme
1.834	0.153	7.081204	6.30E−05	Glioblastoma_Multiforme
1.493	0.119	9.073154	6.08E−05	Glioblastoma_Multiforme
1.805	0.117	9.101025	4.95E−05	Glioblastoma_Multiforme
1.842	0.146	7.392203	6.00E−05	Glioblastoma_Multiforme
1.829	0.143	7.539754	5.92E−05	Glioblastoma_Multiforme
1.424	0.132	8.278005	7.04E−05	Glioblastoma_Multiforme
1.828	0.141	7.639245	5.84E−05	Glioblastoma_Multiforme
1.797	0.122	8.753204	5.17E−05	Glioblastoma_Multiforme
1.671	0.151	7.220961	6.83E−05	Glioblastoma_Multiforme
1.441	0.132	8.26972	6.95E−05	Glioblastoma_Multiforme
1.821	0.131	8.182737	5.46E−05	Glioblastoma_Multiforme
1.475	0.122	8.874687	6.30E−05	Glioblastoma_Multiforme
1.82	0.132	8.125208	5.51E−05	Glioblastoma_Multiforme
1.701	0.119	8.991251	5.33E−05	Glioblastoma_Multiforme
1.824	0.133	8.067043	5.54E−05	Glioblastoma_Multiforme
1.713	T_2 156	6.994028	6.88E−05	Glioblastoma_Multiforme
1.822	0.138	7.795224	5.74E−05	Glioblastoma_Multiforme

(continued)

Table 5.3 (continued)

T_1 relaxation time (s)	T_2 relaxation time (s)	T_0 relaxation rate (s^{-1})	MR signal (S)	T_1 relaxation time (s)
1.442	0.144	7.637926	7.56E−05	Glioblastoma_Multiforme
1.502	0.151	7.288296	7.60E−05	Glioblastoma_Multiforme
1.688	0.116	9.213107	5.24E−05	Glioblastoma_Multiforme
1.59	0.134	8.091617	6.40E−05	Glioblastoma_Multiforme
1.408	0.154	7.203734	8.26E−05	Glioblastoma_Multiforme
1.429	0.127	8.573806	6.76E−05	Glioblastoma_Multiforme
1.667	0.117	9.146889	5.35E−05	Glioblastoma_Multiforme
1.67	0.132	8.17456	6.00E−05	Glioblastoma_Multiforme
1.47	0.133	8.199069	6.87E−05	Glioblastoma_Multiforme
1.68	0.155	7.046851	6.97E−05	Glioblastoma_Multiforme
1.852	0.118	9.014533	4.86E−05	Glioblastoma_Multiforme
1.446	0.135	8.09897	7.08E−05	Glioblastoma_Multiforme
1.662	0.122	8.798406	5.59E−05	Glioblastoma_Multiforme
1.857	0.124	8.603019	5.08E−05	Glioblastoma_Multiforme
1.442	0.15	7.360148	7.86E−05	Glioblastoma_Multiforme
1.879	0.131	8.165786	5.30E−05	Glioblastoma_Multiforme
1.611	0.159	6.910041	7.45E−05	Glioblastoma_Multiforme
1.202	0.094	11.47024	6.03E−05	Metastasis
1.552	0.082	12.83945	4.10E−05	Metastasis
1.579	0.106	10.06727	5.14E−05	Metastasis
1.238	0.119	9.211116	7.33E−05	Metastasis
1.348	0.103	10.45058	5.86E−05	Metastasis
1.308	0.107	10.11032	6.26E−05	Metastasis
1.535	0.132	8.227223	6.53E−05	Metastasis
1.351	0.084	12.64495	4.82E−05	Metastasis
1.554	0.123	8.773582	6.03E−05	Metastasis
1.38	0.082	12.91976	4.61E−05	Metastasis
1.211	0.086	12.45367	5.50E−05	Metastasis
1.522	0.079	13.31526	4.04E−05	Metastasis
1.363	0.111	9.742685	6.23E−05	Metastasis
1.213	0.095	11.35072	6.03E−05	Metastasis
1.325	0.131	8.388305	7.51E−05	Metastasis
1.1	0.107	10.25489	7.45E−05	Metastasis
1.176	0.108	10.1096	7.03E−05	Metastasis
1.32	0.132	8.333333	7.59E−05	Metastasis
1.192	0.083	12.88712	5.40E−05	Metastasis
1.375	0.106	10.16123	5.91E−05	Metastasis
1.246	0.096	11.21923	5.93E−05	Metastasis
1.456	0.107	10.03261	5.63E−05	Metastasis
1.323	0.116	9.376548	6.69E−05	Metastasis
1.086	0.107	10.2666	7.54E−05	Metastasis
1.362	0.132	8.309972	7.36E−05	Metastasis

(continued)

Table 5.3 (continued)

T_1 relaxation time (s)	T_2 relaxation time (s)	T_0 relaxation rate (s^{-1})	MR signal (S)	T_1 relaxation time (s)
1.596	0.13	8.318874	6.19E−05	Metastasis
1.08	0.124	8.990442	8.74E−05	Metastasis
1.264	0.117	9.338148	7.06E−05	Metastasis
1.592	0.091	11.61715	4.41E−05	Metastasis
1.505	0.127	8.538468	6.42E−05	Metastasis
1.218	0.087	12.31527	5.53E−05	Metastasis
1.188	0.079	13.49998	5.17E−05	Metastasis
1.509	0.097	10.97197	4.95E−05	Metastasis
1.409	0.08	13.20972	4.42E−05	Metastasis
1.475	0.104	10.29335	5.41E−05	Metastasis
1.167	0.103	10.56564	6.77E−05	Metastasis
1.226	0.101	10.71665	6.32E−05	Metastasis
1.239	0.122	9.003824	7.50E−05	Metastasis
1.428	0.116	9.32097	6.20E−05	Metastasis
1.508	0.079	13.32136	4.08E−05	Metastasis
1.335	0.094	11.38736	5.43E−05	Metastasis
1.124	0.088	12.25332	6.05E−05	Metastasis
1.26	0.091	11.78266	5.57E−05	Metastasis
1.274	0.126	8.721437	7.52E−05	Metastasis
1.15	0.118	9.344141	7.82E−05	Metastasis
1.174	0.104	10.46717	6.79E−05	Metastasis
1.205	0.127	8.703891	8.01E−05	Metastasis
1.452	0.088	12.05234	4.69E−05	Metastasis
1.136	0.092	11.74985	6.25E−05	Metastasis
1.362	0.124	8.798731	6.93E−05	Metastasis
1.476	0.105	10.20132	5.45E−05	Metastasis
1.218	0.112	9.749589	7.03E−05	Metastasis
1.585	0.079	13.28914	3.88E−05	Metastasis
1.41	0.115	9.404872	6.23E−05	Metastasis
1.13	0.099	10.98597	6.73E−05	Metastasis
1.316	0.106	10.19384	6.17E−05	Metastasis
1.319	0.115	9.453802	6.66E−05	Metastasis
1.2	0.117	9.380342	7.44E−05	Metastasis
1.182	0.079	13.50425	5.20E−05	Metastasis
1.597	0.123	8.756255	5.86E−05	Metastasis
1.528	0.1	10.65445	5.03E−05	Metastasis
1.361	0.084	12.63952	4.79E−05	Metastasis
1.241	0.104	10.42119	6.42E−05	Metastasis
1.541	0.091	11.63794	4.56E−05	Metastasis
1.288	0.108	10.03566	6.42E−05	Metastasis
1.077	0.12	9.261838	8.49E−05	Metastasis
1.122	0.093	11.64395	6.39E−05	Metastasis

(continued)

Table 5.3 (continued)

T_1 relaxation time (s)	T_2 relaxation time (s)	T_0 relaxation rate (s^{-1})	MR signal (S)	T_1 relaxation time (s)
1.359	0.109	9.910147	6.14E−05	Metastasis
1.449	0.125	8.690131	6.56E−05	Metastasis
1.131	0.122	9.080895	8.21E−05	Metastasis
1.456	0.091	11.67582	4.82E−05	Metastasis
1.544	0.107	9.993463	5.31E−05	Metastasis
1.246	0.079	13.4608	4.93E−05	Metastasis
1.413	0.082	12.90284	4.51E−05	Metastasis
1.174	0.101	10.75278	6.60E−05	Metastasis
1.199	0.12	9.167362	7.63E−05	Metastasis
1.422	0.121	8.967698	6.48E−05	Metastasis
1.519	0.102	10.46225	5.15E−05	Metastasis
1.568	0.125	8.637755	6.07E−05	Metastasis
1.355	0.084	12.64277	4.81E−05	Metastasis
1.471	0.083	12.728	4.38E−05	Metastasis
1.419	0.086	12.33263	4.69E−05	Metastasis
1.415	0.118	9.18129	6.36E−05	Metastasis
1.34	0.096	11.16294	5.51E−05	Metastasis
1.104	0.131	8.539385	9.01E−05	Metastasis
1.466	0.125	8.682128	6.49E−05	Metastasis
1.255	0.128	8.609313	7.75E−05	Metastasis
1.148	0.127	8.745096	8.41E−05	Metastasis
1.533	0.099	10.75333	4.96E−05	Metastasis
1.47	0.13	8.37258	6.72E−05	Metastasis
1.166	0.119	9.260994	7.78E−05	Metastasis
1.324	0.085	12.51999	4.97E−05	Metastasis
1.566	0.127	8.512585	6.17E−05	Metastasis
1.561	0.131	8.274203	6.37E−05	Metastasis
1.368	0.094	11.36929	5.29E−05	Metastasis
1.378	0.091	11.7147	5.10E−05	Metastasis
1.065	0.095	11.46528	6.87E−05	Metastasis
1.07	0.098	11.13866	7.04E−05	Metastasis
1.128	0.088	12.25016	6.03E−05	Metastasis
1.077	0.124	8.993021	8.76E−05	Metastasis
1.547	0.151	7.268929	7.38E−05	Lower_Grade_Glioma
1.701	0.169	6.505049	7.48E−05	Lower_Grade_Glioma
1.758	0.193	5.750175	8.24E−05	Lower_Grade_Glioma
1.759	0.157	6.937932	6.74E−05	Lower_Grade_Glioma
1.785	0.181	6.085086	7.62E−05	Lower_Grade_Glioma
1.535	0.139	7.84571	6.86E−05	Lower_Grade_Glioma
1.674	0.205	5.47542	9.18E−05	Lower_Grade_Glioma
1.786	0.154	7.053417	6.51E−05	Lower_Grade_Glioma
1.748	0.138	7.818459	5.99E−05	Lower_Grade_Glioma

(continued)

Table 5.3 (continued)

T_1 relaxation time (s)	T_2 relaxation time (s)	T_0 relaxation rate (s^{-1})	MR signal (S)	T_1 relaxation time (s)
1.796	0.162	6.729632	6.80E−05	Lower_Grade_Glioma
1.649	0.198	5.656933	9.01E−05	Lower_Grade_Glioma
1.458	0.219	5.252081	0.000112	Lower_Grade_Glioma
1.579	0.212	5.350293	0.000101	Lower_Grade_Glioma
1.73	0.147	7.380756	6.43E−05	Lower_Grade_Glioma
1.712	0.204	5.486073	8.93E−05	Lower_Grade_Glioma
1.664	0.203	5.52707	9.15E−05	Lower_Grade_Glioma
1.449	0.194	5.84477	0.0001	Lower_Grade_Glioma
1.689	0.164	6.689627	7.32E−05	Lower_Grade_Glioma
1.58	0.145	7.529463	6.95E−05	Lower_Grade_Glioma
1.589	0.19	5.892485	8.98E−05	Lower_Grade_Glioma
1.443	0.149	7.40441	7.81E−05	Lower_Grade_Glioma
1.569	0.202	5.587844	9.65E−05	Lower_Grade_Glioma
1.672	0.217	5.206381	9.72E−05	Lower_Grade_Glioma
1.741	0.167	6.562406	7.23E−05	Lower_Grade_Glioma
1.796	0.189	5.847798	7.90E−05	Lower_Grade_Glioma
1.555	0.182	6.137592	8.80E−05	Lower_Grade_Glioma
1.782	0.205	5.439216	8.62E−05	Lower_Grade_Glioma
1.736	0.135	7.983444	5.90E−05	Lower_Grade_Glioma
1.483	0.191	5.909911	9.67E−05	Lower_Grade_Glioma
1.572	0.219	5.202342	0.000104	Lower_Grade_Glioma
1.543	0.14	7.790945	6.87E−05	Lower_Grade_Glioma
1.554	0.159	6.932809	7.72E−05	Lower_Grade_Glioma
1.74	0.152	7.15366	6.60E−05	Lower_Grade_Glioma
1.763	0.144	7.511659	6.18E−05	Lower_Grade_Glioma
1.42	0.21	5.46613	0.000111	Lower_Grade_Glioma
1.728	0.217	5.186999	9.40E−05	Lower_Grade_Glioma
1.642	0.222	5.113518	0.000101	Lower_Grade_Glioma
1.6	0.167	6.613024	7.87E−05	Lower_Grade_Glioma
1.548	0.172	6.459948	8.37E−05	Lower_Grade_Glioma
1.665	0.149	7.31201	6.77E−05	Lower_Grade_Glioma
1.51	0.174	6.409378	8.67E−05	Lower_Grade_Glioma
1.576	0.149	7.345927	7.15E−05	Lower_Grade_Glioma
1.784	0.225	5.004983	9.43E−05	Lower_Grade_Glioma
1.486	0.178	6.290925	9.01E−05	Lower_Grade_Glioma
1.551	0.132	8.220503	6.46E−05	Lower_Grade_Glioma
1.707	0.21	5.347728	9.22E−05	Lower_Grade_Glioma
1.438	0.148	7.452167	7.78E−05	Lower_Grade_Glioma
1.455	0.119	9.090647	6.23E−05	Lower_Grade_Glioma
1.526	0.217	5.263603	0.000106	Lower_Grade_Glioma
1.499	T_0.157	7.036538	7.91E−05	Lower_Grade_Glioma
1.567	0.137	7.937432	6.63E−05	Lower_Grade_Glioma

(continued)

Table 5.3 (continued)

T_1 relaxation time (s)	T_2 relaxation time (s)	T_0 relaxation rate (s^{-1})	MR signal (S)	T_1 relaxation time (s)
1.717	0.135	7.989819	5.97E−05	Lower_Grade_Glioma
1.762	0.129	8.319475	5.56E−05	Lower_Grade_Glioma
1.604	0.141	7.71564	6.66E−05	Lower_Grade_Glioma
1.663	0.13	8.293631	5.94E−05	Lower_Grade_Glioma
1.79	0.214	5.231556	8.95E−05	Lower_Grade_Glioma
1.601	0.208	5.432302	9.74E−05	Lower_Grade_Glioma
1.49	0.12	9.004474	6.14E−05	Lower_Grade_Glioma
1.546	0.14	7.789688	6.86E−05	Lower_Grade_Glioma
1.443	0.145	7.589552	7.60E−05	Lower_Grade_Glioma
1.588	0.217	5.238018	0.000102	Lower_Grade_Glioma
1.714	0.154	7.076937	6.79E−05	Lower_Grade_Glioma
1.653	0.213	5.299796	9.65E−05	Lower_Grade_Glioma
1.431	0.222	5.203317	0.000116	Lower_Grade_Glioma
1.59	0.153	7.164879	7.27E−05	Lower_Grade_Glioma
1.695	0.19	5.853128	8.42E−05	Lower_Grade_Glioma
1.41	0.195	5.837425	0.000104	Lower_Grade_Glioma
1.538	0.214	5.323092	0.000104	Lower_Grade_Glioma
1.495	0.206	5.523265	0.000103	Lower_Grade_Glioma
1.775	0.161	6.77456	6.84E−05	Lower_Grade_Glioma
1.617	0.187	5.966023	8.69E−05	Lower_Grade_Glioma
1.754	0.167	6.558149	7.17E−05	Lower_Grade_Glioma
1.466	0.199	5.707254	0.000102	Lower_Grade_Glioma
1.656	0.205	5.481914	9.28E−05	Lower_Grade_Glioma
1.625	0.179	6.201977	8.28E−05	Lower_Grade_Glioma
1.657	0.178	6.221478	8.08E−05	Lower_Grade_Glioma
1.694	0.204	5.49228	9.03E−05	Lower_Grade_Glioma
1.734	0.143	7.569708	6.24E−05	Lower_Grade_Glioma
1.625	0.2	5.615385	9.23E−05	Lower_Grade_Glioma
1.477	0.188	5.996197	9.56E−05	Lower_Grade_Glioma
1.674	0.127	8.471387	5.77E−05	Lower_Grade_Glioma
1.407	0.22	5.256187	0.000117	Lower_Grade_Glioma
1.547	0.154	7.139919	7.52E−05	Lower_Grade_Glioma
1.472	0.14	7.822205	7.21E−05	Lower_Grade_Glioma
1.499	0.155	7.118724	7.81E−05	Lower_Grade_Glioma
1.409	0.172	6.523677	9.19E−05	Lower_Grade_Glioma
1.585	0.211	5.370251	9.97E−05	Lower_Grade_Glioma
1.552	0.12	8.977663	5.89E−05	Lower_Grade_Glioma
1.521	0.181	6.182324	8.95E−05	Lower_Grade_Glioma
1.719	0.154	7.07524	6.77E−05	Lower_Grade_Glioma
1.594	0.199	5.652478	9.36E−05	Lower_Grade_Glioma
1.728	0.191	5.814306	8.30E−05	Lower_Grade_Glioma
1.406	0.122	8.907959	6.61E−05	Lower_Grade_Glioma

(continued)

Table 5.3 (continued)

T_1 relaxation time (s)	T_2 relaxation time (s)	T_0 relaxation rate (s^{-1})	MR signal (S)	T_1 relaxation time (s)
1.545	0.122	8.843971	6.01E−05	Lower_Grade_Glioma
1.689	0.175	6.306352	7.80E−05	Lower_Grade_Glioma
1.436	0.192	5.904712	0.0001	Lower_Grade_Glioma
1.62	0.135	8.024691	6.32E−05	Lower_Grade_Glioma
1.772	0.17	6.446687	7.23E−05	Lower_Grade_Glioma
1.783	0.122	8.757574	5.21E−05	Lower_Grade_Glioma
1.604	0.127	8.497457	6.02E−05	Lower_Grade_Glioma

5.7.2   Support Vector Machine

Support Vector Machine (SVM) is a supervised machine learning algorithm which can be used for both classification or regression problems. However, it is mostly used in classification problems. In this algorithm, each data item is plotted as a point in n-dimensional space (where n is number of features available) with the value of each feature being the value of a particular coordinate (Ozbas et al., 2019). Classification is then performed by finding the hyper-plane that differentiates the two classes very well. Support vectors are simply the co-ordinates of individual observation. SVM is a frontier which best segregates the two classes (hyper-plane/line). In the dataset, the SVM model produced 78.89% accuracy for Table 5.2 and 80% accuracy for Table 5.3.

5.7.3   Naive Bayes

Naive Bayes is a simple, yet effective and commonly-used, machine learning classifier. It is a probabilistic classifier that makes classifications using the Maximum A Posteriori decision rule in a Bayesian setting (Archana & Elangovan, 2014). It can also be represented using a very simple Bayesian network. In the neural network developed for the dataset of this study, Naive Bayes model produced 84.44% accuracy for Table 5.2 and 83.33% for Table 5.3.

5.7.4   Decision Tree

Decision tree is a type of supervised learning algorithm (having a pre-defined target variable) used in classification problems. It works for both categorical and continuous input and output variables (Lakshmi & Kavila, 2018). In this technique, we

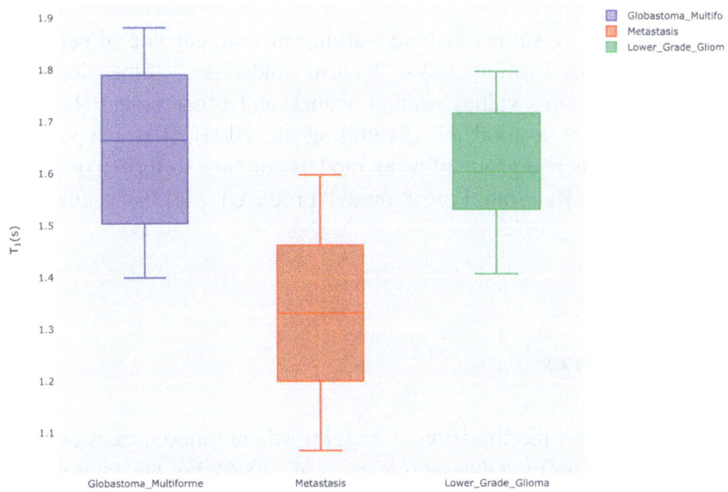

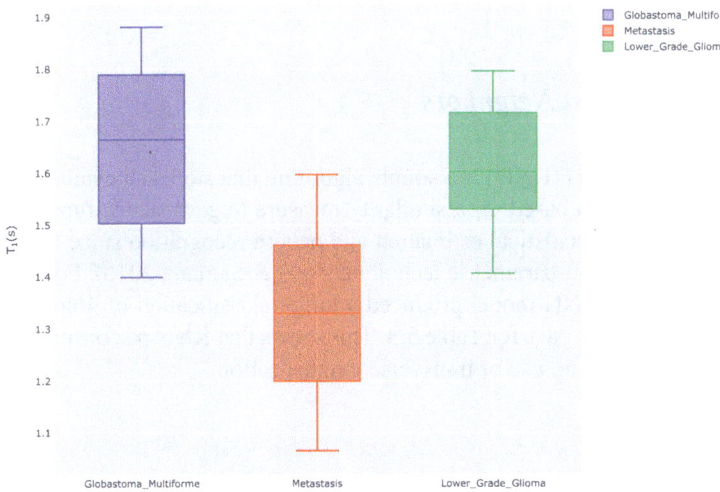

Fig. 5.4 Box plots showing maximum, minimum and the median values of T_1 using the datasets of (**a**) Table 5.2, (**b**) Table 5.3

split the population or sample into two or more homogeneous sets (or sub-populations) based on most significant splitter/differentiator in input variables. In this study, the decision tree model produced a classification accuracy of 75.56% both for Tables 5.2 and 5.3.

5.7.5   Random Forest

Random Forest is a versatile machine learning method capable of performing both regression and classification tasks. It also undertakes dimensional reduction methods, treats missing values, outlier values and other essential steps of data exploration, and does a good job (Svetnik et al., 2003). It is a type of ensemble learning method, where a group of weak models combine to form a powerful model. In the dataset, the Random Forest model produced 81.11% accuracy for both Tables 5.2 and 5.3.

5.7.6   Extra Trees

Extra Trees is another modification of bagging where random trees are constructed from samples of the training dataset (Geurts et al., 2006). For the dataset employed in this study, the Extra Treesmodel produced 75.56% tumor classification accuracy for Table 5.2 and 73.33% for Table 5.3.

5.7.7   K Nearest Neighbors

K Nearest Neighbors (KNN) is a simple algorithm that stores all available cases and classifies new cases based on a similarity measure (e.g., distance functions). KNN had been in use in statistical estimation and pattern recognition since the beginning of the 1970s as a non-parametric technique (Naik & Samant, 2016). For the analysis in this study, the KNN model produced a tumor classification of 46.67% accuracy for Table 5.2 and 77.78% for Table 5.3. This shows that KNN performed very poorly for the dataset making use of transverse magnetization.

5.7.8   XGBoost

XGBoost is a useful algorithm for machine learning; its biggest advantage is its scalability, which drives fast learning through parallel and distributed computing and offers efficient memory usage (Sundar, 2018). For the dataset in this study, XGBoost model produced 80% accuracy for Table 5.2 and 76.67% for Table 5.3.

5.8   Shallow Deep Learning

Deep learning is a component of machine learning in which the developed algorithms function like machine learning algorithms but with various layers, such that each algorithm provides different interpretations of the data employed. This network of algorithms is called artificial neural networks.They resemble the neural connections in the human brain. The datasets of Tables 5.2 and 5.3 were divided into two parts such that 30% were used as a test set and 70% as a training set. Binary cross entropy loss was then used for classification of the data (with Adam optimizer). The artificial neural network has two layers and was trained over 150 epochs. The accuracy and loss for the test/training set are given in Figs. 5.5 and 5.6. As shown in these plots, the final accuracy for the test set is around 84% for Table 5.2 and 89.5% for Table 5.3. The behaviour of the training set produced a value of 83% for Table 5.2 and 84.5% for Table 5.3. As shown in Fig. 5.6, the final loss values for both test and training sets were found to be below 0.35 for Table 5.2 and 0.30 for Table 5.3. Accuracy and loss values were obtained to determine the performance of the classification of tumors on each iteration and often, the accuracy could be used to show optimistic responses if any bias is noticed in the data. We see from the results of these profiles that shallow neural network produced improved training accuracies when compared to classical machine learning techniques presented in Sect. 5.7.

5.9   Deep Neural Network

Deep learning is useful for extraction of high-level abstractions from data. This machine learning technique is executed by making use of hierarchical architectures, an emerging approach that has found various applications in traditional artificial intelligence (Khalifa et al., 2019). To extract high-level information from the datasets of Tables 5.2 and 5.3, we have made use of a nine-layer neural network which uses categorical_crossentropy loss for the classification of datasets (with Adam optimizer). Just like in the earlier section, the dataset was also divided into 30% test set and 70% training set. The model accuracy and loss for the test/training sets are shown in Figs. 5.7 and 5.8. Figure 5.7a shows that for the dataset of Table 5.2, the accuracy of the test set is 82% while the accuracy of the training set is 85%. In Fig. 5.7b, the dataset of Table 5.2 produced a final accuracy is 82% for test set while the final accuracy for the training set is 86%. In Fig. 5.8a, the final losses for both test and training sets are almost the same at 0.41 while in Fig. 5.8b, the final loss is 0.45 for test set and 0.35 for training set. These results demonstrated an improvement on the shallow neural network and the classical machine learning models we have employed in earlier sections. The advantage of the analysis here is that deep learning results can accommodate much more data as normally expected in clinical settings so that the models developed in this study can be seemlessly moved into real-time clinical diagnosis of brain tumors.

(A)

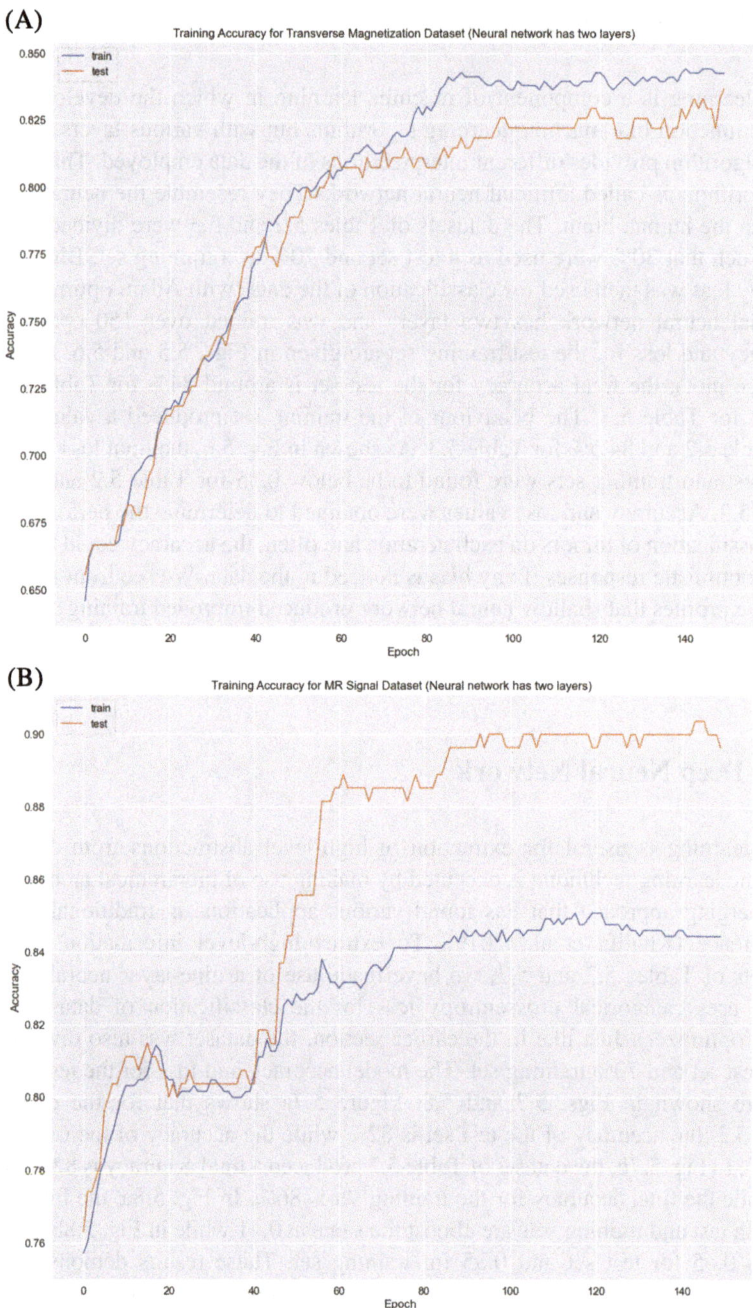

(B)

Fig. 5.5 Profile of the training accuracy of a shallow neural network executed on the datasets of (**a**) Table 5.2, (**b**) Table 5.3

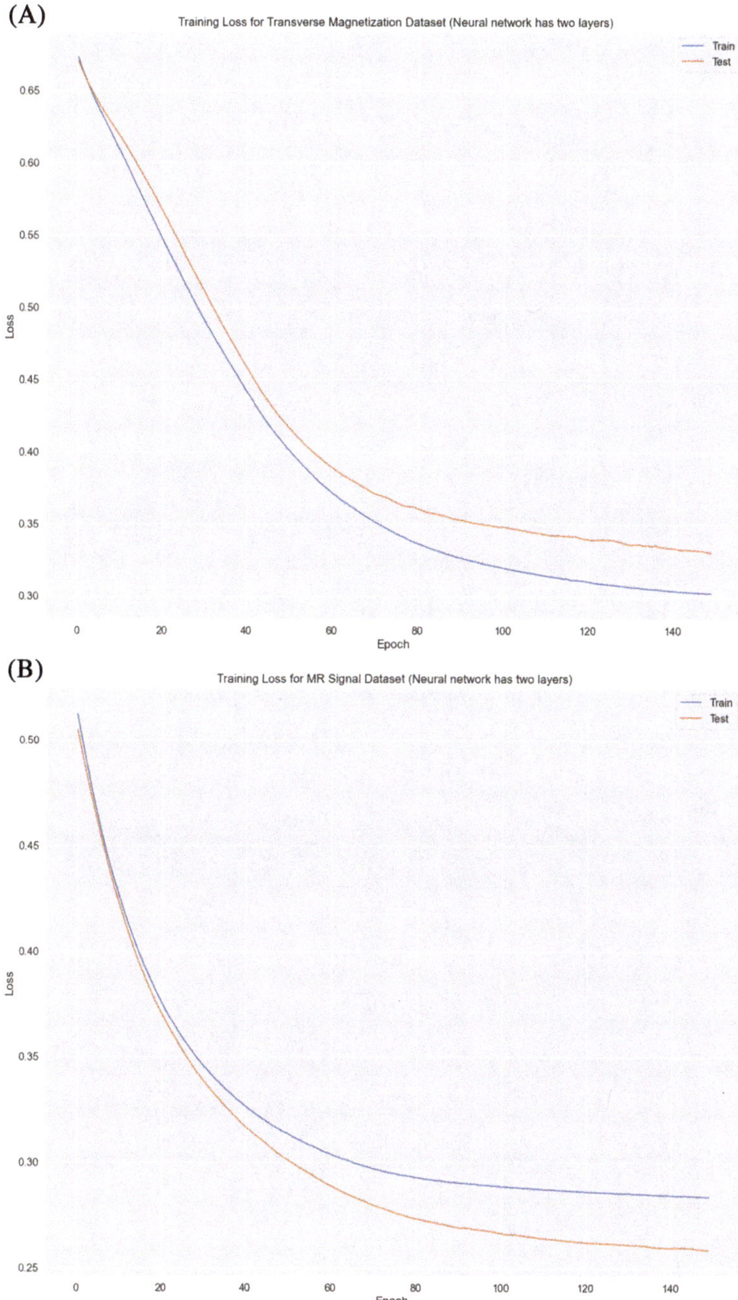

Fig. 5.6 Profile of the training loss of a shallow neural network executed on the datasets of (**a**) Table 5.2, (**b**) Table 5.3

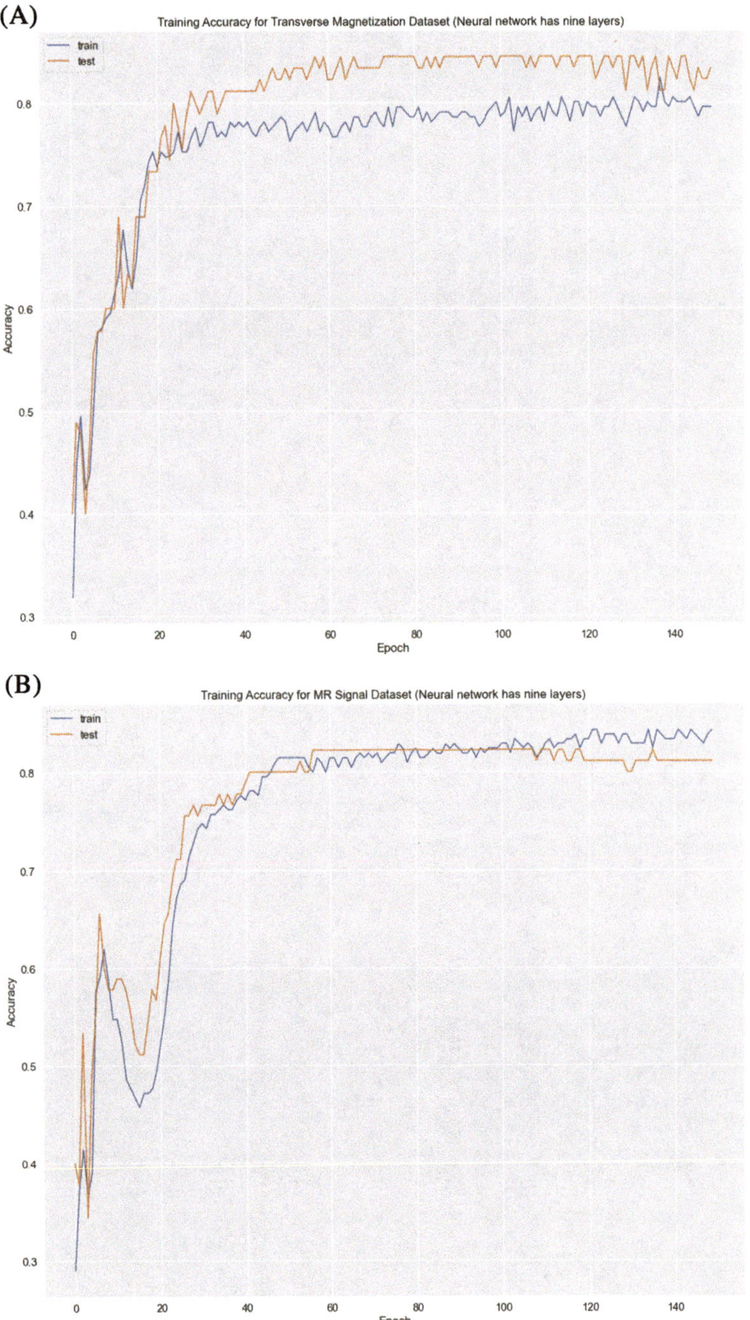

Fig. 5.7 Profile of the training accuracy of a deep neural network executed on the datasets of (**a**) Table 5.2, (**b**) Table 5.3

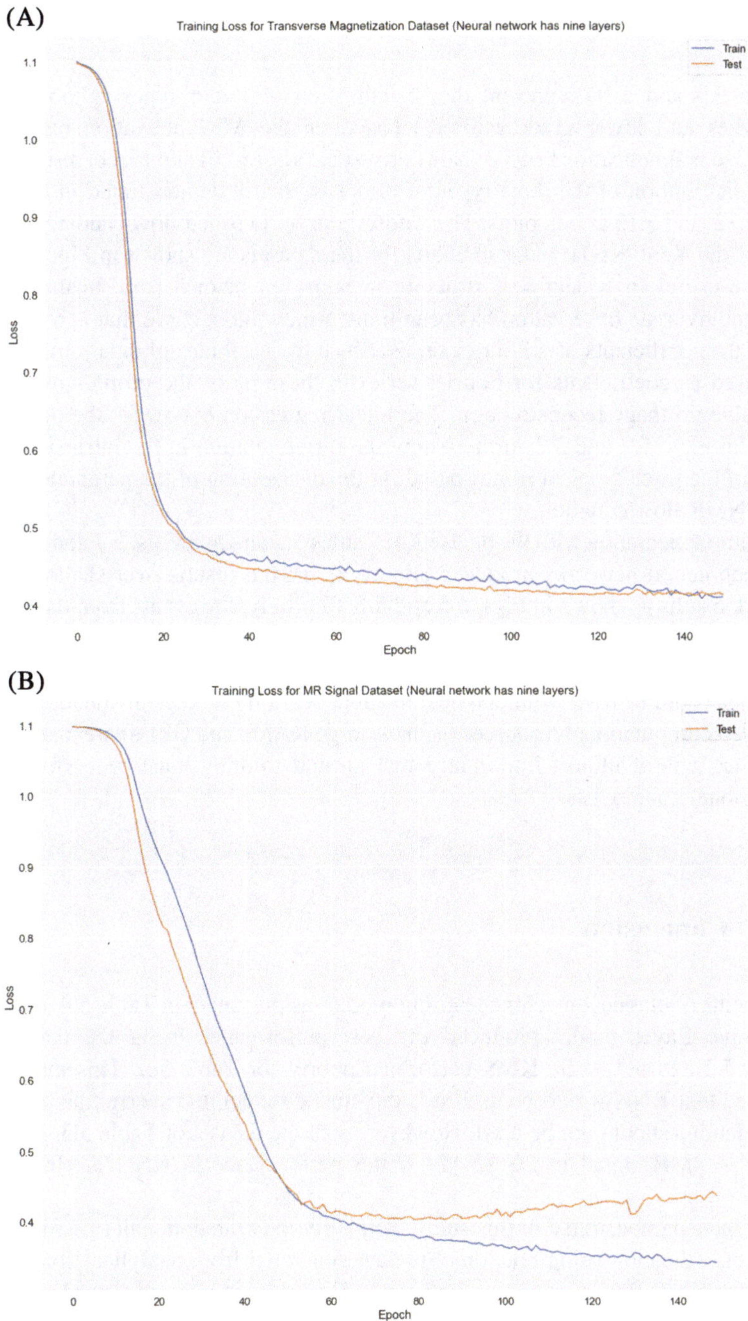

Fig. 5.8 Profile of the training loss of a deep neural network executed on the datasets of (**a**) Table 5.2, (**b**) Table 5.3

5.10   Discussions

Figures 5.9 and 5.10 represent the visualization of tumor types (glioblastomas, metastases and lower grade gliomas) based on the MRI relaxation parameters, transverse magnetizations and signals derived in Eqs. (5.10) and (5.11) respectively. Value distribution of the three types of tumor is clearly demonstrated in Figs. 5.4, 5.11, 5.12, and 5.13 as box plots. The Andrew curves (a rolled-down and non-integer form of the Kent–Kiviat radar m chart) for the datasets are shown in Fig. 5.14 and they are useful for visualizing structure in high-dimensional data. In these plots, every multivariate observation has been transformed into a curve that is representative of the coefficients of a Fourier series. Since the attributes of data samples have been used as coefficients for Fourier series in these plots, the profiles may prove applicable in image reconstruction. The linear regression curves for the datasets as shown in Figs. 5.2 and 5.3 further show the differentiation in the intrinsic mechanisms of the three types of tumor based on the distribution of the parameters of the Bloch NMR flow equation.

On implementation with the datasets of Tables 5.2 and 5.3, Figs. 5.7 and 5.8 show that deep neural networks produced slightly improved results over shallow neural network results as shown in Figs. 5.5 and 5.6. Similarly, the results from the shallow neural network demonstrated some improvements over classical machine learning models as shown in Table 5.4. These results show that computational brain tumor diagnosis could be done with classical machine learning models in situations where advanced computational resources (such as large RAMs and GPUs) are unavailable. However, it is of utmost importance that ground truthing must be performed by professional oncologists.

5.11   Conclusion

The results of classical machine learning models as presented in Table 5.4 show that the Naive Bayes model produced the best performance using the datasets of Tables 5.2 and 5.3 while KNN performed poorly for Table 5.2. This means that provided that KNN would be used in a diagnostic model, transverse magnetization computations should not be used. However, with the dataset of Table 5.3 generated from the NMR signal in Eq. (5.11), better performance in KNN is observed at 77.78%.

We have demonstrated in this study the ability to computationally perform brain tumor classification using relaxometric data generated from analytical solutions to the Bloch NMR flow equation in the form of Bessel functions and Bessel properties as shown in Eqs. (5.10) and (5.11) respectively. The results of this study have motivated data-driven predictions required for statistical learning in the clinic and can lead to the development of interactive clinical apps for classifying brain tumors for faster accurate delineation of tumor margins as illustrated in Tables 5.1, 5.2, and

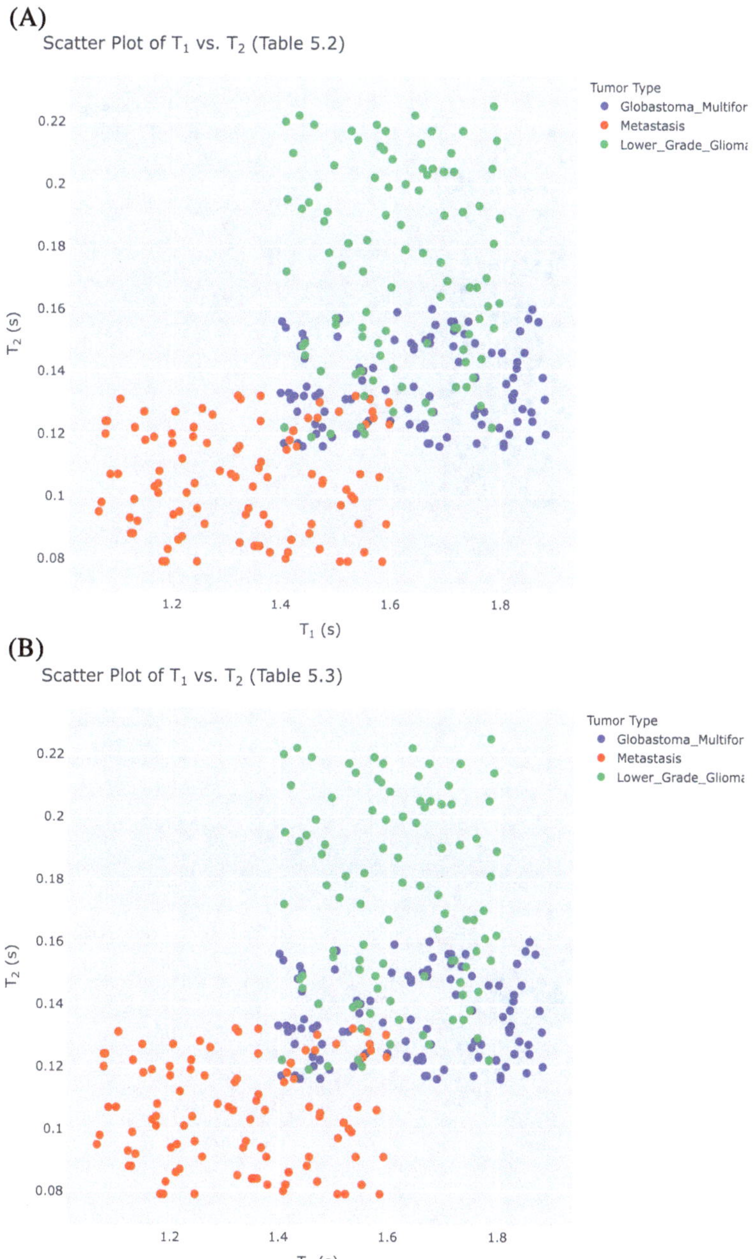

Fig. 5.9 Scatter plots of T1 against T2 for the dataset of (**a**) Table 5.2, (**b**) Table 5.3

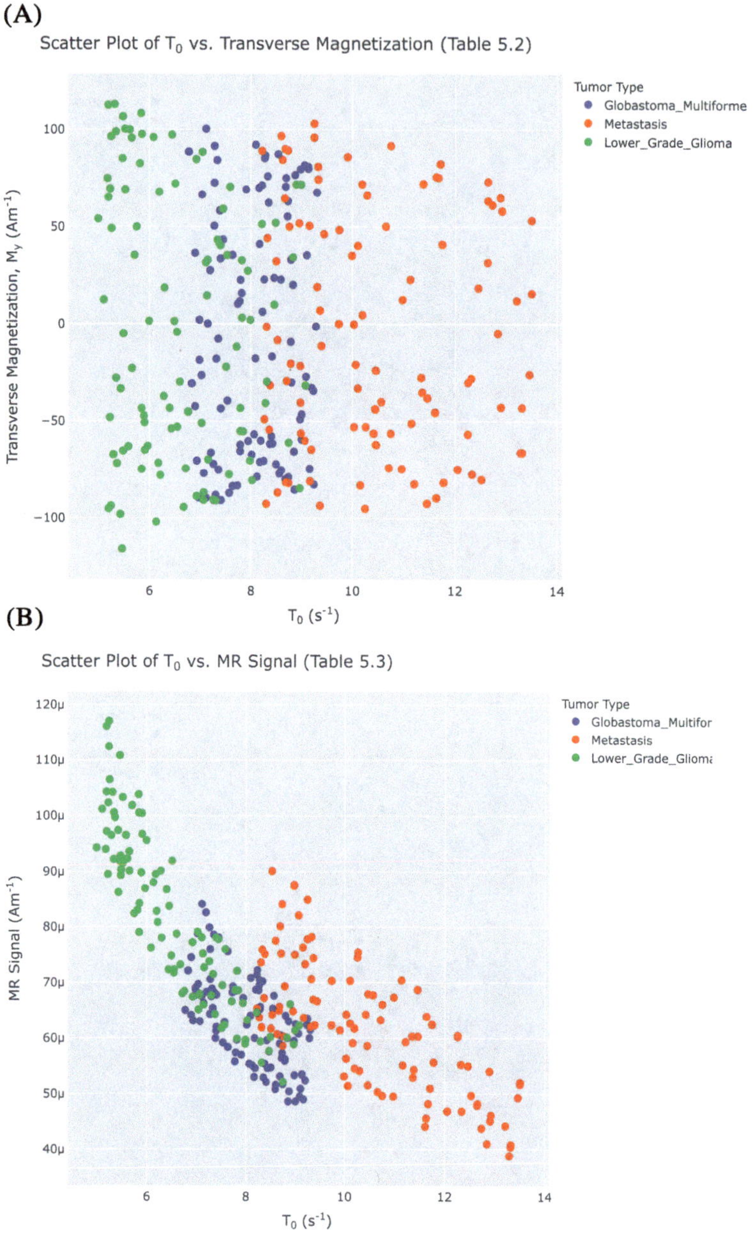

(A)

(B)

Fig. 5.10 Scatter plots of T0 against (**a**) transverse magnetization, (**b**) NMR signal using the data sets of Tables 5.2 and 5.3

(A)

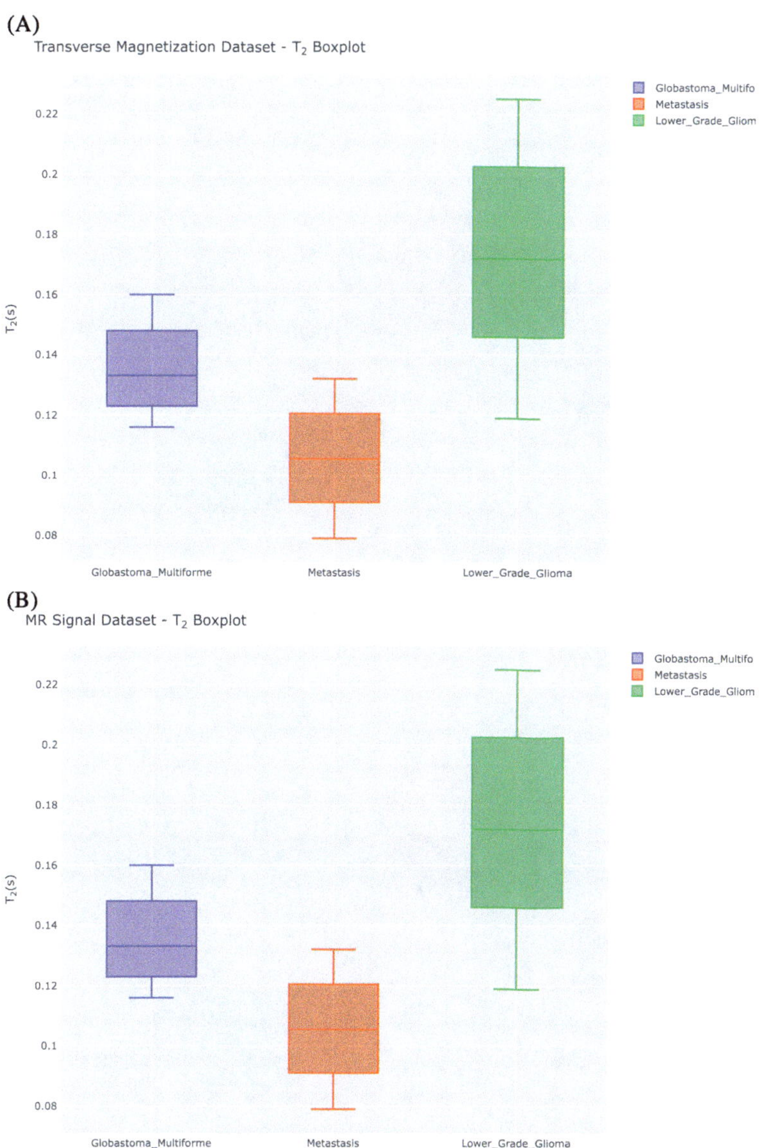

(B)

Fig. 5.11   Box plots showing maximum, minimum and the median values of T2 using the datasets of (**a**) Table 5.2, (**b**) Table 5.3

(A)

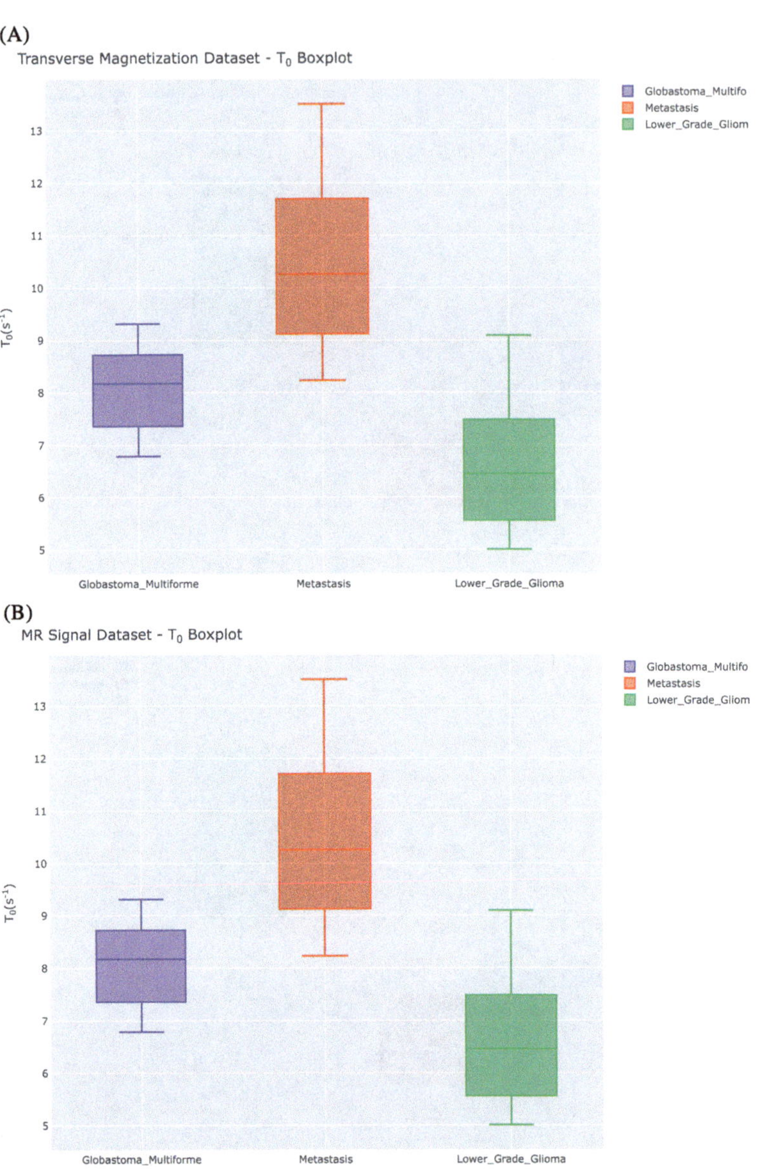

(B)

Fig. 5.12 Box plots showing maximum, minimum and the median values of T0 using the datasets of (**a**) Table 5.2, (**b**) Table 5.3

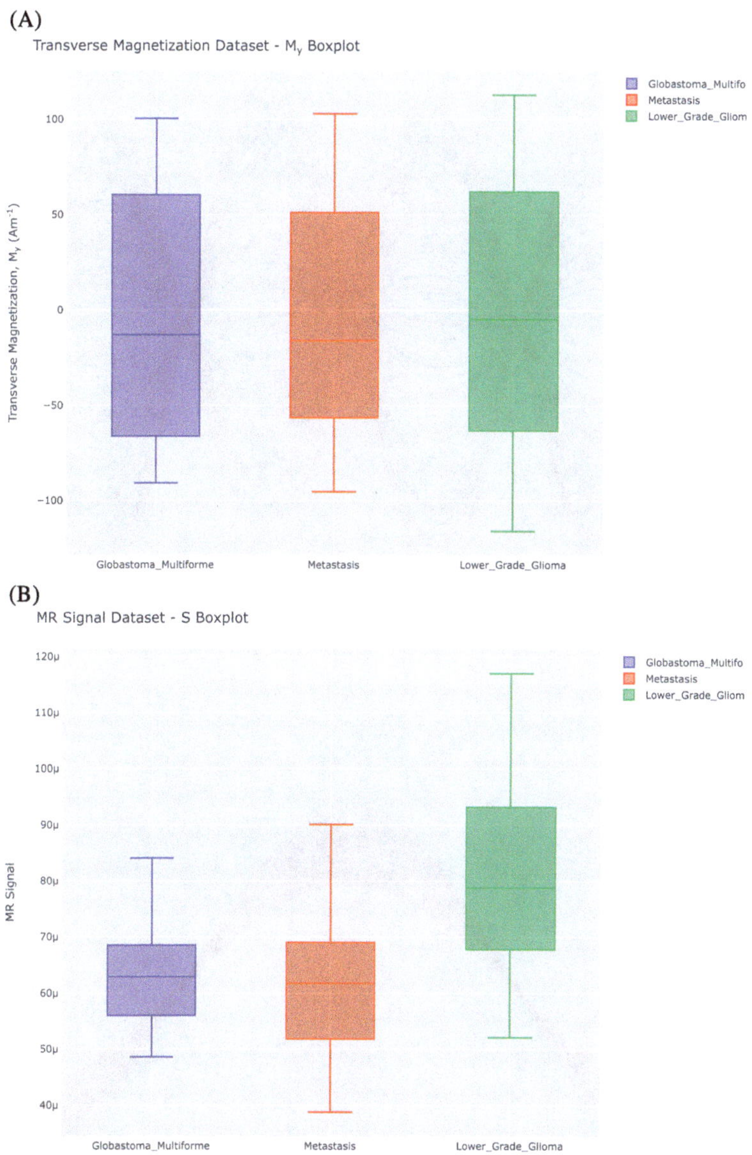

Fig. 5.13 Box plots showing maximum, minimum and the median values of (**a**) transverse magnetization, (**b**) MR signal using the datasets of Tables 5.2 and 5.3

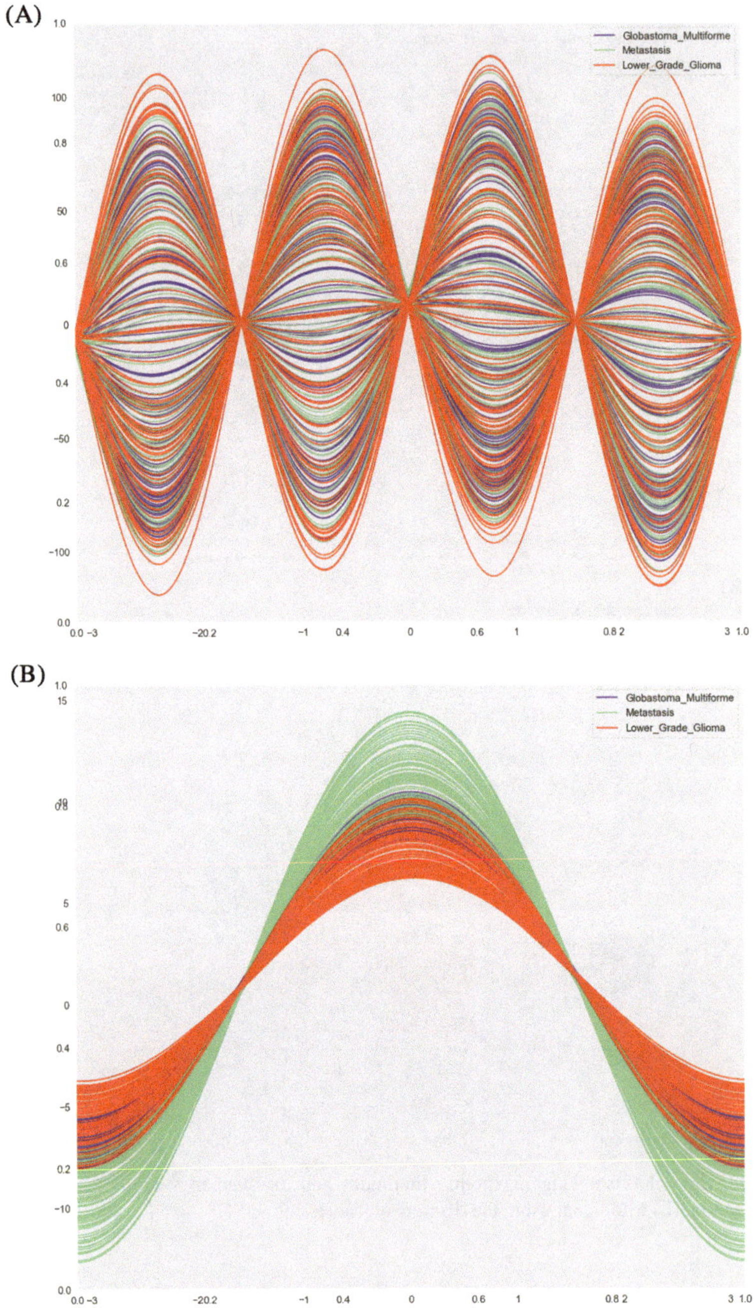

(A)

(B)

Fig. 5.14 Andrew curves for the datasets of (**a**) Table 5.2, (**b**) Table 5.3

Table 5.4 Classification accuracies of different machine learning models from the datasets of Tables 5.2 and 5.3

	Classification accuracy	
Machine learning models	Dataset of Table 5.2 (%)	Dataset of Table 5.3 (%)
Logistic regression	80	78.89
SVM	78.89	80
Naïve Bayes	84.44	83.33
KNN	46.67	77.78
XG boost	80	76.67
Decision tree	75.56	75.56
Random forest	81.11	81.11
Extra tree classifier	75.56	73.33

5.3 and Figs. 5.9, 5.10, 5.4, 5.11, 5.12, 5.13, 5.14, 5.2, and 5.3. The application of these results may raise opportunities and offer the flexibility for the development of new diagnostic and treatment tools for cancer care. This could potentially help in the development of new wearable devices that could revolutionize the diagnosis of brain tumors, heart diseases and neurodegenerative diseases based on the analytical solutions of the Bloch NMR flow equation.

Chapter 6
Analysis of the Hydrogen-Like Atom for Neuro-Oncology Based on Bloch's NMR Flow Equation

Abstract Analytical solutions of Bloch's nuclear magnetic resonance (NMR) flow equations for the hydrogen atom constitute a formidable mathematical problem. It is of such fundamental importance that it merits a lot of attention. This chapter presents classical and quantum mechanical analyses of the Yukawa potential based on a differential equation developed from Bloch's NMR flow equations so that the wave function of hydrogen atoms and hydrogen-like ions can be represented in terms of NMR parameters. The procedure leads to three equations for three spatial variables, and their solutions give rise to the usual quantum numbers associated with the hydrogen energy levels. This allows us to quantitatively evaluate the effect of the hydrogen-like atom with a nucleus charge +Ze and an electron charge –e on the NMR transverse magnetization. The effects of the screening length (defined in the Yukawa potential) on the NMR transverse magnetization at various fluid velocities as discussed in this chapter can provide an excellent model for the NMR study of atomic structure of hydrogen-like particles in general and neuro-oncology specifically. The formulation of a weighting parameter is an interesting physical mechanism that provides a switch between classical and quantum domains. This parameter can be used to tailor laboratory RF pulse to quantum RF energy absorptions for quantum neuro computing.

Keywords Bloch NMR flow equation · Hydrogen-like atom · Yukawa potential · Fluid velocities · Neurocomputing

6.1 Background to Quantum Mechanical Treatment of Bloch Flow Equation

Analytical solutions of Bloch's nuclear magnetic resonance (NMR) flow equations for the hydrogen atom constitute a formidable mathematical problem. It is of such fundamental importance that it merits a lot of attention. This chapter presents quantum and classical mechanical analyses of the Yukawa potential based on a differential equation developed from Bloch's NMR flow equations so that the wave

© The Author(s), under exclusive license to Springer Nature Switzerland AG 2021   219
M. O. Dada, B. O. Awojoyogbe, *Computational Molecular Magnetic Resonance Imaging for Neuro-oncology*, Biological and Medical Physics, Biomedical Engineering, https://doi.org/10.1007/978-3-030-76728-0_6

function of hydrogen atoms and hydrogen-like ions can be represented in terms of NMR parameters. The procedure leads to three equations for three spatial variables, and their solutions give rise to the usual quantum numbers associated with the hydrogen energy levels. For the classical analysis, it is assumed that the NMR transverse magnetization M_y, measures the Yukawa potential. This allows us to quantitatively evaluate the effect of the hydrogen-like atom with a nucleus charge $+Ze$ and an electron charge $-e$ on the NMR transverse magnetization for both low and high instantaneous fluid velocities. The effects of the screening length (defined in the Yukawa potential) on the NMR transverse magnetization at various fluid velocities as discussed in this chapter can provide an excellent model for the NMR study of atomic structure of hydrogen-like particles in general and neuro-oncology specifically.

6.2   Quantum Mechanical Understanding of Classical NMR/MRI

NMR is widely used in physics, chemistry, biology and medicine to characterize materials. It is a microscopic method, in the sense that it probes nuclei and their immediate surroundings. NMR is capable of measuring the local fields at atomic nuclei. The microscopic nature of the NMR measurement makes it extremely useful and sometimes unique. Of course, in order to have a signal of detectable magnitude, we need many of the same configurations to occur in the sample. The nuclei of general interest, hydrogen (proton) and fluorine are very abundant, and they have particularly strong signals. Spins subjected to a combination of external static magnetic field and an RF oscillating field of magnitude B_1 and frequency ω_1, have a rather complex response. The situation is further complicated by the interactions the spins are subjected to. Fortunately, we do not need to consider every detail of the spin response to understand the experiments. A convenient phenomenological theory incorporates most of the basic ideas we need to know. Bloch treated the spins as classical objects, and represented the effect of the interactions by relaxation times.

In this investigation, we consider the flowing fluid particle (on the atomic scale) which either initially or in some average sense is in steady rotation. The particle is assumed to have a behaviour similar to a hydrogen-like atom with a nucleus charge $+Ze$ and an electron charge $-e$ interacting by a columbic force of attraction. Two types of motions are involved here: the first is the translational motion of the atom as a whole, and the second, the internal motion of the electron and the nucleus relative to each other. Our goal is to describe the two motions in fluids through Yukawa potential in terms of NMR flow parameters based on Bloch's NMR flow equations. This is why we pursue the general idea of recasting the Bloch NMR flow equations into a single differential equation based on the fact that any displacement of the elements of a fluid in rigid-body rotation which leads to a non-zero expansion in a lateral plane is accompanied by Coriolis forces which tend to eliminate this expansion.

In this application, it is easy to follow the transition from the Bloch NMR equation (Awojoyogbe, 2002, 2003, 2004) to the Schrödinger equation, up to the quantum numbers, the radial wave functions and the energy levels. The solution of the modelled NMR wave equation and the Schrödinger's equation are, of course, similar (Awojoyogbe, 2007).

Finally, the expression for energy is the same as Bohr's expression for the energy levels. This implies that acceptable solutions of the Bloch NMR wave equation often lead to energy values which are the same as those derived from Bohr's theory. It should also be noted that the whole attitude and philosophy regarding the usual solution of the Schrödinger's equation for the hydrogen atom involving quantum states can also be applied in the theory of NMR physics. This is one of the main problems which have to be clarified in the near future if one intends to go ahead with NMR physics of the hydrogen atom and hydrogen-like ions.

In this section, we present a mathematical algorithm to describe in detail the dynamical state of the hydrogen atom starting from the NMR flow equation, derived from the previous studies (Awojoyogbe, 2007). The hydrogen atom is assumed to be magnetized by the static B_o field to an equilibrium magnetization, M_0, before entering the excitor coil. The z axis in the rotating frame coincides with the laboratory Z axis; the x axis makes an angle 'ωt' at any instant of time 't' with laboratory x-axis. The position x = 0 could be taken such that the transverse magnetic field at the end of the detector coil is negligible.

6.3   Mathematical Representation of NMR/MRI in Quantum Mechanical Domain

We study the flow properties of the modified time independent Bloch NMR flow equations which describe the dynamics of the hydrogen atom under the influence of RF magnetic field as follows (Awojoyogbe, 2002, 2003, 2004, 2007):

$$\frac{d^2 M_y}{dx^2} + \frac{1}{v}\left(\frac{1}{T_1} + \frac{1}{T_2}\right)\frac{dM_y}{dx} + \frac{1}{v^2}\left(\gamma^2 B_1^2(x) + \frac{1}{T_1 T_2}\right)M_y = \frac{M_0 \gamma B_1(x)}{v^2 T_1} \tag{6.1}$$

It is convenient to use as dependent variable the departure of the stream function from its classical form and write:

$$M_y(x) = \psi(x)e^{\zeta x} \tag{6.2}$$

where $\zeta = \frac{1}{2vT_A}$, v is the instantaneous velocity of the fluid and $\psi(x)$ is a special function of the transverse magnetization. However, when the RF $B_1(x)$ field is applied, M_y has a maximum value when RF $B_1(x)$ is maximum and $M_0 = 0$. At the point when maximum NMR signal is received (maximum values of M_y and $B_1(x)$ respectively), Eq. (6.1) becomes:

$$\frac{d^2\psi}{dx^2} + \frac{1}{v^2}\left[\gamma^2 B_1^2(x) + T_g - T_R\right]\psi = 0 \tag{6.3}$$

subject to the following two conditions:

1.

$$e^{\zeta x} \neq 0 \tag{6.4}$$

2. Resonance condition exists at Larmor frequency

$$f_0 = \gamma B - \omega = 0 \tag{6.5}$$

$$T_g = \frac{1}{T_1 T_2}, \quad T_R = \frac{1}{4T_A^2} \quad and \quad \frac{1}{T_A} = \frac{1}{T_1} + \frac{1}{T_2} \tag{6.6}$$

where γ denotes the gyromagnetic ratio of fluid spins; $\omega/2\pi$ is the RF excitation frequency; f_0/γ is the off-resonance field in the rotating frame of reference. T_1 and T_2 are the spin-lattice and spin-spin relaxation times respectively, the reciprocals of T_1 and T_2 are defined as relaxation rates. RF B_1 is treated as constant and of the order of 1G. Details of how Eq. (6.3) has been obtained from Eq. (6.1) can be found in Chap. 3.

The exponential function in Eq. (6.2) can be defined as follows:

$$e^x = \sum_{n=0}^{\infty} \frac{(\zeta x)^n}{n!} = F(x) \tag{6.7}$$

Equation (6.6) is extremely useful in obtaining approximations to complicated formulas, valid when x is small (especially when $x = 4vT_A$). Equation (6.3) may be referred to as the NMR wave equation. This equation completely describes the wave properties of the Bloch NMR equations in terms of the mechanical wave function $\psi(x)$, the velocity v, radio-frequency $\omega_1 = \gamma B_1(x)$, T_1 and T_2 relaxation parameters. The uniqueness of Eq. (6.3) is apparent from the RF B_1 field which can be a constant or a variable. The relaxation parameters are intrinsic properties of the NMR system and are constants. In order to learn how to solve the NMR wave equation and to acquire a feeling for its peculiarities, we proceed to treat a few applications. The novel quantity appearing in the NMR wave equation $\psi(x)$, expresses the mathematical form of particle wave properties associated with the fluid particles of the flow system. We now examine some important features of the wave equation by considering the values of the radio frequency relative to the relaxation parameters.

When the RF B_1 field spatially varies with x. Equation (6.3) can be written as (Awojoyogbe, 2007):

$$\frac{d^2\psi}{dx} + G^2(x)\psi = 0 \tag{6.8}$$

where

$$G^2(x) = \frac{\gamma^2 B_1^2(x)}{v^2} + \frac{T_g - T_R}{v^2} \tag{6.9}$$

The following assumptions will be made:

$$G^2(x) = \frac{2\mu\varsigma}{\hbar^2}\left[E - E_p(x)\right] \tag{6.10}$$

where $\hbar$ is the reduced Planck's constant, μ is particle mass, ς is a dimensionless weighting parameter while E and $E_p(x)$ are the total energy and potential energy of the fluid particle respectively. For simplicity, let us suppose that the RF B_1 field is designed such that $G^2(x)$ vanishes at a single point $x = x_o$ only. Then

$$\begin{aligned} G^2(x) > 0; & \quad x > x_0 \\ G^2(x) < 0; & \quad x < x_0 \end{aligned} \tag{6.11}$$

The implication of this condition according to Eq. (6.9) is:

$$\begin{aligned} \gamma^2 B_1^2(x) > T_R - T_g; & \quad x > x_0 \\ \gamma^2 B_1^2(x) < T_R - T_g; & \quad x < x_0 \end{aligned} \tag{6.12}$$

Equation (6.12) is related to the linearity condition in nuclear magnetic resonance.

6.4   Formulation of the Yukawa Potential for NMR Wave Equation

Equation (6.8) can be written as follows:

$$\frac{d^2\psi}{dx^2} + \frac{2\mu\varsigma}{\hbar^2}\left[E - E_p(x)\right]\psi = 0 \tag{6.13}$$

$$\frac{d^2\psi}{dx^2} - \frac{2\mu\varsigma}{\hbar^2}E_p(x)\psi = -\frac{2\mu\varsigma}{\hbar^2}E\psi$$

$$-\frac{\hbar^2}{2\mu\varsigma}\frac{d^2\psi}{dx^2} + E_p(x)\psi = E\psi \tag{6.14}$$

In the generalized coordinate, Eq. (6.14) can be assumed to have the form (Shankar, 2012; Griffiths, 2019):

$$-\frac{\hbar^2}{2\mu\varsigma}\nabla^2\psi\left(\vec{r}\right) + E_p\left(\vec{r}\right)\psi = E\psi\left(\vec{r}\right) \tag{6.15}$$

If we consider a spherical voxel, $\vec{r} = r\widehat{e}_r + \theta\widehat{e}_\theta + \phi\widehat{e}_\phi$, Eq. (6.15) becomes:

$$\left[-\frac{\hbar^2}{2\mu\varsigma}\left(\frac{\partial}{r^2\partial r}r^2\frac{\partial}{\partial r} + \frac{1}{r^2\sin\theta}\frac{\partial}{\partial\theta}\sin\theta\frac{\partial}{\partial\theta} + \frac{1}{r^2\sin^2\theta}\frac{\partial^2}{\partial\phi^2}\right) + E_p(r)\right]\psi(r,\theta,\phi)$$
$$= E\psi(r,\theta,\phi)$$

$$\tag{6.16}$$

where $\nabla^2 = \frac{\partial}{r^2\partial r}r^2\frac{\partial}{\partial r} + \frac{1}{r^2\sin\theta}\frac{\partial}{\partial\theta}\sin\theta\frac{\partial}{\partial\theta} + \frac{1}{r^2\sin^2\theta}\frac{\partial^2}{\partial\phi^2}$ is the Laplace's operator.

If we set the function ψ as follows (Shankar, 2012):

$$\psi(r,\theta,\phi) = R(r)Y_l^m(\theta,\phi) \tag{6.17}$$

where $Y_l^m(\theta,\phi)$ represents spherical harmonics (Zwillinger, 1997). Using Eq. (6.17) in Eq. (6.16),

$$\left[-\frac{\hbar^2}{2\mu\varsigma}\left(\frac{\partial}{r^2\partial r}r^2\frac{\partial}{\partial r} + \frac{1}{r^2\sin\theta}\frac{\partial}{\partial\theta}\sin\theta\frac{\partial}{\partial\theta} + \frac{1}{r^2\sin^2\theta}\frac{\partial^2}{\partial\phi^2}\right) + E_p(r)\right]R(r)Y_l^m(\theta,\phi)$$
$$= ER(r)Y_l^m(\theta,\phi)$$

$$\left[-\frac{\hbar^2}{2\mu\varsigma}\left(Y_l^m\frac{\partial}{r^2\partial r}r^2\frac{\partial R}{\partial r} + \frac{R}{r^2\sin\theta}\frac{\partial}{\partial\theta}\sin\theta\frac{\partial Y_l^m}{\partial\theta} + \frac{R}{r^2\sin^2\theta}\frac{\partial^2 Y_l^m}{\partial\phi^2}\right)\right] + E_p(r)RY_l^m$$
$$= ERY_l^m$$

$$\left[-\frac{\hbar^2}{2\mu\varsigma}\left(\frac{1}{r^2R}\frac{d}{dr}r^2\frac{dR}{dr} + \frac{1}{Y_l^m r^2\sin\theta}\frac{\partial}{\partial\theta}\sin\theta\frac{\partial Y_l^m}{\partial\theta} + \frac{1}{Y_l^m r^2\sin^2\theta}\frac{\partial^2 Y_l^m}{\partial\phi^2}\right) + E_p(r)\right]$$
$$= E$$

$$\tag{6.18}$$

Re-arranging Eq. (6.18) leads to:

$$\frac{1}{R}\frac{d}{dr}\left(r^2\frac{dR}{dr}\right)+\frac{2\mu\varsigma}{\hbar^2}(E-E_p(r))r^2=-\left(\frac{1}{Y_l^m\sin\theta}\frac{\partial}{\partial\theta}\sin\theta\frac{\partial Y_l^m}{\partial\theta}+\frac{1}{Y_l^m\sin^2\theta}\frac{\partial^2 Y_l^m}{\partial\phi^2}\right)$$

$$(6.19)$$

Since both sides of Eq. (6.19) are independent of each other, they must individually be equal to a constant $l(l+1)$. Therefore,

$$\frac{1}{R}\frac{d}{dr}\left(r^2\frac{dR}{dr}\right)+\frac{2\mu\varsigma}{\hbar^2}(E-E_p(r))r^2=-\left(\frac{1}{Y_l^m\sin\theta}\frac{\partial}{\partial\theta}\sin\theta\frac{\partial Y_l^m}{\partial\theta}+\frac{1}{Y_l^m\sin^2\theta}\frac{\partial^2 Y_l^m}{\partial\phi^2}\right)$$

$$=l(l+1)$$

$$(6.20)$$

This leads to the following equations:

$$\frac{d}{dr}\left(r^2\frac{dR}{dr}\right)+\frac{2\mu\varsigma}{\hbar^2}(E-E_p(r))r^2R-l(l+1)R=0 \tag{6.21}$$

$$\frac{1}{\sin\theta}\frac{\partial}{\partial\theta}\sin\theta\frac{\partial Y_l^m}{\partial\theta}+\frac{1}{\sin^2\theta}\frac{\partial^2 Y_l^m}{\partial\phi^2}+l(l+1)Y_l^m=0 \tag{6.22}$$

Equation (6.21) can be resolved further,

$$\frac{d^2R}{dr^2}+\frac{2}{r}\frac{dR}{dr}+\frac{2\mu\varsigma}{\hbar^2}(E-E_p(r))R-\frac{l(l+1)}{r^2}R=0 \tag{6.23}$$

$$\frac{d^2R}{dr^2}+\frac{2}{r}\frac{dR}{dr}+\frac{2\mu\varsigma}{\hbar^2}\left[E-E_p(r)-\frac{l(l+1)\hbar^2}{2\mu\varsigma r^2}\right]R=0 \tag{6.24}$$

Setting:

$$R=\frac{y}{r} \tag{6.25}$$

Equations (6.26) and (6.27) can easily be obtained:

$$\frac{dR}{dr}=\frac{1}{r}\frac{dy}{dr}-\frac{y}{r^2} \tag{6.26}$$

$$\frac{d^2R}{dr^2}=\frac{1}{r}\frac{d^2y}{d^2r}-\frac{2}{r^2}\frac{dy}{dr}+\frac{2y}{r^3} \tag{6.27}$$

Equation (6.24) becomes:

$$\frac{1}{r}\frac{d^2y}{d^2r} - \frac{2}{r^2}\frac{dy}{dr} + \frac{2y}{r^3} + \frac{2}{r}\left(\frac{1}{r}\frac{dy}{dr} - \frac{y}{r^2}\right) + \frac{2\mu\varsigma}{\hbar^2}\left[E - E_p(r) - \frac{l(l+1)\hbar^2}{2\mu\varsigma r^2}\right]\frac{y}{r} = 0$$

$$(6.28)$$

$$\frac{d^2y}{d^2r} + \frac{2\mu\varsigma}{\hbar^2}\left[E - E_p(r) - \frac{l(l+1)\hbar^2}{2\mu\varsigma r^2}\right]y = 0 \qquad (6.29)$$

If we consider E_p to be a Yukawa potential (screened Coulomb potential), then we have (Yukawa, 1935):

$$E_p(r) = -g_p^2\frac{e^{-\alpha_p\mu r}}{r} \qquad (6.30)$$

where g_p is amplitude of the potential, α_p is a scaling constant while μ is the fluid particle mass. Equation (6.30) is a screened version of the Coulomb potential in which g_p^2 describes the strength of the interaction and $\frac{1}{\alpha_p\mu}$ its range (Edwards et al., 2017). it should be noted that the Yukawa interaction can be switched off at a distance scale $r \approx \frac{1}{\alpha_p\mu}$. Hence, Eq. (6.29) becomes:

$$\frac{d^2y}{d^2r} + \frac{2\mu\varsigma}{\hbar^2}\left[E + g_p^2\frac{e^{-\alpha_p\mu r}}{r} - \frac{l(l+1)\hbar^2}{2\mu\varsigma r^2}\right]y = 0 \qquad (6.31)$$

To resolve Eq. (6.31), we set:

$$\sigma = g_p\lambda_0\sqrt{\frac{e^{-\alpha_p\mu r}}{r}} = g_p\lambda_0 e^{-\frac{\alpha_p\mu r}{2}}r^{-\frac{1}{2}} \qquad (6.32)$$

where r_0 is a dimensionless weighting parameter.

$$\frac{d\sigma}{dr} = -g_p\lambda_0\frac{\alpha_p\mu}{2}r^{-\frac{1}{2}}e^{-\frac{\alpha_p\mu r}{2}} - \frac{g_p\lambda_0}{2}r^{-\frac{3}{2}}e^{-\frac{\alpha_p\mu r}{2}} \qquad (6.33)$$

$$\frac{d^2\sigma}{dr^2} = \frac{g_p\lambda_0\alpha_p^2\mu^2}{4}r^{-\frac{1}{2}}e^{-\frac{\alpha_p\mu r}{2}} + \frac{2g_p\lambda_0\alpha_p\mu}{4}r^{-\frac{3}{2}}e^{-\frac{\alpha_p\mu r}{2}} - \frac{3g_p\lambda_0}{4}r^{-\frac{5}{2}}e^{-\frac{\alpha_p\mu r}{2}}$$

$$= \left(\frac{\alpha_p^2\mu^2}{4} + \frac{2\alpha_p\mu}{4r} + \frac{3}{4r^2}\right)g_p\lambda_0 r^{-\frac{1}{2}}e^{-\frac{\alpha_p\mu r}{2}}$$

$$(6.34)$$

Hence,

$$\frac{dy}{dr} = \frac{dy}{d\sigma}\frac{d\sigma}{dr}$$

$$\frac{d^2y}{dr^2} = \frac{d^2y}{d\sigma^2}\left(\frac{d\sigma}{dr}\right)^2 + \frac{dy}{d\sigma}\frac{d^2\sigma}{dr^2} \qquad (6.35)$$

This leads to:

$$\left(\frac{d\sigma}{dr}\right)^2 = \left(-\frac{g_p\lambda_0\alpha_p\mu}{2}r^{-\frac{1}{2}}e^{-\frac{\alpha_p\mu r}{2}} - \frac{g_p\lambda_0}{2}r^{-\frac{3}{2}}e^{-\frac{\alpha_p\mu r}{2}}\right)^2 = \left(\frac{\alpha_p\mu}{2}+\frac{1}{2r}\right)^2\left(-g_p\lambda_0 r^{-\frac{1}{2}}e^{-\frac{\alpha_p\mu r}{2}}\right)^2$$

$$= \left(\frac{\alpha_p\mu r+1}{2r}\right)^2 \left(g_p\lambda_0\right)^2\frac{e^{-\alpha_p\mu r}}{r}$$

$$(6.36)$$

Equation (6.31) becomes:

$$\left(\frac{\alpha_p\mu r+1}{2r}\right)^2\left(g_p\lambda_0\right)^2 e^{-\alpha_p\mu r}\frac{d^2y}{d\sigma^2} + \left(\frac{\alpha_p^2\mu^2}{4}+\frac{2\alpha_p\mu}{4r}+\frac{3}{4r^2}\right)\left(g_p\lambda_0\right)r^{-\frac{1}{2}}e^{-\frac{\alpha_p\mu r}{2}}\frac{dy}{d\sigma}$$

$$+\frac{2\mu\varsigma}{\hbar^2}\left[E+\left(g_p\lambda_0\right)^2\frac{e^{-\alpha_p\mu r}}{r}-\frac{l(l+1)\hbar^2}{2\mu\varsigma r^2}\right]y = 0$$

$$(6.37)$$

$$\frac{\left(\alpha_p\mu r+1\right)^2}{4r^2}\sigma^2\frac{d^2y}{d\sigma^2}+\left(\frac{\alpha_p^2\mu^2 r^2+2\alpha_p\mu r+3}{4r^2}\right)\sigma\frac{dy}{d\sigma}$$

$$+\left(\frac{2\mu\varsigma}{\hbar^2\lambda_0^2}\sigma^2+\frac{2\mu\varsigma E}{\hbar^2}-\frac{l(l+1)}{r^2}\right)y = 0 \qquad (6.38)$$

$$\sigma^2\frac{d^2y}{d\sigma^2}+\left(\frac{\alpha_p^2\mu^2 r^2+2\alpha_p\mu r+3}{\alpha_p^2\mu^2 r^2+2\alpha_p\mu r+1}\right)\sigma\frac{dy}{d\sigma}$$

$$+\left(\frac{8\mu\varsigma r^2\sigma^2}{\left(\alpha_p\mu r+1\right)^2\lambda_0^2\hbar^2}+\frac{8\mu\varsigma E r^2}{\left(\alpha_p\mu r+1\right)^2\hbar^2}-\frac{4l(l+1)}{\left(\alpha_p\mu r+1\right)^2}\right)y = 0 \qquad (6.39)$$

An equation is transformable to Bessel differential equation has the form (Wylie & Barrett, 1982):

$$x^2\frac{d^2y}{dx^2}+x(a+2\varepsilon x^{p_0})\frac{dy}{dx}+\left(c+d_0 x^{2q}+\varepsilon(a+p_0-1)x^{p_0}+\varepsilon^2 x^{2p_0}\right)y = 0 \quad (6.40)$$

Equation (6.40) has a complete solution

$$y = x^{p_0}e^{-\beta x p_0}\left[C_1 J_v(\lambda x^q)+C_2 Y_v(\lambda x^q)\right] \qquad (6.41)$$

where

$$\rho_0 = \frac{1-a}{2}, \quad \beta = \frac{\varepsilon}{q}, \quad \lambda = \frac{\sqrt{|d_0|}}{q}, \quad \upsilon = \frac{\sqrt{(1-a)^2 - 4c}}{2q} \tag{6.42}$$

Comparing Eq. (6.39) with Eq. (6.40)

$$a = \frac{\alpha_p^2 \mu^2 r^2 + 2\alpha_p \mu r + 3}{\alpha_p^2 \mu^2 r^2 + 2\alpha_p \mu r + 1}, \quad \varepsilon = 0, \quad q = 1, \quad d_0 = \frac{8\mu\varsigma r^2}{(\alpha_p \mu r + 1)^2 \lambda_0^2 \hbar^2},$$

$$c = \frac{8\mu\varsigma E r^2}{(\alpha_p \mu r + 1)^2 \hbar^2} - \frac{4l(l+1)}{(\alpha_p \mu r + 1)^2} \tag{6.43}$$

$$\rho_0 = \frac{1 - \left(\frac{\alpha_p^2 \mu^2 r^2 + 2\alpha_p \mu r + 3}{\alpha_p^2 \mu^2 r^2 + 2\alpha_p \mu r + 1}\right)}{2} = -\frac{1}{\alpha_p^2 \mu^2 r^2 + 2\alpha_p \mu r + 1} = -\frac{1}{(\alpha_p \mu r + 1)^2} \tag{6.44}$$

$$\beta = 0, \quad \lambda = \sqrt{\frac{8\mu\varsigma r^2 r_0^2}{(\alpha_p \mu r + 1)^2 \hbar^2}} = \frac{r\sqrt{8\mu\varsigma}}{(\alpha_p \mu r + 1)\lambda_0 \hbar} \tag{6.45}$$

$$\upsilon = \frac{1}{2}\sqrt{\left(\frac{\alpha_p^2 \mu^2 r^2 + 2\alpha_p \mu r + 1 - \alpha_p^2 \mu^2 r^2 - 2\alpha_p \mu r - 3}{\alpha_p^2 \mu^2 r^2 + 2\alpha_p \mu r + 1}\right)^2 - \frac{32\mu\varsigma E r^2}{(\alpha_p \mu r + 1)^2 \hbar^2} + \frac{16l(l+1)}{(\alpha_p \mu r + 1)^2}}$$

$$= \frac{1}{2}\sqrt{\frac{4}{(\alpha_p \mu r + 1)^2} - \frac{32\mu\varsigma E r^2}{(\alpha_p \mu r + 1)^2 \hbar^2} + \frac{16l(l+1)}{(\alpha_p \mu r + 1)^2}}$$

$$= \frac{1}{(\alpha_p \mu r + 1)}\sqrt{1 + 4l(l+1) - \frac{8\mu\varsigma E r^2}{\hbar^2}}$$

$$\tag{6.46}$$

Finally,

$$y(\sigma) = \sigma^{-\frac{1}{(\alpha_p \mu r + 1)^2}}\left[C_1 J_\upsilon\left(\frac{r\sqrt{8\mu\varsigma}}{(\alpha_p \mu r + 1)\lambda_0 \hbar}\sigma\right) + C_2 Y_\upsilon\left(\frac{r\sqrt{8\mu\varsigma}}{(\alpha_p \mu r + 1)\lambda_0 \hbar}\sigma\right)\right] \tag{6.47}$$

Using Eq. (6.31) in Eq. (6.46),

$$y(r) = \left(g_p \lambda_0 \sqrt{\frac{e^{-\alpha_p \mu r}}{r}}\right)^{-\frac{1}{(\alpha_p \mu r + 1)^2}}\left[C_1 J_\upsilon\left(\frac{r\sqrt{8\mu\varsigma}}{(\alpha_p \mu r + 1)\hbar}g_p\sqrt{\frac{e^{-\alpha_p \mu r}}{r}}\right) + \right.$$

$$\left. \times C_2 Y_\upsilon\left(\frac{r\sqrt{8\mu\varsigma}}{(\alpha_p \mu r + 1)\hbar}g_p\sqrt{\frac{e^{-\alpha_p \mu r}}{r}}\right)\right]$$

$$y(r) = \left(g_p \lambda_0 \sqrt{\frac{e^{-\alpha_p \mu r}}{r}} \right)^{-\frac{1}{(\alpha_p \mu r + 1)^2}} \left[C_1 J_v \left(g_p \frac{\sqrt{8\mu\varsigma r} e^{-\alpha_p \mu r}}{(\alpha_p \mu r + 1)\hbar} \right) + C_2 Y_v \left(g_p \frac{\sqrt{8\mu\varsigma r} e^{-\alpha_p \mu r}}{(\alpha_p \mu r + 1)\hbar} \right) \right]$$

$$(6.48)$$

According to Eq. (6.25),

$$R(r) = \frac{1}{r} \left(g_p \lambda_0 \sqrt{\frac{e^{-\alpha_p \mu r}}{r}} \right)^{-\frac{1}{(\alpha_p \mu r + 1)^2}} \left[C_1 J_v \left(g_p \frac{\sqrt{8\mu\varsigma r} e^{-\alpha_p \mu r}}{(\alpha_p \mu r + 1)\hbar} \right) + C_2 Y_v \left(g_p \frac{\sqrt{8\mu\varsigma r} e^{-\alpha_p \mu r}}{(\alpha_p \mu r + 1)\hbar} \right) \right]$$

$$(6.49)$$

where C_1 and C_2 are both constants, J_v is Bessel function of the first kind while Y_v is Bessel function of the second kind.

Since the MRI signals should be measurable at all radial points, we set $C_2 = 0$. Hence,

$$R(r) = \frac{C_1}{r} \left(g_p \lambda_0 \sqrt{\frac{e^{-\alpha_p \mu r}}{r}} \right)^{-\frac{1}{(\alpha_p \mu r + 1)^2}} J_v \left(g_p \frac{\sqrt{8\mu\varsigma r} e^{-\alpha_p \mu r}}{(\alpha_p \mu r + 1)\hbar} \right) \qquad (6.50)$$

According to Eq. (6.17),

$$\psi(r, \theta, \phi) = \frac{C_1}{r} \left(g_p \lambda_0 \sqrt{\frac{e^{-\alpha_p \mu r}}{r}} \right)^{-\frac{1}{(\alpha_p \mu r + 1)^2}} J_v \left(g_p \frac{\sqrt{8\mu\varsigma r} e^{-\alpha_p \mu r}}{(\alpha_p \mu r + 1)\hbar} \right) Y_l^m(\theta, \phi) \quad (6.51)$$

The Bohr radius a_0, which characterizes atomic dimensions, is given as (Peleg et al., 1998):

$$a_0 = \frac{\hbar^2}{\mu e_c^2} = 0.5 \times 10^{-10} m \qquad (6.52)$$

where e_c is electronic charge while the reduced mass μ is defined as:

$$\mu = \frac{m_p m_e}{m_p + m_e} \qquad (6.53)$$

m_p is proton mass while m_e is electron mass. Equation (6.51) becomes:

$$\psi(r,\theta,\phi) = \frac{C_1}{r}\left(g_p\lambda_0\right)^{-\frac{1}{(\alpha_p\mu r+1)^2}}\left(\frac{e^{-\alpha_p\mu r}}{r}\right)^{-\frac{1}{2(\alpha_p\mu r+1)^2}}J_v\left(\frac{g_p\sqrt{8\mu\varsigma r}e^{-\alpha_p\mu r}}{(\alpha_p\mu r+1)\hbar}\right)Y_l^m(\theta,\phi)$$

$$(6.54)$$

The spherical harmonics function is given as (Zwillinger, 1997):

$$Y_l^m(\theta,\phi) = (-1)^m\sqrt{\frac{(2l+1)}{4\pi}\frac{(l-m)!}{(l+m)!}}P_l^m(\cos\theta)e^{im\phi};\quad\begin{cases}l=0,1,2,3,\dots\\ m=-l,-l+1,\dots,l-1,l\end{cases}$$

$$(6.55)$$

In spherical coordinates, the parameter from Eq. (6.2) may be modified to:

$$\zeta x \rightarrow \vec{\zeta}\cdot\vec{r} = \vec{\zeta}\cdot r\hat{e}_r + \theta\hat{e}_\theta + \phi\hat{e}_\phi$$

$$(6.56)$$

If the fluid velocity is not significantly dependent on θ and ϕ, Eq. (6.2) becomes:

$$M_y(r,\theta,\phi) = \psi(r,\theta,\phi)e^{-\frac{r}{2vT_A}}$$

$$(6.57)$$

Therefore, from Eq. (6.54),

$$M_y(r,\theta,\phi) = \frac{C_1}{r}\left(g_p\lambda_0\right)e^{-\frac{r}{2vT_A}-\frac{1}{(\alpha_p\mu r+1)^2}}\left(\frac{e^{-\alpha_p\mu r}}{r}\right)^{-\frac{1}{2(\alpha_p\mu r+1)^2}}J_v\left(\frac{g_p\sqrt{8\mu\varsigma r}e^{-\alpha_p\mu r}}{(\alpha_p\mu r+1)\hbar}\right)Y_l^m(\theta,\phi)$$

$$(6.58)$$

The total energy, E in Eq. (6.46) is dependent on the quantum numbers and using supersymmetric approach, this energy is expressed as (Onate & Ojonubah, 2016):

$$E_{n,l} = \frac{(\alpha_p\mu)^2\hbar^2 l(l+1)}{2\varsigma\mu} - \frac{(\alpha_p\mu)^2\hbar^2}{2\varsigma\mu}\left[\frac{-\frac{2\varsigma\mu g_p^2}{(\alpha_p\mu)\hbar^2}-(n+l+1)^2-l(l+1)}{2(n+l+1)}\right]^2$$

$$(6.59)$$

However, the amplitude of the Yukawa potential can be given as (Garavelli & Oliveira, 1991):

$$g_p = e_c$$

$$(6.60)$$

$$E_{n,l} = \frac{(\alpha_p\mu)^2\hbar^2 l(l+1)}{2\varsigma\mu} - \frac{(\alpha_p\mu)^2\hbar^2}{2\varsigma\mu}\left[\frac{-\frac{2\varsigma\mu e_c^2}{(\alpha_p\mu)\hbar^2} - (n+l+1)^2 - l(l+1)}{2(n+l+1)}\right]^2$$

$$(6.61)$$

$$\frac{8\mu\varsigma E_{n,l}r^2}{\hbar^2} = (2\alpha_p\mu)^2 l(l+1)r^2 - (2\alpha_p\mu)^2 r^2\left[\frac{-\frac{2\varsigma\mu e_c^2}{(\alpha_p\mu)\hbar^2} - (n+l+1)^2 - l(l+1)}{2(n+l+1)}\right]^2$$

$$(6.62)$$

Equation (6.58) becomes:

$$M_y(r,\theta,\phi) = \frac{C_1}{r}(e_c\lambda_0)e^{-\frac{r}{2vT_A} - \frac{1}{(\alpha_p\mu r+1)^2}}\left(\frac{e^{-\alpha_p\mu r}}{r}\right)^{-\frac{1}{2(\alpha_p\mu r+1)^2}}J_v\left(\frac{e_c\sqrt{8\mu\varsigma}re^{-\alpha_p\mu r}}{(\alpha_p\mu r+1)\hbar}\right)Y_l^m(\theta,\phi)$$

$$(6.63)$$

Hence, Eq. (6.46) is given as:

$$v = \frac{1}{(\alpha_p\mu r+1)}$$

$$\times\sqrt{1+4l(l+1) - (2\alpha_p\mu)^2 r^2\left(l(l+1) - \left[\frac{-\frac{2\varsigma\mu e_c^2}{(\alpha_p\mu)\hbar^2} - (n+l+1)^2 - l(l+1)}{2(n+l+1)}\right]^2\right)}$$

$$(6.64)$$

6.5 Formulation of the Coulomb Potential for NMR Wave Equation

Screening is the damping of electric fields due to the presence of mobile charge carriers. Section 6.4 presented a case of screened Coulomb potential (Yukawa potential). Fluid flow within biological systems often reduces the effective interaction between fluid particles to a short-range "screened" Coulomb interaction. However, in the absence of electron screen, we set $\alpha_p = 0$ and Yukawa potential reduces to Coulomb potential. Since the potential energy of the electron in the Coulomb potential $\Phi = \frac{e}{r}$ due to the NMR proton is, hydrogen atom in N-space dimensions has the potential defined as (Nieto, 1979):

$$E_p(r) = -\frac{e_c^2}{r} \tag{6.65}$$

Hence, Eq. (6.29) becomes:

$$\frac{d^2 y}{d^2 r} + \frac{2\mu\varsigma}{\hbar^2}\left[E + \frac{e_c^2}{r} - \frac{l(l+1)\hbar^2}{2\mu\varsigma r^2}\right] y = 0 \tag{6.66}$$

where:

$$\psi(r, \theta, \phi) = \frac{y(r)}{r} Y_l^m(\theta, \phi) \tag{6.67}$$

where $Y_l^m(\theta, \phi)$ represents spherical harmonics (Zwillinger, 1997). Due to the nature of Eq. (6.66), we can assume an asymptotic behaviour (Shankar, 2012) such that:

$$y \underset{r\to\infty}{\approx} e^{-i\frac{\sqrt{2\mu\varsigma E}}{\hbar}r} \tag{6.68}$$

$$y \underset{r\to 0}{\approx} r^{l+1} \tag{6.69}$$

It is important to note that the energy required to liberate the electron (binding energy) is given as:

$$E_B = -E \tag{6.70}$$

Equation (6.68) then takes the form:

$$y \underset{r\to\infty}{\approx} e^{-\frac{\sqrt{2\mu\varsigma E_B}}{\hbar}r} \tag{6.71}$$

Setting:

$$\varepsilon = \frac{\sqrt{2\mu\varsigma E_B}}{\hbar} r \tag{6.72}$$

We can then assume an auxiliary function y_1 such that,

$$y = e^{-\varepsilon} y_1 \tag{6.73}$$

Hence, from Eqs. (6.72) and (6.73),

$$\frac{d\varepsilon}{dr} = \frac{\sqrt{2\mu\varsigma E_B}}{\hbar}, \quad r^2 = \frac{\hbar^2}{2\mu\varsigma E_B}\varepsilon^2 \tag{6.74}$$

$$\frac{dy}{dr} = \frac{dy}{d\varepsilon}\frac{d\varepsilon}{dr} = \frac{\sqrt{2\mu\varsigma E_B}}{\hbar}\frac{d(e^{-\varepsilon}y_1)}{d\varepsilon} = \frac{\sqrt{2\mu\varsigma E_B}}{\hbar}\left(e^{-\varepsilon}\frac{dy_1}{d\varepsilon} - e^{-\varepsilon}y_1\right) \tag{6.75}$$

$$\frac{d^2y}{dr^2} = \frac{d^2y}{d\varepsilon^2}\left(\frac{d\varepsilon}{dr}\right)^2 + \frac{dy}{d\varepsilon}\frac{d^2\varepsilon}{dr^2} = \frac{2\mu\varsigma E_B}{\hbar^2}\frac{d^2(e^{-\varepsilon}y_1)}{d\varepsilon^2} + 0 = \frac{2\mu\varsigma E_B}{\hbar^2}\frac{d}{d\varepsilon}\left(e^{-\varepsilon}\frac{dy_1}{d\varepsilon} - e^{-\varepsilon}y_1\right)$$

$$= \frac{2\mu\varsigma E_B}{\hbar^2}e^{-\varepsilon}\left(\frac{d^2y_1}{d\varepsilon^2} - 2\frac{dy_1}{d\varepsilon} + y_1\right) \tag{6.76}$$

Equation (6.66) then becomes:

$$\frac{2\mu\varsigma E_B}{\hbar^2}e^{-\varepsilon}\left(\frac{d^2y_1}{d\varepsilon^2} - 2\frac{dy_1}{d\varepsilon} + y_1\right)$$

$$+ \left[-\frac{2\mu\varsigma E_B}{\hbar^2} + \frac{2\mu\varsigma}{\hbar^2}\frac{e_c^2}{r} - \frac{2\mu\varsigma}{\hbar^2}\frac{l(l+1)\hbar^2}{2\mu\varsigma r^2}\right]e^{-\varepsilon}y_1$$

$$= 0 \tag{6.77}$$

$$\frac{2\mu\varsigma}{\hbar^2}\left(\frac{d^2y_1}{d\varepsilon^2} - 2\frac{dy_1}{d\varepsilon} + y_1\right) + \left[-1 + \frac{1}{E_B}\frac{e_c^2}{r} - \frac{\hbar^2}{2\mu\varsigma E_B}\frac{l(l+1)}{r^2}\right]y_1 = 0$$

$$\left(\frac{d^2y_1}{d\varepsilon^2} - 2\frac{dy_1}{d\varepsilon} + y_1\right) + \left[-1 + \frac{\sqrt{2\mu\varsigma E_B}}{\hbar E_B}\frac{e_c^2}{\varepsilon} - \frac{\hbar^2}{2\mu\varsigma E_B}\frac{2\mu\varsigma E_B}{\hbar^2}\frac{l(l+1)}{\varepsilon^2}\right]y_1 = 0$$

$$\frac{d^2y_1}{d\varepsilon^2} - 2\frac{dy_1}{d\varepsilon} + \left(\frac{e_c^2}{\varepsilon}\sqrt{\frac{2\mu\varsigma}{\hbar^2 E_B}} - \frac{l(l+1)}{\varepsilon^2}\right)y_1 = 0 \tag{6.78}$$

A power series solution to Eq. (6.78) can be assumed as follows (Zwillinger, 1997; Shankar, 2012):

$$y_1(\varepsilon) = \varepsilon^{l+1}\sum_{k=0}^{\infty}B_k\varepsilon^k \tag{6.79}$$

Expanding Eq. (6.78) in terms of Eq. (6.79),

$$\frac{B_{k+1}}{B_k} = \frac{-e_c^2\sqrt{\frac{2\mu\varsigma}{\hbar^2 E_B}} + 2(k+l+1)}{(k+l+2)(k+l+1) - l(l+1)} \tag{6.80}$$

As $\varepsilon \to \infty$, a series of the form $\varepsilon^{\mu\varsigma}e^{2\varepsilon}$ leads to:

$$y \approx e^{-\varepsilon} y_1 \approx e^{-\varepsilon} \left(\varepsilon^{\mu\varsigma} e^{2\varepsilon} \right) \approx \varepsilon^{\mu\varsigma} e^{\varepsilon} \tag{6.81}$$

and

$$\frac{B_{k+1}}{B_k} \underset{k \to \infty}{\longrightarrow} \frac{2}{k} \tag{6.82}$$

It follows that the series of Eq. (6.79) will termite at some k provided that:

$$e_c^2 \sqrt{\frac{2\mu\varsigma}{\hbar^2 E_B}} = 2(k + l + 1) \tag{6.83}$$

$$\frac{\hbar^2 E_B}{\mu\varsigma} = \frac{e_c^4}{2(k + l + 1)^2} \tag{6.84}$$

Hence, the particle energy becomes:

$$\frac{\hbar^2 E_B}{\mu\varsigma} = \frac{e_c^4}{2(k + l + 1)^2} \tag{6.85}$$

Hence, the particle energy becomes:

$$E = -E_B = -\frac{\mu\varsigma e_c^4}{2\hbar^2 (k + l + 1)^2} \tag{6.86}$$

If we set the principle quantum number as:

$$n = k + l + 1 \tag{6.87}$$

Equation (6.86) becomes:

$$E_n = -\frac{\mu\varsigma e_c^4}{2\hbar^2 n^2} \tag{6.88}$$

For each value of n, the allowed values of l are then given as (Eisberg & Resnick, 1985):

$$l = n - k - 1 = n - 1, n - 2, \ldots, 1, 0 \tag{6.89}$$

It is important to note that if the Hamiltonian operator contains more symmetries beside rotational invariance, states of different l should be degenerate. The degeneracy at each n is given as:

Table 6.1 Spectroscopic notations and quantum states

Notation	State
1s	$n = 1, l = 0$
2s and 2p	$l = 0, (l = 1$ at $n = 2)$
3s, 3p and 3d	$l = 0, l = 1$ and two states $n = 3$

$$\sum_{l=0}^{n-1} (2l + 1) = n^2 \tag{6.90}$$

These states are usually denoted as s ($l = 0$), p ($l = 1$), d ($l = 2$), f ($l = 3$), g ($l = 4$), h ($l = 5$), ... states. In this spectroscopic notation, Table 6.1 gives the first three states (Shankar, 2012):

In terms of a natural unit of energy, known as a Rydberg (Ry), the energy levels of NMR proton are given as:

$$Ry = \frac{\mu \varsigma e_c^4}{2\hbar^2} \tag{6.91}$$

or

$$E_n = -\frac{Ry}{n^2}; \quad n \neq 0 \tag{6.92}$$

Since the series in Eq. (6.79) terminates at $k = n - l - 1$,

$$y_1(\varepsilon) = \varepsilon^{l+1} L_{n-l-1}^{2l+1}(2\varepsilon) \tag{6.93}$$

Consequently,

$$y(\varepsilon) = C_2 e^{-\varepsilon} \varepsilon^{l+1} L_{n-l-1}^{2l+1}(2\varepsilon) \tag{6.94}$$

where C_2 is a constant.

According to Eq. (6.25),

$$R(\varepsilon) = \frac{y(\varepsilon)}{\varepsilon} \tag{6.95}$$

$$R(\varepsilon) = C_2 \frac{e^{-\varepsilon} \varepsilon^{l+1}}{\varepsilon} L_{n-l-1}^{2l+1}(2\varepsilon) = C_2 e^{-\varepsilon} \varepsilon^l L_{n-l-1}^{2l+1}(2\varepsilon) \tag{6.96}$$

Transforming back into r-coordinate,

$$R(r) \approx C_2 e^{-\left(\frac{\sqrt{2\mu\varsigma E_B}}{\hbar}r\right)} \left(\frac{\sqrt{2\mu\varsigma E_B}}{\hbar}r\right)^l L_{n-l-1}^{2l+1}\left(2\frac{\sqrt{2\mu\varsigma E_B}}{\hbar}r\right) \tag{6.97}$$

Note that from Eqs. (6.86) and (6.87),

$$E_B = \frac{\mu\varsigma e_c^4}{2\hbar^2 n^2} \tag{6.98}$$

Equation (6.60) becomes:

$$R(r) \approx C_2 e^{-\left(\frac{\mu\varsigma e_c^2}{\hbar^2 n}r\right)} \left(\frac{\mu\varsigma}{\hbar^2}\frac{e_c^2}{n}r\right)^l L_{n-l-1}^{2l+1}\left(2\frac{\mu\varsigma}{\hbar^2}\frac{e_c^2}{n}r\right) \tag{6.99}$$

Using Eq. (6.52),

$$R(r) \approx C_2 e^{-\left(\frac{\varsigma r}{na_0}\right)} \left(\frac{\varsigma r}{na_0}\right)^l L_{n-l-1}^{2l+1}\left(\frac{2\varsigma r}{na_0}\right) \tag{6.100}$$

Hence,

$$\psi_{n,l,m}(r,\theta,\phi) = C_2 e^{-\left(\frac{\varsigma r}{na_0}\right)} \left(\frac{\varsigma r}{na_0}\right)^l L_{n-l-1}^{2l+1}\left(\frac{2\varsigma r}{na_0}\right) Y_l^m(\theta,\phi) \tag{6.101}$$

Provided that the condition leading to Eq. (6.57) holds in this formulation also, we write:

$$M_y(r,\theta,\phi) = C_2 e^{-\frac{r}{2vT_A}} e^{-\left(\frac{\varsigma r}{na_0}\right)} \left(\frac{\varsigma r}{na_0}\right)^l L_{n-l-1}^{2l+1}\left(\frac{2\varsigma r}{na_0}\right) Y_l^m(\theta,\phi) \tag{6.102}$$

The first few normalized transverse magnetizations are given as follows:

$$M_y\big|_{1,0,0} = C_2 e^{-\frac{r}{2vT_A}} e^{-\left(\frac{\varsigma r}{a_0}\right)} \sqrt{\frac{1}{\pi a_0^3}} \tag{6.103}$$

$$M_y\big|_{2,0,0} = C_2 \left(2 - \frac{\varsigma r}{a_0}\right) e^{-\frac{r}{2vT_A}} e^{-\left(\frac{\varsigma r}{2a_0}\right)} \sqrt{\frac{1}{32\pi a_0^3}} \tag{6.104}$$

$$M_y\big|_{2,1,0} = C_2\left(\frac{\varsigma r}{a_0}\right) e^{-\frac{r}{2vT_A}} e^{-\left(\frac{\varsigma r}{2a_0}\right)} \sqrt{\frac{1}{32\pi a_0^3}}\cos\theta \tag{6.105}$$

$$M_y\big|_{2,1,\pm1} = \mp\left(\frac{\varsigma r}{a_0}\right) C_2 e^{-\frac{r}{2vT_A}} e^{-\left(\frac{\varsigma r}{2a_0}\right)} \sqrt{\frac{1}{64\pi a_0^3}}\sin\theta e^{\pm i\phi} \tag{6.106}$$

6.6 Quantum Neuro-Oncology

In order to analyse the results obtained in Sects. 6.4 and 6.5, we shall consider the electrostatic interaction (Peleg et al., 1998) between a proton and electron moving in a fluid with velocity of v. The results obtained in the earlier sections are valid for hydrogen and hydrogen-like atoms. It is worthy of note that nearly half of the atoms in proteins are made up of hydrogen atoms (Engler et al., 2003). These atoms mediate hydrogen bridges while ensuring non-bonding interactions such as electrostatic and van der Waals forces. In fact, they are crucial for the stabilization of defined three-dimensional structure of proteins through which they make contributions to the complicated energy landscape of proteins (Frauenfelder et al., 1991). Hydrogen atoms found in polarized bonds have essential roles in enzymatic catalysis, play important roles in hydrogen bonds within the substrate-binding process and are crucial to proton-transfer reactions during the catalysis. Knowledge of the protonation states of amino acid residues within active centre of proteins is important to the understanding of reaction mechanisms. In most situations, identifying protonated amino acids in the catalytic centre can help in making decisions between two or more competing reaction models (Schmidt et al., 1996). Furthermore, the knowledge of hydrogen positions in non-polarized bonds (COH bonds) can help in better understanding of the dynamic properties of proteins (Engler et al., 2003). Also, the conformational flexibility of the protein matrix is essential for enzymatic catalysis and for general biological molecular activity. A protein (Receptor-tyrosine-kinase-like Orphan Receptor 1) which is used in early developmental process of embryo cells have been recently found in various types of cancer, in which they act as a switch regulating the spread of cancer cells (Cui et al., 2013). Understanding the electrostatic reactions involving these proteins may add to knowledge of carcinogenesis and development of appropriate protein-based anti-cancer remedies. Currently, various existing proteins (Fusco & Fedele, 2007) are being examined as biomarkers in carcinogenesis and the contribution of these proteins in anti-cancer activity are also under evaluation for therapeutic strategies (Zarogoulidis et al., 2015). Cytochrome complex or Cytochrome C is another protein that has been implicated in apoptosis, a controlled form of cell death used to eliminate harmful cells in the process of development or in response to infection or DNA damage (Liu et al., 1996). Cytochrome C is a small globular protein with an important role in the transfer of electrons in the respiratory chain of mitochondria. It has a c-type heme which is covalently attached by the two cysteine residues of the peptide via thioether bonds and mediates electron transfer (Sakamoto et al., 2010).

The 'two-Yukawa (2Y) fluid' model which assumes an interparticle potential consisting of a hard core plus an attractive and a repulsive Yukawa tail has found applications in explaining the structural properties of various colloidal materials as evaluated in small-angle neutron scattering (SANS) experiments (Chen et al., 2016). In fact, Yukawa models have helped in improved understanding of the behaviour of cytochrome C protein solutions such that effective potential of this protein were modelled as a hard-sphere in addition to a short-range attraction and a weaker screened electrostatic repulsion. In current research studies, the effective inter-particle potential is known to determine the structure and the phase behaviour of a colloidal solution and a solution of cytochrome C protein is an example of Yukawa fluid (Liu et al., 2005).

Therefore, a small particle like a cytochrome C protein or lysozyme protein molecule could provide a very interesting scenario which can be used to simulate the results obtained in Sects. 6.4 and 6.5 (in the presence of screening and absence of screening). Since the attractive range of the van der Waals potential is a few angstroms cytochrome C, the diameter is about 33 Å (Liu et al., 2005). This implies that the proton-electron system in this molecule will be distributed over a radial distance (r) of 16.5 Å. Cytochrome C binds to cardiolipin within the inner mitochondrial membrane to ensure that it is not released from the mitochondria to initiate apoptosis (Liu et al., 1996). Hence, in this simulation, the fluid velocity (v) will be taken as the average velocity of the mitochondria (Chen et al., 2016) which varies across a wide range (0.05–1.0 μm/s). The NMR relaxation times of cytochrome C have been determined in a related study (Sakamoto et al., 2010) and these measurements are given in Table 6.2.

Furthermore, other parameters used in the simulation are given in Table 6.3

Table 6.2 Relaxation rates of human cytochrome C

	Reduced cytochrome C (s^{-1})	Oxidized cytochrome C (s^{-1})
$R_1 = 1/T_1$	1.43 ± 0.02	1.84 ± 0.02
$R_2 = 1/T_2$	11.2 ± 0.52	9.42 ± 0.52

Table 6.3 Parameters used for simulations of Yukawa potential formulation and Coulomb potential formulation

Parameter	Yukawa potential formulation	Coulomb potential formulation
C_1	4×10^6	–
C_2	–	4×10^6
ς	100,000	0.4
α_p	9.0×10^{38}	–
λ_0	1.8×10^{24}	–
μ	$\approx 9.095 \times 10^{-31}$	$\approx 9.095 \times 10^{-31}$
a_0	–	0.52×10^{-10} m

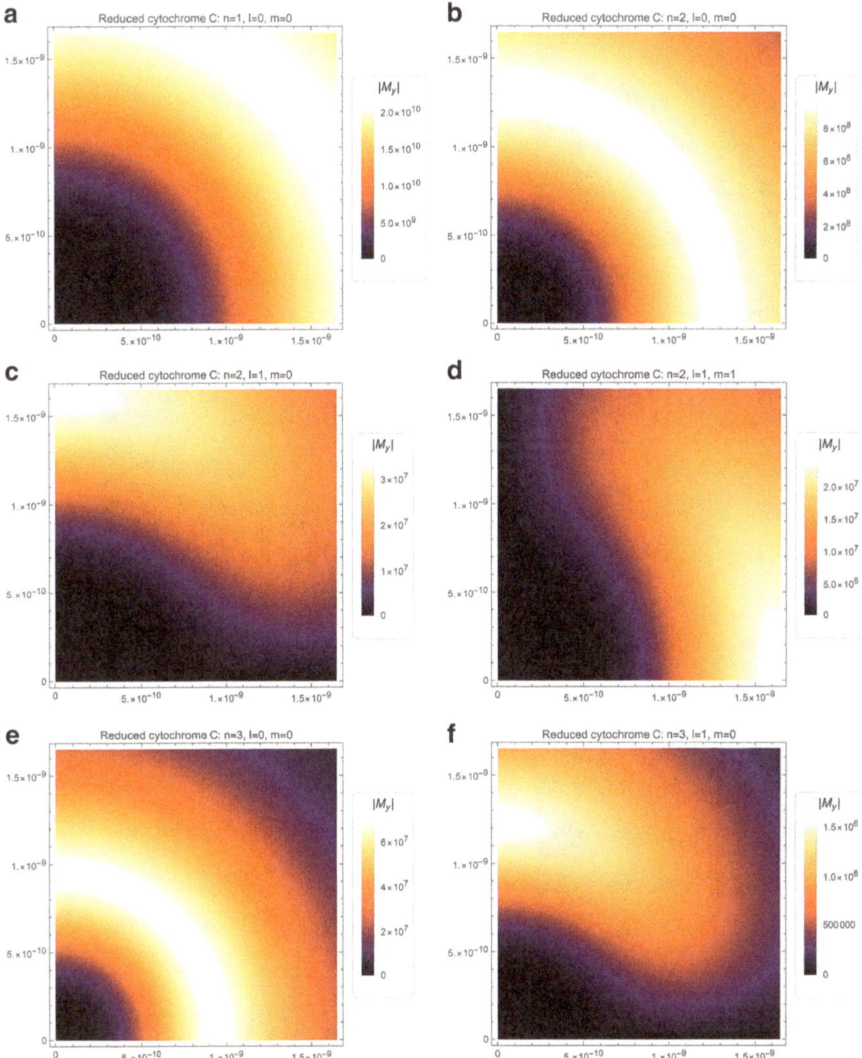

Fig. 6.1 NMR transverse magnetization profiles of cytochrome C molecule under the influence of Yukawa potential for (**a**) 1s, (**b**) 2s, (**c**) $2p_x$, (**d**) $2p_y$, (**e**) 3s, (**f**) $3p_x$ orbitals

The profiles of the NMR transverse magnetization for cytochrome C molecule in Yukawa and Coulomb potentials are given in Figs. 6.1 and 6.2 respectively.

It is important at this point to investigate the nature of the RF pulse needed to excite the cytochrome C molecules while exhibiting both Yukawa and Coulomb potentials. In order to achieve this, we need to generalize Eqs. (6.9) and (6.10) into spherical coordinates.

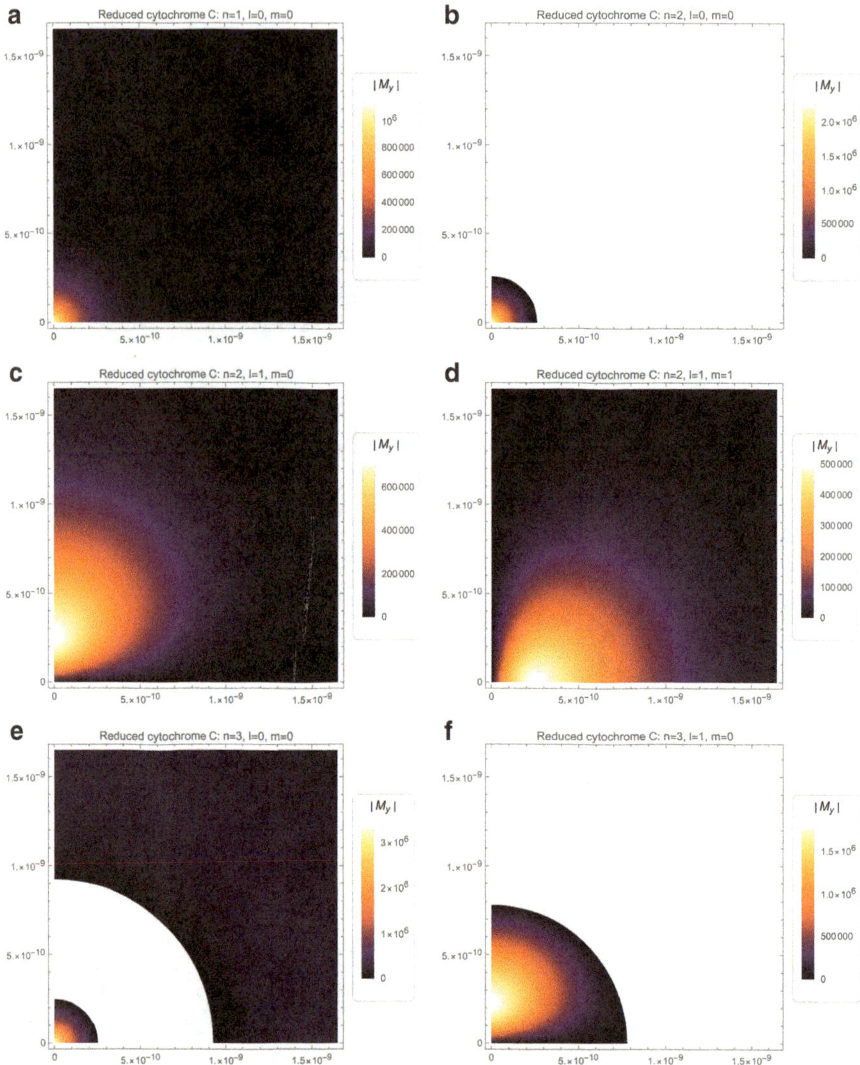

Fig. 6.2 NMR transverse magnetization profiles of cytochrome C molecule under the influence of Coulomb potential for (**a**) 1s, (**b**) 2s, (**c**) 2p$_x$, (**d**) 2p$_y$, (**e**) 3s, (**f**) 3p$_x$ orbitals

6.7   Analysis of Radiofrequency Pulse as a Function of Radial Distance of the Atoms

Assuming radiofrequency pulse is only dependent on the radial distance of the atoms, Eqs. (6.9) and (6.10) can be written as follows:

$$G^2(r) = \frac{\omega_1(r)}{v^2} + \frac{T_g - T_R}{v^2} \tag{6.107}$$

$$G^2(r) = \frac{2\mu\varsigma}{\hbar^2}\left[E - E_p(r)\right] \tag{6.108}$$

These equations lead to:

$$\omega_1(r) = \frac{2\mu\varsigma v^2}{\hbar^2}\left[E - E_p(r)\right] + T_R - T_g \tag{6.109}$$

Hence, the excitation frequencies are given as:

$$\omega_1(r) = \frac{2\mu\varsigma v^2}{\hbar^2}\left[E + e_c^2 \frac{e^{-\alpha_p \mu r}}{r}\right] + T_R - T_g \tag{6.110}$$

for Yukawa scenario while,

$$\omega_1(r) = \frac{2\mu\varsigma v^2}{\hbar^2}\left[E + \frac{e_c^2}{r}\right] + T_R - T_g \tag{6.111}$$

Spatial distributions of the angular frequency ω_1 are given in Figures 6.5 and 6.6.

6.8 Discussion

Based on the derived NMR differential Eq. (6.1), the Yukawa potential has been analysed in the domain of weak and strong nuclear charges. The Yukawa potential is usually obtained in terms of (a) the potential energy, due to the coulomb field of the nucleus and (b) the potential energy, of the averaged-out interactions with the other electrons. We considered an atom or ion with a single electron and nuclear charge Ze when the spin of the electron is ignored. $Z = 1$ refers to a hydrogen atom, $Z = 2$ to a He^+ ion, $Z = 3$ to a Li^{++} ion, and so on. Based on Eq. (6.3) or Eq. (6.5), the transverse magnetizations are uniquely specified by the eigen functions which are totally dependent on the three quantum numbers n, ℓ, m, and hence on the spherical harmonics $Y_l^m(\theta, \phi)$. The quantum number n is a positive integer, and, for a fixed value of n, ℓ takes integral values from zero to $n - 1$. It may be noted that if the potential $E_p(r)$ is a function only of the magnitude of r which represents a central force field, the angular part of the solution of the NMR wave equation is a spherical harmonic $Y_l^m(\theta, \phi)$. This is true whatever the functional form of $E_p(r)$. The function $E_p(r)$ determines the radial part of the solution.

The electron traveling around the nucleus generally has a probability of being found in any direction from the nucleus, so it tends to form a spherically shaped distribution of the possible positions around the nucleus. For this reason, spherical coordinates are used when describing the position of the electron relative to the nucleus instead of rectangular coordinates (x, y, z).

The quantity of interest, energy (E) of the hydrogen-like atom, appears as an eigenvalue. Thus, we conclude that acceptable solutions of the Bloch NMR wave equation leads to energy values which are the same as those derived from Bohr's theory. The radial part of the wave function, $R_{n\ell}(r)$, describes how far from the nucleus the electron might be found with respect to the accompanying proton. In particular, the absolute-square of the wave function gives information about a probability function. In the case of the wave function, its absolute-square gives information about the probability of finding the electron around the proton in a hydrogen-like atom as a function of the (3D) position and time. The radial part of the probability function depends on the principal quantum number n and the orbital angular momentum quantum number l, $P_{n\ell}(r)$. This function tells us the probability of detecting an electron in a hydrogen-like atom at some distance r from the nucleus (the proton).

Classically, we obtained the NMR transverse magnetization as a measure of the Yukawa and Coulomb potential so as to demonstrate the presence or absence of screening. Despite the difficulty in solving this problem, this study has presented a solution that is easy to follow as well as making them into computer codes for simulation. The proton-electron system has been taken as a single quantum particle whose mass is the reduced mass. As an example, this proton-electron system which can be found in proteins has been used to simulate the results obtained in this study (given in Eqs. (6.63) and (6.102)). Specifically, the particle dynamics of cytochrome C protein has been used. The simulations in this study have been achieved with wolfram Mathematica software package.

Using the NMR relaxation properties given in Table 6.2 and simulation parameters in Table 6.3, we have simulated NMR signal in terms of the transverse magnetization in Figs. 6.1 and 6.2. It is interesting to note that the quantum approach used in this study produced a measurable NMR signal for cytochrome C protein molecule which is quantum in nature. Although C_1 and C_2 serve as amplification factors, the simulation in Figs. 6.1 and 6.2 still shows a significantly large NMR signal for quantum molecule. This means that the model developed in this study may prove to be quite useful in molecular imaging especially in single-atom imaging. This could help improve real-time imaging of molecular processes in proteins and related solutions. In fact, we may be able to observe cancer cells in real-time as well as observation of the performance of molecular anti-cancer remedies. For example, 1s and 2s (Fig. 6.1a, b) orbitals showed very high NMR signals when compared to the other selected orbitals. This may prove beneficial as MR imaging and NMR spectroscopy in one single method. It would be observed that significant differences exist in the profiles of the transverse magnetization of Yukawa potential formulation and Coulomb potential formulation. This is probably due to the screening of the particles. The implication of this is that at the molecular level in cytochrome C proteins, screening increases NMR signal and such signal is quantifiable at all points. Figure 6.2b, e, f show regions (in white) for which the NMR signal could not be quantified. Figure 6.2c, d show unique patterns (belonging to the $2p_x$ and $2p_z$ orbitals) in which the NMR transverse magnetization has the highest magnitudes close to the origin. Interesting profile patterns are possible at higher energy levels.

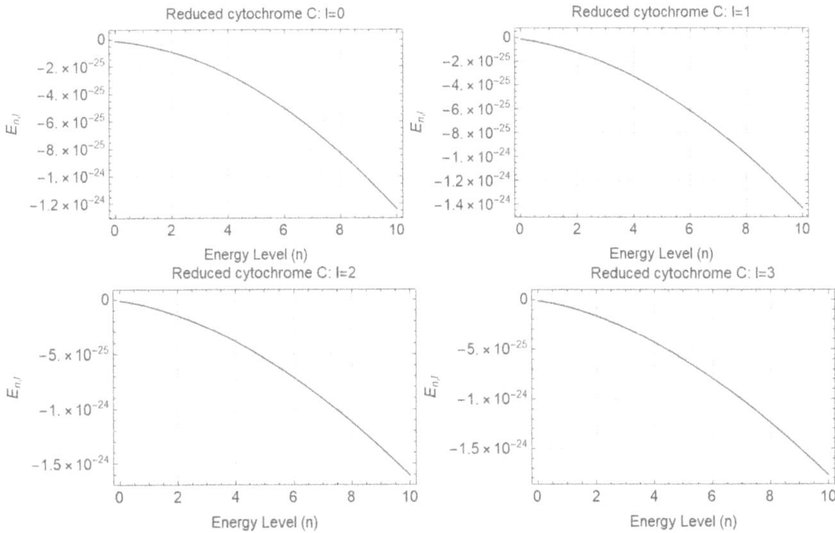

Fig. 6.3 Plots of particle energy in the presence of Yukawa potential as a function of energy level at different angular momentum states

We observed almost identical profiles in oxidized cytochrome C and this is due to less contribution of the flow part (containing T_1 and T_2 relaxation times). This flow contribution improves as the fluid velocity becomes smaller. This shows that it is still possible to improved image contrasts for any protein whose flow velocity is lower than those of the mitochondria. The energy of the molecule in Yukawa potential as given by Eq. (6.61) is shown in Fig. 6.3. These plots show that the energy is decreasing with increase in the quantum numbers n and l (Fig. 6.3). However, energy increases with increasing energy level and magnitude of the weighting parameter, ς (Fig. 6.4). The RF excitation pulses necessary for these situations to be achieved are given in Eq. (6.110) and (6.111). These equations interestingly demonstrate the connection between particle geometry, its flow properties and NMR relaxation properties. Figures 6.5 and 6.6 show the variation of the angular frequency with respect to atomic radial distributions. In all situations, angular frequency decreased with increase in r.

Basically, normal concentrations of electrolytes are essential for physiologic processes such as nerve conduction, muscle contraction, and blood clotting. The main electrolytes (Na^+), (K^+), (Ca^{++}), (Cl^-) may be treated like hydrogen atoms. Hydrogen, being the simplest of all chemical elements, provides a model for the study of atomic structure of elements in general. It constitutes one of the basic examples of chemical systems exhibiting rotational motion. The spatial rotation of an electron around the nucleus in the hydrogen atom can be described in terms of the electron's two angular variables θ and ϕ, and its radial distance r from the centre of the nucleus. This system is equivalent to a particle moving in an infinite number of concentric spheres.

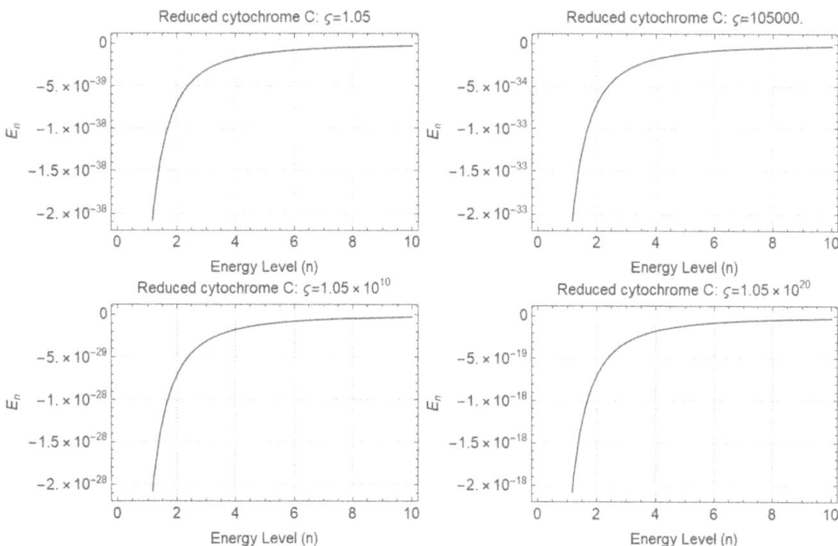

Fig. 6.4 Plots of particle energy in the presence of Coulomb potential as a function of energy level at different values of ς

Fig. 6.5 Profiles of applied radiofrequency in the presence of Yukawa potential ($\varsigma = 700$) for (**a**) reduced cytochrome C, (**b**) oxidized cytochrome C

(A)

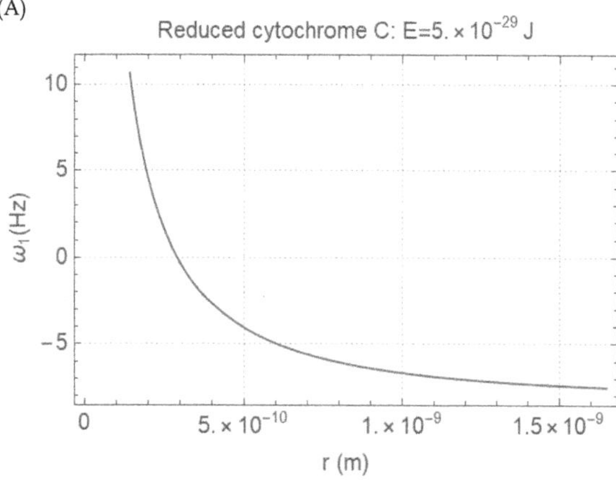

(B)

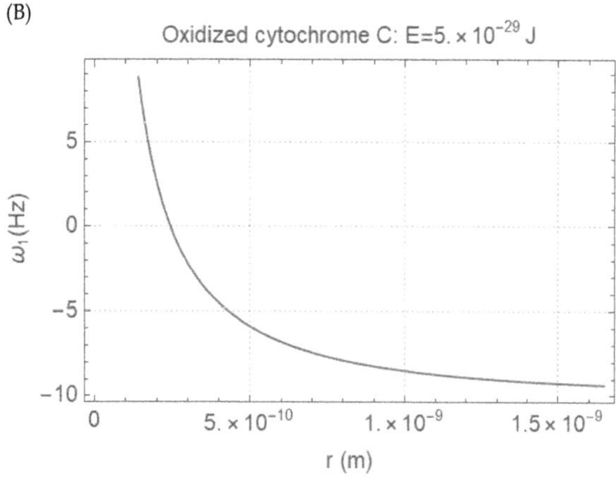

Fig. 6.6 Profiles of applied radiofrequency in the presence of Coulomb potential ($\varsigma = 500$) for (**a**) reduced cytochrome C, (**b**) oxidized cytochrome C

(A)

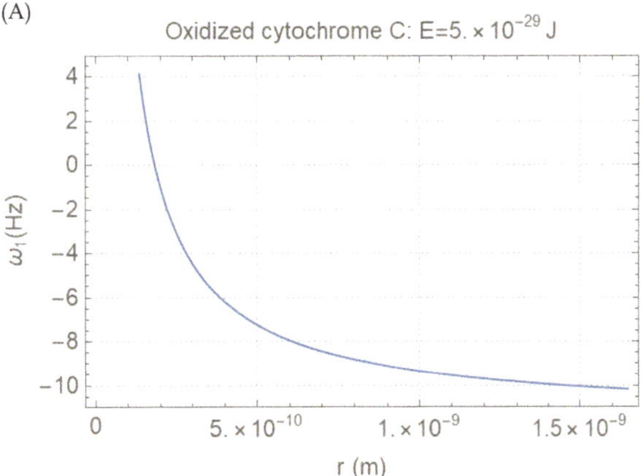

(B)

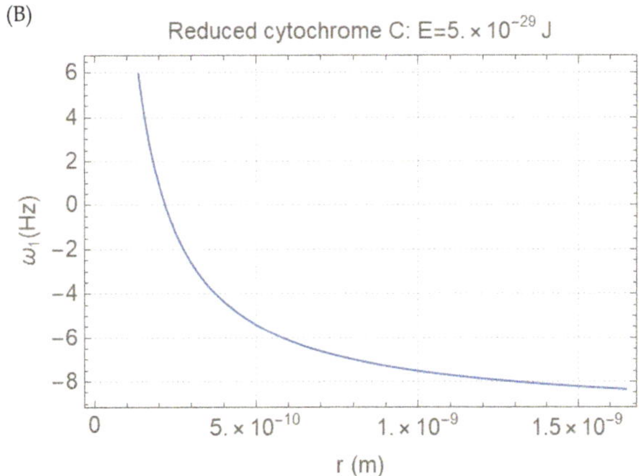

In comparison with some other commonly used numerical and approximation techniques for the analyses of hydrogen atom and hydrogen-like ions, this model may have three main advantages. First, it is based on the analytical solution of the fundamental Bloch NMR flow equations which is easier to visualize. Second, the quantum and the classical models can complement each other to provide an excellent model for the NMR study of atomic structure of hydrogen-like particles in general. Similarly, the formulation of the parameter ς is an interesting physical mechanism that provides a switch between classical and quantum mechanics. This is possible because if an atom other than hydrogen is ionized to an extent that it only has one electron left, then it behaves like a hydrogen atom and all of the hydrogen atom equations will still hold true for that ionized atom if one replaces the quantity e in the equations by Ze, where Z is the number of protons in the nucleus.

6.9   Conclusion

We have developed in this chapter the classical and quantum mechanical analyses of the Yukawa potential based on Bloch's NMR flow equation so that the wave function of hydrogen atoms and hydrogen-like ions can be represented in terms of NMR parameters. The formulation of the parameter ς is an interesting physical mechanism that provides a switch between classical domains and quantum domains. This parameter can be used to tailor laboratory RF pulse to quantum RF energy absorptions. A future study exploring the properties for experimental application of this parameter may be exciting and revealing.

Chapter 7
Quantum Mechanical Model of the Bloch NMR Flow Equations for Transport Analysis of Quantum-Drugs in Microscopic Blood Vessels Applicable in Nanomedicine

Abstract In the human cardiovascular system, there are more than 10^{10} capillary blood vessels the diameter of which is about the same size as that of blood cells. When blood flows through the capillaries, the blood cells have to be squeezed, deformed and move in single files. This makes it very difficult to analyse blood flow in these microscopic blood vessels in terms of classical mechanical quantities. The focus of this chapter is the application of known quantum mechanical formulations and models to the Bloch NMR flow equations; it provides a theoretical foundation that may enhance accurate understanding of the transport of nanodevices in microscopic blood vessels used in nanomedicine.

Keywords Bloch NMR flow equations · Quantum mechanical model · Microscopic blood vessels · Quantum-drugs · Nano-medicine

7.1 Introduction

Nanomedicine is the application of nanotechnology, the science of the manipulation and creation of microscopic objects, to the medical field. The unit of measurement in nanotechnology is the nanometre (symbol nm), which is one billionth of a meter. To put such measurements in perspective, a red blood cell is about 2500 nm in diameter. A bacterium is about 1000 nm long and a virus about 100 nm. Human DNA measures about 2.5 nm in diameter (Jain, 2008).

The ultimate goal of nanomedicine is to build microscopic devices designed to perform medical procedures inside the human body through what may be referred to as *nanomachines*. Nanomedicine may trigger research towards the design of nanodevices that may be able

1. to travel through tiny capillaries and deliver oxygen to anaemic tissues, remove obstruction from blood vessels and plaque from brain cells, and even hunt down and destroy viruses, bacteria, and other infectious agents.

© The Author(s), under exclusive license to Springer Nature Switzerland AG 2021   247
M. O. Dada, B. O. Awojoyogbe, *Computational Molecular Magnetic Resonance Imaging for Neuro-oncology*, Biological and Medical Physics, Biomedical Engineering, https://doi.org/10.1007/978-3-030-76728-0_7

2. to deliver drugs directly to specifically targeted cells. Cancer detection will improve dramatically with the aid of nanomachines. This will create possibilities for finding very small cancers far earlier than ever before and treating them with powerful drugs at the tumor site alone, while at the same time reducing any harmful side effects.

Through nanomachines, new gene-based designer drugs may be developed for diabetes, heart disease, Alzheimer's disease, schizophrenia, and many other conditions that exert a high toll on human society. The goal of this chapter is to present realistic mathematical models for the application of NMR/MRI in nanomedicine based on the fundamental Bloch NMR flow equations using basic quantum mechanical tools.

Although this may sound like futuristic fantasy, nanomedicine is real in the minds of some scientists. Even now some scientists are reporting good success in the use of nanoparticles in modern medicine (Allen & Cullis, 2004; Moghimi et al., 2005; Boisseau & Loubaton, 2011; Jorgensen, 2004; Jiang et al., 2010). As medical knowledge continues to grow, cutting-edge technology is revolutionizing the medical field. It has been said that medical knowledge doubles every 8 years. In the study reported here, we utilize the classical and quantum mechanical principles of physics based on Bloch's NMR flow equations to provide theoretical foundations that may enhance further application of nanotechnology to modern medicine by providing analytical and modelling tools which can incite, motivate and encourage further investigations in nanomedicine.

This is through the development of a mathematical algorithm to describe in detail the dynamical state of the hydrogen atom starting from the NMR flow equation, derived from the previous studies (Awojoyogbe, 2002, 2003, 2004, 2007; Awojoyogbe & Boubaker, 2009; Kundu & Cohen, 2004). The hydrogen atom is assumed to be magnetized by the static B_0 field to an equilibrium magnetization, M_0, before entering the excitor coil. The z-axis in the rotating frame coincides with the laboratory z-axis; the x axis makes an angle 'ωt' at any instant of time 't' with laboratory x-axis. $x = 0$ position could be such that the transverse magnetic field at the end of the detector coil is negligible.

7.2 Quantum Mechanical Model of Bloch NMR Flow Equations

We study the flow properties of the modified time-independent Bloch NMR flow equations which describe the dynamics of the hydrogen atom under the influence of a radio frequency (RF) magnetic field as follows (Awojoyogbe, 2003, 2007; Awojoyogbe & Boubaker, 2009):

$$\frac{d^2M_y}{dx^2} + \frac{1}{v}\left(\frac{1}{T_1} + \frac{1}{T_2}\right)\frac{dM_y}{dx} + \frac{1}{v^2}\left(\gamma^2 B_1^2(x) + \frac{1}{T_1 T_2}\right)M_y = \frac{M_o \gamma B_1(x)}{v^2 T_1} \tag{7.1}$$

It is convenient to use the departure of the stream function in its classical form as the dependent variable and write:

$$M_y(x) = \phi(x)e^{\varepsilon x} \tag{7.2}$$

where $\varepsilon = -\frac{1}{2vT_A}$, v is the instantaneous velocity of the fluid and $\phi(x)$ is a special function of the transverse magnetization M_y. However, when the RF $B_1(t)$ field is applied, M_y has a maximum value when RF $B_1(t)$ is maximum and $M_0 = 0$. At the point when the maximum NMR signal is received (maximum values of M_y and $B_1(t)$ respectively), Eq. (7.1) becomes:

$$\frac{d^2\phi}{dx^2} + \frac{\gamma^2 B_1^2}{v^2}\phi = 0 \tag{7.3}$$

subject to the following two conditions:

1.
$$e^{\lambda x} \neq 0 \tag{7.4}$$

2. Resonance condition exists at Larmor frequency $f_0 = \gamma B - \omega = 0$
3.
$$\gamma^2 B_1^2(x) \gg \left(T_g - T_R\right) \tag{7.5}$$

$$T_g = \frac{1}{T_1 T_2}, \quad T_R = \frac{1}{4T_A^2} \quad and \quad \frac{1}{T_A} = \frac{1}{T_1} + \frac{1}{T_2}$$

where γ denotes the gyromagnetic ratio of fluid spins; $\omega/2\pi$ is the RF excitation frequency; and f_0/γ is the off- resonance field in the rotating frame of reference. T_1 and T_2 are the spin-lattice and spin-spin relaxation times respectively, as defined in Chap. 6. RF B_1 is treated as constant and of the order of 1G.

The exponential function in Eq. (7.2) can be defined as follows,

$$e^x = \sum_{n=0}^{\infty} \frac{(\zeta x)^n}{n!} = F(x) \tag{7.6}$$

Equation (7.6) is extremely useful in obtaining approximations to complicated formulas, valid when x is small. Equation (7.3) is known as the NMR wave equation (Awojoyogbe, 2003, 2007). It completely describes the wave properties of the signal obtainable from the Bloch NMR equations in terms of the mechanical wave function $\phi(x)$, the velocity v, radio-frequency $\omega_1 = \gamma B_1(x)$, as well as T_1 and T_2 relaxation parameters. The uniqueness of Eq. (7.3) is apparent from the RF B_1 field which can be a constant or a variable. The relaxation parameters are intrinsic properties of the NMR system and are constants. In order to learn how to solve the NMR wave equation and to acquire a feeling for its peculiarities and nature, we proceed to treat a few applications. The novel quantity appearing in the NMR wave equation $\phi(x)$ expresses the mathematical form of particle wave properties associated with the fluid particles of the flow system. We now examine some important features of the wave equation by considering the value of the radio frequency relative to the relaxation parameters.

When the RF B_1 field spatially varies with x, Eq. (7.3) can be written as (Awojoyogbe & Boubaker, 2009; Dada et al., 2009):

$$\frac{d^2\phi}{dx} + G^2(x)\phi = 0 \tag{7.7}$$

where

$$G^2(x) = \frac{\gamma^2 B_1^2(x)}{v^2} = \frac{8\pi^2 m}{h^2}(E - V(x)) \tag{7.8}$$

E and V(x) are the total energy and potential energy (in the magnetic resonance system) of the nanoparticle. Assuming the RF B_1 field is designed such that $G^2(x)$ vanishes at a single point x = a only (Thirring, 2013). Then,

$$\begin{aligned} G^2(x) > 0 \quad for \ \ x > a \\ G^2(x) < 0 \quad for \ \ x < a \end{aligned} \tag{7.9}$$

If Eq. (7.3) is applied to matter waves (Aruldhas, 2009), it must satisfy the de Broglie relation and uncertainty principle where the total energy of the system is the sum of the kinetic energy E_k and the potential energy V(x).

$$E = E_k + V(x) \quad or \quad p = [2m(E - V(x))]^{1/2}$$

Using de Broglie equation, the wavelength λ of the nanoparticle is given as:

$$\lambda = \frac{h}{p} = \frac{h}{[2m(E - V(x))]^{1/2}} \tag{7.10}$$

where p and m refer to the momentum and the mass of the nanoparticle respectively. Equation (7.8) becomes:

$$G^2(x) = \frac{\gamma^2 B_1^2(x)}{v^2} = \frac{1}{\lambda^2} = \frac{8\pi^2 m}{h^2}(E - V(x)) \tag{7.11}$$

7.3 Application to Nanotechnology

In an attempt to build nanodevices designed to perform medical procedures inside the human body, we consider a quantum drug confined to a nanomachine of length 1 nm in a blood vessel of cross-sectional area d^2 (Aruldhas, 2009; Shankar, 2012), where d is assumed to be the diameter of the blood vessel. Based on Eq. (7.7), the energy eigenvalue for the quantum drug is

$$E_n = \frac{n^2 \hbar^2}{8md^2} \tag{7.12}$$

The minimum energy of the quantum drug at the ground state (n = 1) is 36.28 kJ mol^{-1} and the minimum excitation energy from the ground state is 108.84 kJ mol^{-1} (Aruldhas, 2009). The probability that the quantum drug is in the range x = 0 to x = a is:

$$\int \phi^* \phi dx \tag{7.13}$$

and the wave function is:

$$\phi(x) = \begin{cases} \left(\frac{2}{d}\right)^{1/2} \cos\left(n\pi \frac{x}{d}\right), & \text{for } n = 1, 3, 5, \ldots \ldots \\ \left(\frac{2}{d}\right)^{1/2} \sin\left(n\pi \frac{x}{d}\right), & \text{for } n = 2, 4, 6, \ldots \ldots \end{cases} \tag{7.14}$$

Equations (7.12)–(7.14) are obtained from the assumption that for a constant B_1 field, the potential V(x) becomes constant, such that

$$G^2 = \frac{\gamma^2 B_1^2}{v^2} = \frac{1}{\lambda^2} = \frac{8\pi^2 m}{h^2}(E - V) \tag{7.15}$$

The total energy delivered to the system by the constant B_1 field is

$$E = \frac{\gamma^2 B_1^2 \hbar^2}{2mv^2} + V \tag{7.16}$$

This implies that the kinetic energy of any quantum drug being administered is given as

$$\frac{\gamma^2 B_1^2 \hbar^2}{2mv^2} \tag{7.17}$$

It is very interesting to note that if the quantum drugs under consideration are taking part in a fluid flow within a porous cavity, the expression in Eq. (7.17) is equivalent to (Awojoyogbe et al., 2009):

$$\frac{\gamma^2 B_1^2 \hbar^2}{2mv^2} = \frac{p^2}{x^2} \tag{7.18}$$

where p is the porosity of the medium and x is the spatial distance travelled by the quantum drugs. Assuming that the saturation condition holds; we have (Cowan, 1997)

$$\gamma^2 B_1^2 T_1 T_2 = 1 \tag{7.19}$$

The implication of Eqs. (7.16)–(7.19) is that we are able to manipulate the energy of the quantum drugs through small spaces even when a constant potential is present. This is made possible by simply altering the magnitude of the constant B_1 field as the situation requires. From Eqs. (7.2) and (7.14), we have

$$M_y(x)\big|_n = \begin{cases} \left(\frac{2}{d}\right)^{1/2} \cos\left(n\pi\frac{x}{d}\right) e^{\varepsilon x}, & \text{for } n = 1, 3, 5, \ldots\ldots \\ \left(\frac{2}{d}\right)^{1/2} \sin\left(n\pi\frac{x}{d}\right) e^{\varepsilon x}, & \text{for } n = 2, 4, 6, \ldots\ldots \end{cases} \tag{7.20}$$

where $\varepsilon = -\frac{1}{2vT_A}$.

Hence, we have obtained the transverse magnetization in terms of the underlying quantum mechanics (QM) of the fundamental Bloch equation. The discrete number n may represent a single spin or a group of spins with similar quantum drugs such that a sum overall n gives the bulk magnetization. In most NMR systems in which the resonant frequency of the hydrogen atom is used, we see that the hydrogen atom does not exist in isolation but usually as part of a compound. Therefore, our study is based on the fact that any nanomachine designed must be a compound in which proton spin or electron spin is available.

7.4   Quantum Drugs Model

The probability of finding a quantum drug between x = 0 and x = 0.2 nm is 0.0486. If the size of the nanomachine is increased to 2 nm, the minimum energy of the quantum drug becomes 9.07 kJ mol^{-1}. Equation (7.14) can be used to locate the possible positions of the quantum drug. Magnetic resonance becomes useful here because quantum drugs emit energies containing information about what is happening in their vicinity. For example, if a particular disease condition is maintained by the presence of a functional group attached to a protein compound, a nanomachine may be designed in such a way that whenever it gets very close to the protein compound, it becomes very reactive such that it frustrates the chemical bonds (by supplying extra quantum drugs or otherwise) linking the undesirable bond to the compound and continues until the bond is broken. However, while this operation is on, we must be able to exercise as much control on the nanomachine as possible. Using hemodynamic and NMR relaxation properties of human blood vessels at 1.5 T (Glaser, 2012; Tortora & Derrickson, 2012; Marieb & Hoehn, 2013), hemodynamic data is shown in Table 7.1. At a uniform magnetic field $B_0 = 1.5$ T, blood flowing within the blood vessels has $T_1 = 1.2$ s and $T_2 = 0.1$ s (Moratal et al., 2006) (Table 7.1). Using the expression of Eq. (7.20) and the hemodynamic data in Table 7.1, the profiles of the NMR signal for the different blood vessels are presented in Fig. 7.1. For this case, we have evaluated the scenario at n = 1. These profiles unique signal contrasts that can be used for magnetic resonance imaging of capillaries.

Table 7.1 Hemodynamic properties of human blood vessels

Vessel	Average velocity (m/s)	Diameter (m)	Wall shear rate (s^{-1})	Reynolds number (Re)
Aorta	0.48	0.025	155	3400
Artery	0.46	0.004	900	500
Arteriole	0.05	0.00005	8000	0.7
Capillary	0.001	0.000008	1000	0.002

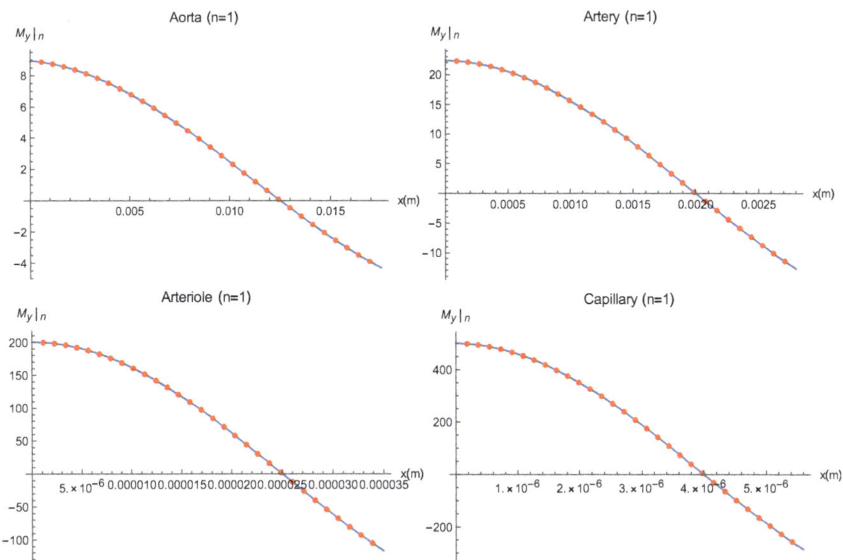

Fig. 7.1 2D profiles of NMR transverse magnetization for different types of human blood vessels at energy level, n = 1

7.5 Application of the WKB Approximation

However, in real life systems, the potential function $V(x)$ is not constant but is very much dependent on the nature and condition of the system being considered. In such cases, we must find an approximate solution to Eq. (7.7), (7.8) and (7.9). An approximate solution to Eq. (7.7) may be assumed to take the form (Aruldhas, 2009; Shankar, 2012):

$$\phi(x) = \frac{1}{\sqrt{p(x)}} \exp\left[\pm \int_{x_0}^{x} p(s)ds\right] \tag{7.21}$$

where we are to determine the function $p(x)$. If we substitute for Eq (7.21) in Eq. (7.7), we obtain:

$$p^2(x) + h(x) = G^2(x) \tag{7.22}$$

where

$$h(x) = \frac{1}{2p}\frac{d^2p}{dx^2} - \frac{3}{4p^2}\left(\frac{dp}{dx}\right)^2 \tag{7.23}$$

In the WKB approximation, $h(x)$ is neglected so that:

$$p^2(x) = G^2(x) \tag{7.24}$$

This approximation is valid only when

$$h(x) \ll \left|G^2(x)\right| \tag{7.25}$$

The general solution to Eq. (7.7) becomes:

$$\phi(x) = \frac{1}{\sqrt{G(x)}}\left[A\exp\left(i\int_{x_0}^{x} G(s)ds\right) + B\exp\left(-i\int_{x_0}^{x} G(s)ds\right)\right] \text{ for } x$$
$$> x_0 \tag{7.26}$$

$$\phi(x) = \frac{1}{\sqrt{|G(x)|}}$$
$$\times \left[C\exp\left(i\int_{x_0}^{x} |G(s)|ds\right) + D\exp\left(-\int_{x_0}^{x} |G(s)|ds\right)\right] \text{ for } x$$
$$< x_0 \tag{7.27}$$

where A, B, C and D are all constants. It is required that the boundary condition holds

$$\phi(x) \to 0 \ \ as \ \ x \to \infty \tag{7.28}$$

Therefore, C must be zero so that within the range $x < x_o$, we have

$$\phi(x) = \frac{D}{\sqrt{|G(x)|}} \exp\left(-\int_{x_0}^{x} |G(s)|ds\right) \tag{7.29}$$

If the WKB approximation fails at the turning point and the neighbourhood of the turning point, we may write A and B as functions of D:

$$\left.\begin{array}{l} A = \dfrac{D}{i} e^{i(\pi/4)} \\[3mm] B = -\dfrac{D}{i} e^{-i(\pi/4)} \end{array}\right\} \tag{7.30}$$

Therefore, we finally have:

$$\phi(x) = \frac{2C}{\sqrt{|G(x)|}} \sin\left(\int_{x_0}^{x} G(s)ds + \frac{\pi}{4}\right) \text{ for } x > x_0 \tag{7.31}$$

$$\phi(x) = \frac{D}{\sqrt{|G(x)|}} \exp\left(-\int_{x_0}^{x} |G(s)|ds\right) \text{ for } x < x_0 \tag{7.32}$$

Since the total magnetic field experienced by a given spin could be expressed as (Haacke et al., 2000):

$$B(x) = B_0 + B_1(x) \tag{7.33}$$

or in the presence of a magnetic field gradient, g

$$B(x) = B_0 + gx \tag{7.34}$$

we may assume the following:

$$\gamma B_1(x) = \gamma g x \tag{7.35}$$

Then, from Eq. (7.8), we may write

$$G^2(x) = \frac{\gamma^2 g^2 x^2}{v^2} = \frac{2m}{\hbar^2}(E - V(x)) \tag{7.36}$$

The kinetic energy of the quantum drug in the MR system is

$$E - V(x) = \frac{\gamma^2 g^2 x^2 \hbar^2}{2mv^2} \tag{7.37}$$

It also follows that the potential is a quadratic function in x, given as:

$$V(x) = E - \frac{\gamma^2 g^2 x^2 \hbar^2}{2mv^2} \tag{7.38}$$

Using Eq. (7.36) in Eqs. (7.26) and (7.27), we have

$$\phi(x) = \frac{2C}{\sqrt{\left|\frac{\gamma g x}{v}\right|}} \sin\left(\int_{x_0}^{x} \left|\frac{\gamma g s}{v}\right| ds + \frac{\pi}{4}\right) \text{ for } x > x_0 \tag{7.39}$$

$$\phi(x) = \frac{D}{\sqrt{\left|\frac{\gamma g x}{v}\right|}} \exp\left(-\int_{x_0}^{x} \left|\frac{\gamma g s}{v}\right| ds\right) \text{ for } x < x_0 \tag{7.40}$$

which is:

$$\phi(x) = \frac{2C}{\sqrt{\left|\frac{\gamma g x}{v}\right|}} \sin\left(\left|\frac{\gamma g x^2}{2v}\right| - \left|\frac{\gamma g x_0^2}{2v}\right| + \frac{\pi}{4}\right) \text{ for } x > x_0 \tag{7.41}$$

$$\phi(x) = \frac{D}{\sqrt{\left|\frac{\gamma g x}{v}\right|}} \exp\left(\left|\frac{\gamma g x_0^2}{2v}\right| - \left|\frac{\gamma g x^2}{2v}\right|\right) \text{ for } x < x_0 \tag{7.42}$$

Finally, using Eq. (7.2), the transverse magnetization becomes:

$$M_y(x) = \frac{2C}{\sqrt{\left|\frac{\gamma g x}{v}\right|}} \sin\left(\left|\frac{\gamma g x^2}{2v}\right| - \left|\frac{\gamma g x_0^2}{2v}\right| + \frac{\pi}{4}\right) \exp\left(\frac{x}{2vT_A}\right) \text{ for } x > x_0 \tag{7.43}$$

$$M_y(x) = \frac{D}{\sqrt{\left|\frac{\gamma g x}{v}\right|}} \exp\left(\left|\frac{\gamma g x_0^2}{2v}\right| - \left|\frac{\gamma g x^2}{2v}\right| - \frac{x}{2vT_A}\right) \text{ for } x < x_0 \tag{7.44}$$

where C and D are constants. Since x_0 represents the turning points, we require that the spins are always away from these points. At such points, the potential energy (while aligning with the B_0 field) is approximately equal to the total energy; the kinetic energy is also nearly equal to zero. This is possible only if the applied gradient (quantifying the applied RF field) is removed, so that:

$$\frac{\gamma^2 g^2 x^2 \hbar^2}{2mv^2} = 0 \tag{7.45}$$

We can say, therefore, that the point x_0 represents the point at which the NMR magnetization has the highest possible amplitude. Equation (7.43) is therefore not possible under the situation defined by this study. Consequently, as long as x is less than x_0, the transverse magnetization is given as:

$$M_y(x) = \frac{D\sqrt{v}}{\sqrt{\gamma g x}} \exp\left(\left| \frac{\gamma g x_0^2}{2v} \right| - \left| \frac{\gamma g x^2}{2v} \right| - \frac{x}{2vT_A} \right) \tag{7.46}$$

Hence, Eq. (7.46) is very useful for real NMR systems in which there is signal attenuation. This is because the spins display free induction decay immediately after leaving the point x_0. It is worthy of note that a Fourier transform of Eq. (7.46) is:

$$M_y(k) = \frac{iD(2kvT_A + i)^2 \left[i\sqrt{\frac{2ikv}{\gamma g} - \frac{1}{\gamma g T_A}} - \sqrt{\frac{2ikv}{\gamma g} - \frac{1}{\gamma g T_A}} \right]}{(4kvT_A + 2i)\sqrt{\frac{2\pi\gamma g}{v}}}$$
$$\times K_{\frac{1}{4}}\left(-\frac{(2kvT_A + i)^2}{16\gamma g v T_A^2} \right) \exp\left(\frac{8\gamma^2 g^2 x_0^2 T_A^2 - 4k^2v^2 T_A^2 - 4ikvT_A + 1}{16\gamma g v T_A^2} \right) \tag{7.47}$$

where $K_{1/4}$ is the modified Bessel function of the second kind (Watson, 1966).

The profiles of Eqs. (7.46) and (7.47) are presented in the following illustrations. For this analysis, we have made use of the NMR properties (Moratal et al., 2006) of human blood ($T_1 = 1.2$ s and $T_2 = 0.1$ s), g $= 0.002$ T m^{-1}, $\gamma = 42666666.67$ s^{-1} T^{-1} and hemodynamic features of human blood vessels as shown in Table 7.1.

7.6   The Tunnelling Effect of Quantum Drugs

In quantum mechanical terms, drugs can be designed to travel through tiny capillaries and remove obstruction from blood vessels by initiation of controlled chemical reactions as shown in Eq. (7.6) and Figs. 7.2 and 7.3.

$$V(x) = V_1 = V_{III} \quad for \ x < a \quad and \quad x > a$$
$$V(x) = V_I + \Delta V \quad for \ 0 < x < a \tag{7.48}$$

The key feature in this structure is the quantum mechanical tunnelling effect. For the three–region case, the general form of the solution of Eqs. (7.7) and (7.8) is (Tang, 2005):

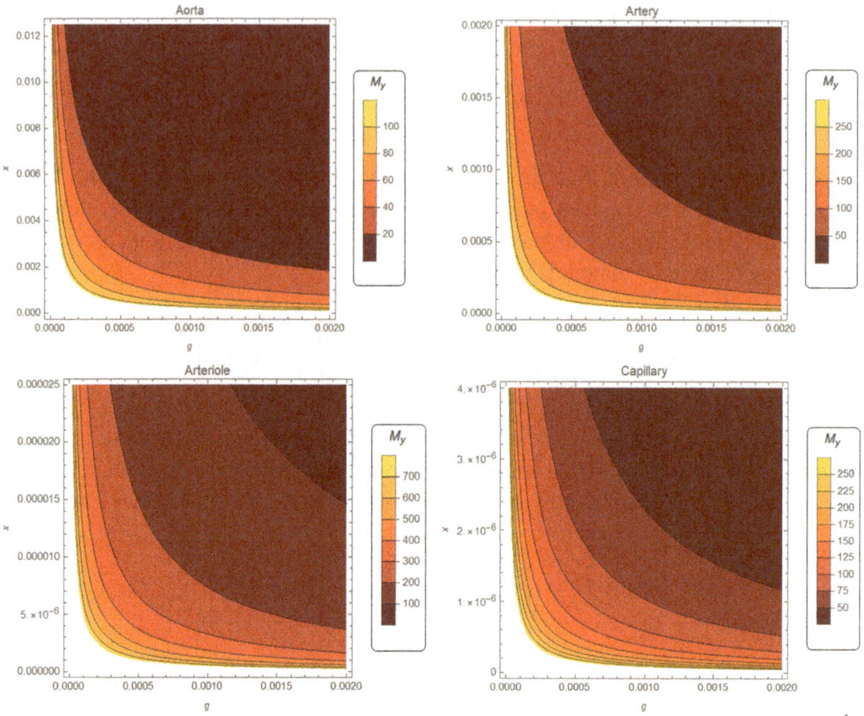

Fig. 7.2 Contour profiles of NMR transverse magnetization for different types of human blood vessels (D = 500)

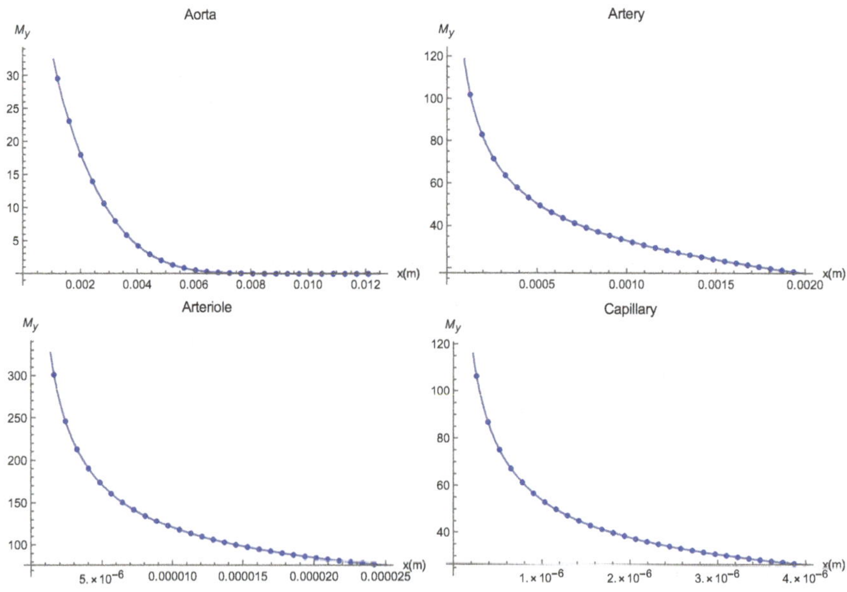

Fig. 7.3 2D profiles of NMR transverse magnetization for different types of human blood vessels (D = 500)

$$\psi_I(x) = Ae^{ik_I x} + Be^{-ik_I x} \quad for \ x < 0$$
$$\psi_{II}(x) = Ce^{ik_{II} x} + De^{-ik_{II} x} \quad for \ 0 < x < d \tag{7.49}$$
$$\psi_{III}(x) = Fe^{ik_{III} x} \quad for \ x > d$$

where

$$k_I = \sqrt{\frac{2m(E - V_I)}{\hbar^2}}, \quad k_{II} = \sqrt{\frac{2m(E - V_{II})}{\hbar^2}} \quad and \quad k_{III}$$

$$= \sqrt{\frac{2m(E - V_{III})}{\hbar^2}} \tag{7.50}$$

Since the fluid velocity is constant in this analysis, we note that:

$$\alpha k_I = \frac{\gamma B_1^I}{v} = \frac{\omega_{1,I}}{v}, \quad \alpha k_{II} = \frac{\gamma B_1^{II}}{v} = \frac{\omega_{1,II}}{v} \quad and \quad \alpha k_{II} = \frac{\gamma B_1^{III}}{v} = \frac{\omega_{1,III}}{v} \tag{7.51}$$

If we consider the case where $E > V_I = V_{III}$, k_I and k_{III} are always real while k_{II} can either be real or imaginary, in which case we define the imaginary part of k_{II} as (Tang, 2005):

$$k_{II} = i\alpha_{II} \quad and \quad \alpha_{II} = \sqrt{\frac{2m(V_{II} - E)}{\hbar^2}} \tag{7.52}$$

$$\psi_{II}(x) = Ce^{-\alpha_{II} x} + De^{\alpha_{II} x} \quad for \quad 0 < x < a \tag{7.53}$$

Applying the usual boundary conditions gives:

$$\frac{F}{A} = \frac{e^{-ik_{III} d}}{\left(\cos k_{II} a - i\frac{k_I^2 + k_{II}^2}{2k_I k_{II}} \sin k_{II} d \right)} \tag{7.54}$$

And the corresponding current transmission coefficient is:

$$TC = \left[1 + \frac{(V_{II} - V_I)^2 \sin^2 k_{II} d}{4(E - V_I)(E - V_{II})} \right]^{-1} \tag{7.55}$$

$$TC = \frac{4(E - V_I)(E - V_{II})}{4(E - V_I)(E - V_{II}) + (V_{II} - V_I)^2 \sin^2 k_{II} d} \tag{7.56}$$

$$TC = \frac{4(E - V_I)(E - V_{II})}{4(E - V_I)(E - V_{II}) + (V_{II} - V_I)^2 \sin^2 \left(\frac{\omega_{1,II} d}{\alpha v} \right)} \tag{7.57}$$

when $E > V_{II}$ or $E < V_{II}$.

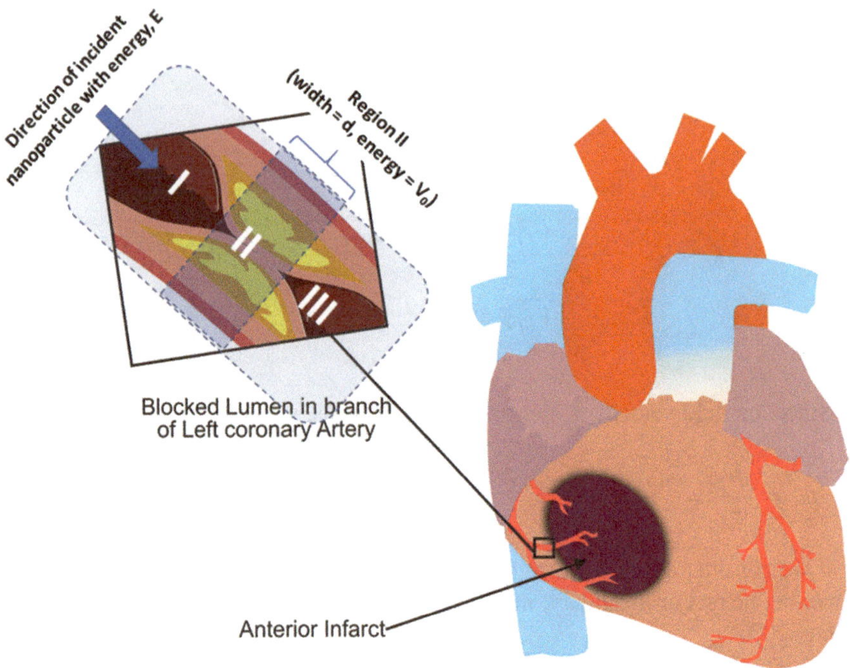

Fig. 7.4 A typical barrier potential from human heart with quantum drug(s) motion from region I to region III nearly impossible. (Adapted from https://www.labroots.com/trending/cardiology/2335/preventing-heart-failure-after-acute-myocardial-infarction)

The transmission coefficient of the quantum drug is given in the above equations. The quantum drug is definitely expected to find its way through the plaque within region II in Fig. 7.4. However, even with the condition $E < V_{II}$, the quantum drug would be able to tunnel through to region III according to the quantum mechanical rules. Therefore, from Eq. (7.52), we have the following

$$TC = \left[1 + \frac{m(V_{II} - V_I)^2 \sin^2 k_{II} a}{2\hbar^2 k_{II}^2 (E - V_I)}\right]^{-1} \tag{7.58}$$

$$TC = \frac{\left(\frac{\hbar \omega_{1,II}}{av}\right)^2 (E - V_I)}{\left(\frac{\hbar \omega_{1,II}}{av}\right)^2 (E - V_I) + (V_{II} - V_I)^2 \frac{m}{2} \sin^2\left(\frac{\omega_{1,II} a}{av}\right)} \tag{7.59}$$

where m is the mass of the quantum drug, and $h = 2\pi\hbar$ is the Planck constant.

For the analysis of the transmission coefficient, we shall assume that the energy of the quantum drug is the same as the left ventricle (connected with the aorta) inflow kinetic energy (Bolger et al., 2007) of the order of 8.2 mJ. The potential energy as a result of slight restriction in blood flow (in which the quantum drug is carried), would be assumed to be $V_I = 2.6$ mJ and the mass the quantum drug is 5.356×10^{-26} kg while the total energy for this particular system is $E = 10.8$ mJ. It is worthy of note that for these assume parameters, the RF field required to drive a nano-particle of mass m through region I is:

$$\omega_{1,I} = \gamma B_{1,I} = \frac{\alpha v}{\hbar} \sqrt{0.0164m} \qquad (7.60)$$

The graphical representations TC as function of quantum and NMR parameters are given in Figs. 7.5 and 7.6. In these profiles, m $= 5.356 \times 10^{-26}$ kg, E $= 10.8$ mJ, v $= 0.48$ m s^{-1}, $\alpha = 1.0 \times 10^{-18}$ kg.

It is interesting to note from Fig. 7.5 that the magnitude of the width of the plague (the quantum barrier) has no significant influence on the probability of the quantum drug finding itself in region III. Transmission of the nano drug is ensured by the use of higher magnitude of the excitation radiofrequency in this region especially barrier potential energy is significant.

In Fig. 7.5, we see that despite increasing barrier width does not have any noticeable influence on particle across the barrier. However, we observe that it is much easier for the quantum drug to overcome the barrier at higher values of RF energy coming through region I. In addition to this, the picks show some selective

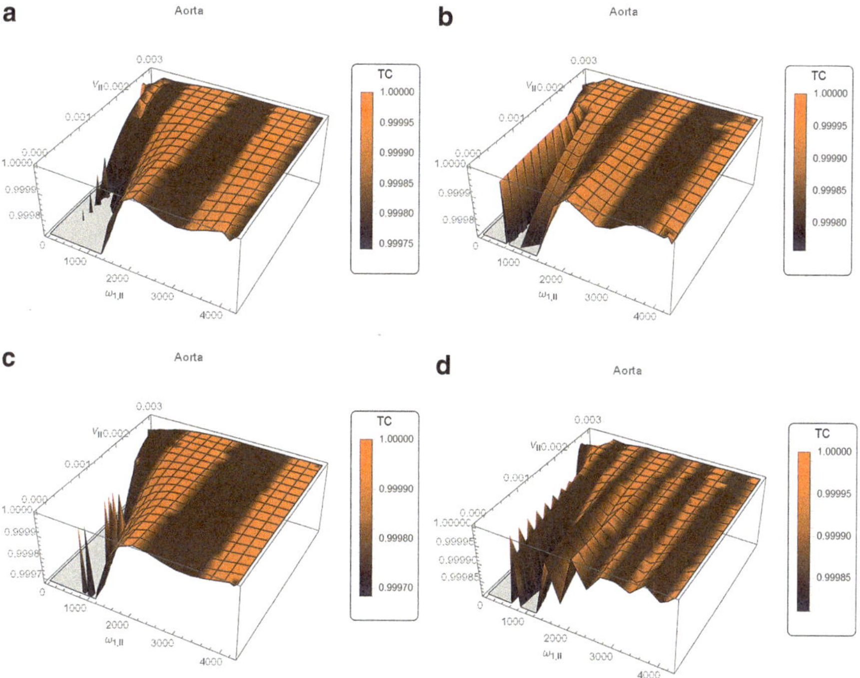

Fig. 7.5 Plots of the transmission coefficient TC against V_{II} and potential $\omega_{1,II}$ based on Eq. (7.59) for barrier width of (**a**) 2.5 nm, (**b**) 3.5 nm, (**c**) 5.5 nm, (**d**) 10.5 nm

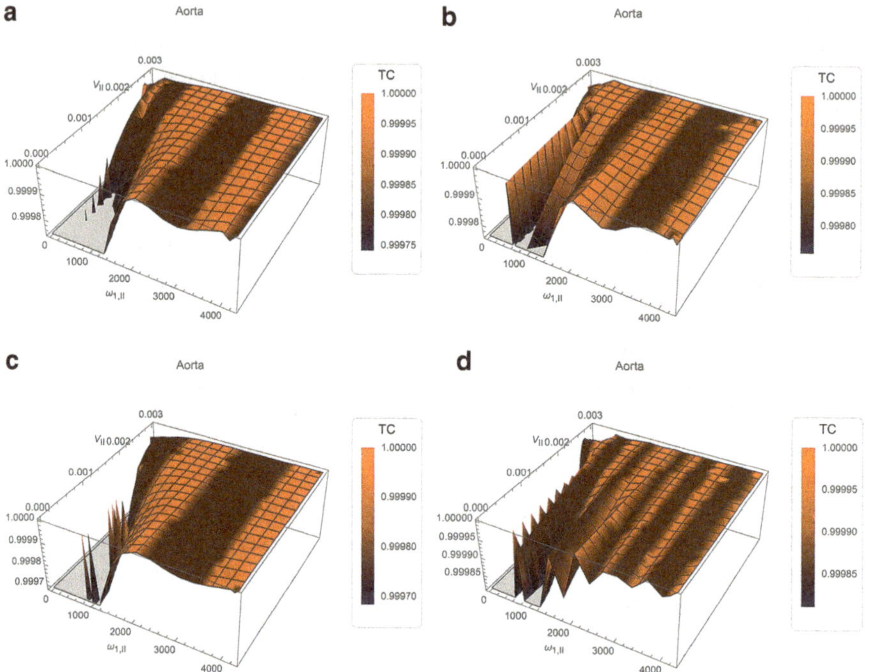

Fig. 7.6 Plots of the transmission coefficient TC against $\omega_{1,II}$ and the barrier width for (**a**) $E < V_{II}$ ($V_I = 2.6$ mJ, $V_{II} = 11.2$ mJ), (**b**) $E = V_{II}$ ($V_I = 2.6$ mJ, $V_{II} = 10.8$ mJ), (**c**) $E > V_{II}$ ($V_I = 2.6$ mJ, $V_{II} = 4.5$ mJ), (**d**) $E > V_{II} < V_I$ ($V_I = 2.6$ mJ, $V_{II} = 1.2$ mJ)

values of potential V_{II} at which the transmission is close to 100%. Figure 7.6 shows that quantum drug transmission across the barrier is possible as long as the condition in Fig. 7.6b is avoided. That is, to ensure quantum drug tunnelling, $E \neq V_I$ and the higher the values of excitation radiofrequency, the better the transmission. This shows that a nano particle may be designed to tunnel through a plague, lose its energy within the barrier and made to execute another task until the plague is dissolved. For example, iron oxide nanoparticles are known to be thermal sensitive and if such particles are made to tunnel to the centre of a fatty deposit, they could be thermally manipulated to dissolve these deposits without affecting the surrounding normal tissues by ensuring magnetic resonance monitoring via real time reporting on T_1 and T_2 within the plague environment.

It has been noted that plaque size is not always the determining factor in patient symptomatology and that other plaque characteristic (such as surface ulceration, subintimal haemorrhage, and composition) may play a significant role in plaque pathogenicity (Morrisett et al., 2003). The results from the transmission coefficient lay credence to this experimental observation. The wave vector k_{II} is an important property of the barrier and its combination with the potential V_{II} may provide the opportunity of understanding the roles of other plague characteristics. The issues emerging from Fig. 7.6a may provide some support to related clinical and pathoanatomic observations (Bolger et al., 2007; Libby, 1995) in which there are indications that plaque type (composition and biology) is a more significant determinant than plaque size (degree of stenosis) in the development of thrombus-mediated acute coronary syndromes. It would be observed that the transmission coefficient produced a similar profile whether the wave energy is lower or higher than the barrier energy. However, the nano-drug is completely blocked from tunnelling through the barrier at the point the wave energy equals the barrier energy of region I (as shown in Fig. 7.6b). Fortunately, MRI provides an opportunity for accessing the complex plague characteristics through quantification of T_1 and T_2 relaxation times for different barrier potentials and then we can tune the wave energy until the nano-drug is able to vanquish the blockage within the blood vessel. This study may be very important in this direction because it provides a means of understanding the quantum drugs according to Eqs. (7.2) and (7.20) in addition to the quantum dynamic information provided on the nanodrug by Eqs. (7.10)–(7.13) and (7.59).

Figure 7.5a, c demonstrate tunnelling effect occurs such that most of the quantum drugs would overcome the barrier even when $E < V_{II}$. This makes the job of the quantum drug very easy in the sense that, irrespective of the composition and size of the plaque, we can get the quantum drugs to the centre of the barrier to accomplish their tasks. In fact, the transmission coefficient becomes almost unity as the magnitude of the radiofrequency increases.

Figures 7.5 and 7.6c show the obvious, because with $E > V_{II}$, we definitely expect the quantum drugs to overcome the barrier. However, even with this condition, the probability of the quantum drug overcoming the barrier is not evenly distributed; and some of them will still not be able to cross as quickly as others. This shows that plaque composition is not always uniform even when hemodynamic pressure is high enough to drive the quantum drug across the barrier. Hence, an appropriate amount of RF energy correctly mapped to the tissue's relaxation properties can be used to control the nano/quantum drug as the situation requires.

Finally, when the kinetic energy $E - V_I$ of the quantum drugs in region I is greater than the potential step ΔV, as shown in Fig. 7.6, the transmission coefficient TC is a damped oscillatory function of E and it asymptotically approaches value 1. It can be very useful to note that there are particular values of E where $k_{II}d$ is an integral multiple of π such that

$$k_{II}d = n\pi; \quad n = 1, 2, 3, 4, 5, 6, \ldots \tag{7.61}$$

At these values, the transmission coefficient is also equal to unity (Fig. 7.4). Even though there is an obstruction present, the quantum drugs have a 100% probability of going through the obstruction as if it were transparent. This process can be faster and better depending on the chemical or biochemical composition of the quantum drugs as well as the RF energy applied.

What happens when the blood cells carrying the nanomachine oscillates as the blood flows through the microscopic blood vessels with a potential energy

$$V(x) = \frac{1}{2}fx^2 \tag{7.62}$$

where f is the force constant.

These equations are well known and need only be applied here. For example, the energy separation $h\upsilon$ is negligible in large blood vessels such as in the aorta as we have employed in this analysis. However, this separation can be very significant in microscopic blood vessels.

7.7 Description of QM-Designed Drugs in Protein-Structured Nanomachines

Generally, the transport mode of nanomachines in this study is based on the assumption that a single red blood cell can carry about a million QM-designed drugs in thousands of protein-structured nanomachines and deliver the drugs to body tissues that need them. When a red blood cell leaves the arteries and moves into the tiny capillaries deep in the body tissues, the environment around the red blood cell changes. The environment is warmer than in the large arteries. These thermal signals which result from increase in energy (for example, nh), tell the nanomachines inside the cell that it is time to release their contents where they are most needed.

We have presented a quantum mechanical analysis of the Bloch NMR flow equations using basic quantum mechanical tools. Two important results hold true for the energy of nanomachines traveling in tiny blood vessels such as capillaries and veins of cross-sectional area d. This mathematical approach may be useful not only in the special cases considered above but also in any general situation of restricted fluid flow in small blood vessels. It is apparent from the results that as the cross-sectional area d of the blood vessel decreases, the kinetic energy of any nanodevice flowing through the vessel increases. Other factors being the same, the freer the nanomachine, the lower the kinetic energy. This is crucial for the transport of nanodevices in very small blood vessels because the more localized the motion, the higher its kinetic energy. Such localization of nanomachines can occur in capillaries and veins. In such sensitive blood vessels where pathologies like brain tumor, cancerous tissue, cerebral edema, conjugated double bond systems and aromatic compounds can develop, the mathematical concepts presented in this chapter can be useful for accurate assessment of these specific cases. For example, Fig. 7.5 may indicate the severity of the biological condition before treatment and Fig. 7.6, may correspond to different levels of treatment as transmission coefficient asymptotically equal to one. Further application of the mathematical tools presented in this chapter especially using basic quantum mechanical tools such as quantum well, quantum wire and quantum dots can provide more insight into the application of nanotechnology to modern medicine.

7.8 Conclusion

It is particularly interesting to note that the separation between the energy levels ΔE is the same for all values of n (thus, the value of n does not influence the NMR signal) as shown in Fig. 7.7, but it decreases as the size of the blood vessels increases. The separation is very small for large blood vessels but is significant for microscopic blood vessels. This indicates that a free nanomachine moving in an unrestricted region of space ($d \rightarrow \infty$), has unquantized translational energy levels. This has even been mapped to transverse magnetization as shown in Fig. 7.8. For this reason, a nanomachine moving in a laboratory-sized vessel may be treated as though their translational energy is not quantized. Furthermore, Eq. (7.47) and Fig. 7.9 show that WKB approximation has provided us with a straight forward means of doing MR image reconstruction for the nanomachines.

Based on the theoretical procedures described in this chapter, it is possible in the near future that scientists and engineers will be able to build nanomachines designed to perform medical procedures inside the human body. These minuscule robots

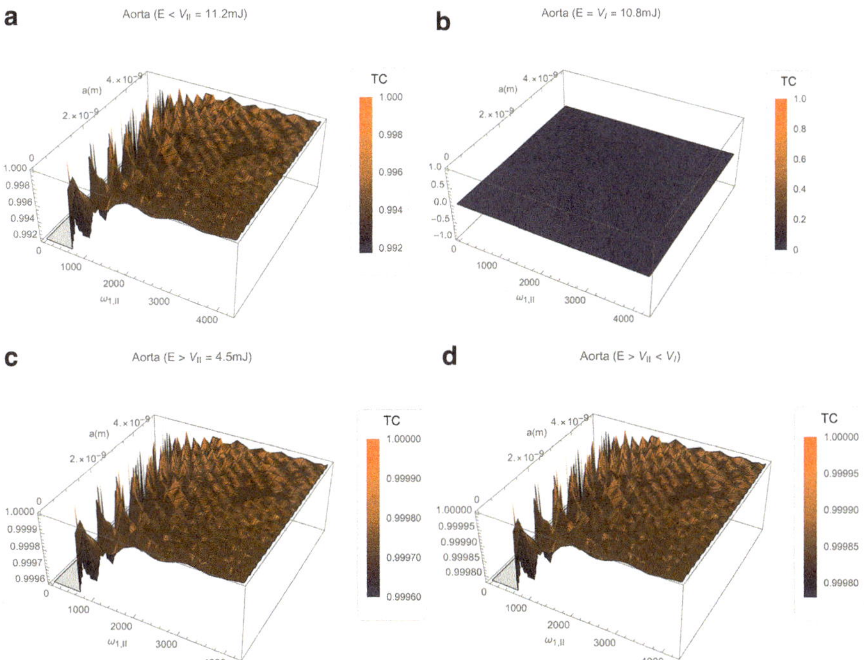

Fig. 7.7 Contour profiles of NMR transverse magnetization for different types of human blood vessels

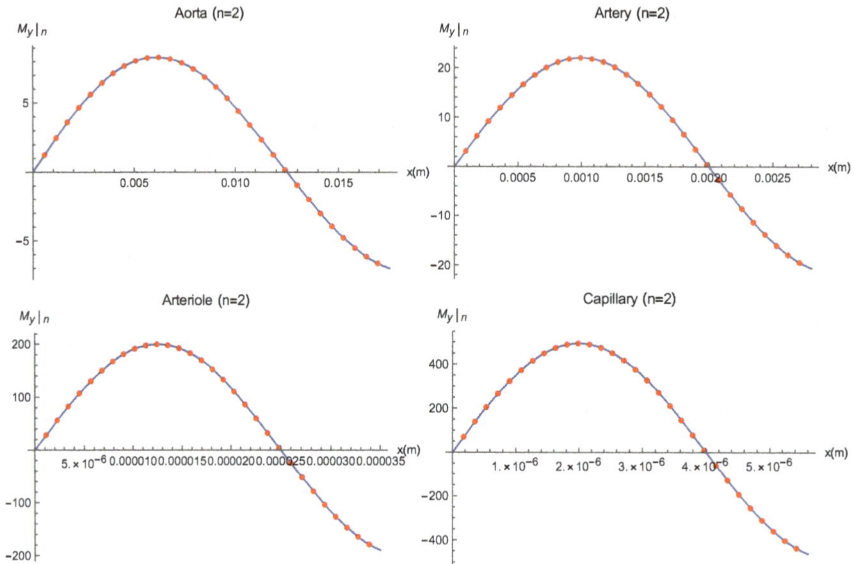

Fig. 7.8 2D profiles of NMR transverse magnetization for different types of human blood vessels at energy level, $n = 2$

could carry quantum computers programmed with very specific instructions. Amazingly, these fairly complex machines will be built with components based on quantum mechanical laws.

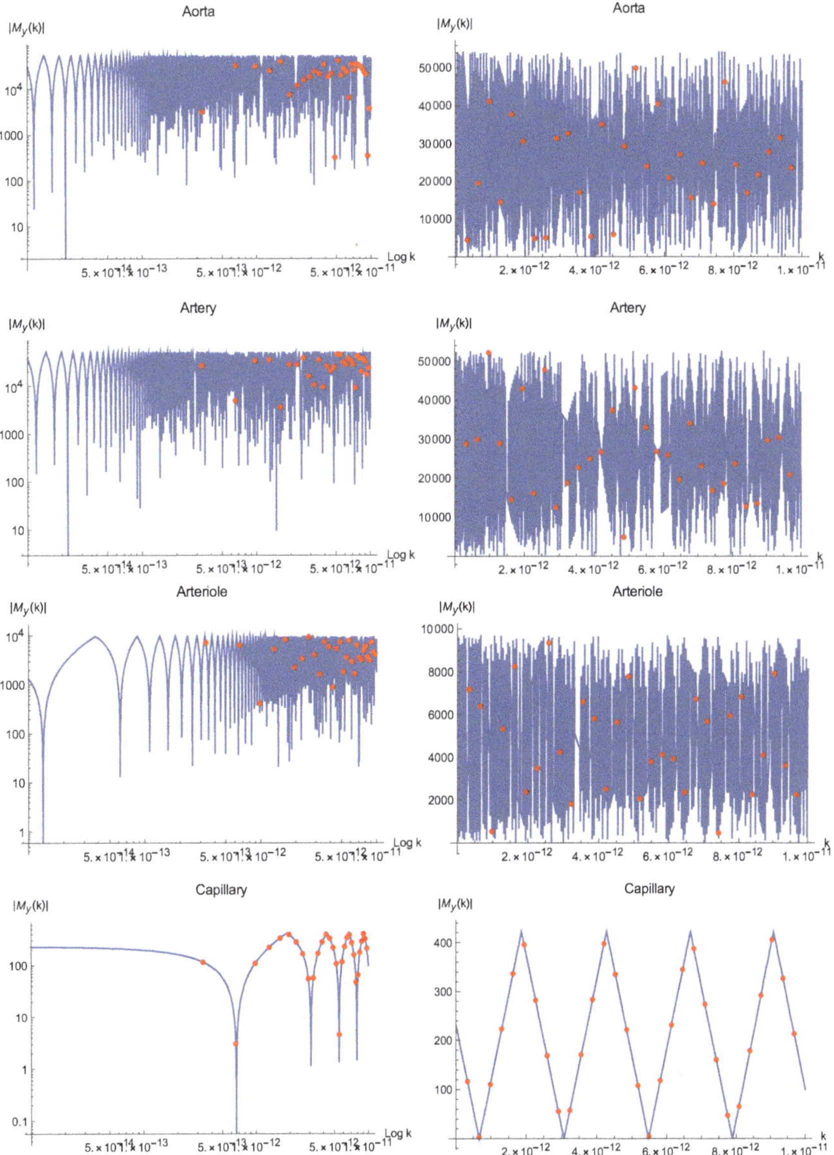

Fig. 7.9 Plots of the Fourier transform of NMR transverse magnetization for different types of human blood vessels at both linear and log scales ($D = 5 \times 10^{24}$)

Chapter 8
Application of "R" Machine Learning for Magnetic Resonance Relaxometry Data Representation and Classification of Human Brain Tumours

Abstract Experimental methods alone are no longer enough for efficient diagnosis. The development of appropriate mathematical models and sophisticated computer simulations is now required to complement laboratory and clinical observations. Such mathematical models have the potential of providing insights into the imaging of the molecular interactions through the analysis of the behaviour of relaxation processes as observed on magnetic resonance (MR) scans and the Bloch NMR flow equation. The aim of this chapter is to explore the analytical features of Bloch's NMR flow equations by using machine learning to develop training software to provide visual training tools. Such tools can facilitate research activities in the application of magnetic resonance imaging (MRI) methods. With the analytical methods and the computer programs presented in this chapter, all that may be needed to perform brain tissue diagnosis is the availability of relaxometric data and some level of computer program proficiency. The programs are easy to use, highly interactive and the data processing is fast and unambiguous. Laboratories which are unable to acquire or maintain MRI equipment can perform computational magnetic resonance diagnosis with straightforward computer programs as long as they find ways of obtaining the required relaxation data (T_1 and T_2). The results can motivate the use of data to produce data-driven predictions required for statistical learning or machine learning that may lead to the application of artificial intelligence (AI) and deep learning in drug delivery and development.

Keywords Bloch NMR flow equation · "R" machine learning · Magnetic resonance relaxometry data · Brain tumor

8.1 Introduction

Abundant research for segmentation and detection of brain tumor using MRI scans has been reported in the literature. This section presents a brief review of some recent works (Mehrotra and Ansari, 2020) on brain tumor segmentation and classification. Joseph et al. (2014) segmented and clustered MRI images of the brain using

M. O. Dada, B. O. Awojoyogbe, *Computational Molecular Magnetic Resonance Imaging for Neuro-oncology*, Biological and Medical Physics, Biomedical Engineering, https://doi.org/10.1007/978-3-030-76728-0_8

K-means clustering and then employed morphological filtering for missed clustered tumors. Bhosale et al. (2017) employed K-Means and Fuzzy C-means (FCM) for segmentation of MRI scan images of brain tumors and then classified the stage of the patient using a support vector machine model.

Dahab et al. (2012) modified the algorithm of Canny edge detection for the detection of contours of brain tumor and then applied probabilistic artificial neural networks (PNN) based on learning vector quantization (LVQ) classification for brain tumor on MRI images. The training set consisted of 64 images while the test set consisted of 16 images. They not only claimed that the algorithm achieved 100% accuracy, but also that the modified PNN based LVQ system decreased the processing time by approximately 79% compared to conventional PNN.

Mustaqeem et al. (2012) conducted experiments on the MRI images of 30 male and 30 female subjects aged 20–60. The images were converted into gray scale and different filters were used for noise removal. The images were then sharpened using different high pass filters. For segmentation, thresholding and watershed techniques were employed and the resulting images were then converted into binary for the identification of brain tumor using different morphological operations. Patil and Bhalchandra (2012) also used MRI images for segmentation and detection of brain tumor. Amedian filter was utilized for noise removal and high pass for enhancement in order to gain the high-frequency information. Finally, the segmentation and extraction of tumors was done using Meyer's flooding watershed and morphological operation for a brain tumor. Selkar and Thakare (2014) also used MRI images for segmenting and detecting brain tumors. First the images were converted into gray scale and sharpening and enhancement methods applied. After that, threshold segmentation using histogram was applied to the desired output. Thereafter, extraction of boundary was performed using edge detection methods. Finally, the shape and size of the brain tumors were detected using mathematical operations.

Laddha and Ladhake (2014) utilized different pre-processing techniques for removing noise, enhancing the texture of the images and sharpening the image. Segmentation was accomplished through thresholding, and watershed segmentation was used to intensify the different elements of an image. Finally, morphological operators were applied to shrink and dilate the image for the detection of tumors.

Subramanyam and Raviraj (2015) performed detection of a brain tumor in four steps. The first step involved the elimination of noise using a 3×3 median filter. The second step was the extraction of significant features using shapes, intensities, and texture, etc. The penultimate step was image segmentation, to facilitate the processing of small segments to detect the main region of the brain tumor. The final stage, post processing, involved several operations, including calculating the high-intensity areas. For the best results, mean, variance, sensitivity and specificity are measured. Salam and Deorankar (2015) also presented four steps for the detection of brain tumors. In a pre-processing step, noise is removed by applying a Gaussian filter. In the second step, color-based segmentation is done by histon and histogram for determining the peak and valley points. The third step is image extraction: the abnormal features of an image are first extracted, and then pixel characteristics are identified by thresholding. The final step: approximation or classification includes defining the mass using dilation and shape similarity using grading rules with the age of tumors through specifying grades of Astrocytoma. Fuzzy expert systems are used for comparing before and after results of an image.

Ghare et al. (2015) calculated brain tumor presence using FCM. The proposed system has eight steps. The first step is the inputting of an MRI image. The second step is the transformation of the MRI image to gray scale. The third step is thresholding and the conversion of the gray scale image to binary. The fourth step involves separating the regions of the skull into segments using Sobel edge detection using discrete operations. In the fifth step, FCM soft clustering method is applied to divide the different elements into C-Fuzzy clusters. The sixth step involves calculating white points of the image. Thereafter, the different parts are measured to determine which parts are affected by the tumor. That is the seventh step. The eighth and final step is the display of the regions affected by the tumor.

In the case of Ambily et al. (2015) noise was removed to produce enhanced images which were then transformed and decomposed using a wavelet filter. These decomposed images were then fused by their coefficients. The fused image was then transferred to the segmentation step. In segmentation, different regions were analysed by color segmentation and intensities. Through intensities, threshold segmentation was facilitated. Feed-forward Neural Network technique was used for the extraction of brain tumor regions. Another skull stripping technique was used to identify the abnormal parts of the brain. Naik and Patel (2014) similarly applied noise removal, median filter, morphological operations, and power law transformation for pre-processing. Then histogram and texture-based features extracted. The feature vector fed to a decision tree.

The algorithm by Aggarwal and Kaur (2014a, b) is based on segmentation techniques. A histogram is used for brain tumor segmentation using histogram is used; while for enhancing the results of the segmentation, FCM clustering-based optimized fuzzy logic is applied. Membership functions are used to calculate the neighbour pixels of the images in the spatial domain to calculate statistics and then these statistics transform into the membership function. Angoth et al. (2013) performed MRI and CT analyses to determine the size and location of the tumor. Then wavelet analysis is done by decomposing the image into the low and high frequencies and then comparing their results with the gradient vector flow and segmentation algorithms. They conclude that the proposed algorithm gives better results than other algorithms.

Divya et al. (2020) detect brain tumors even in clinical noisy images and the images are fused. Segmentation is applied to the fused binary images and the images are enhanced. The images are decomposed through wavelet transforms by changing the parameters. The tumor is detected through the intensity of the pixels. Padmavathi and Megal (2015) performed spatial filtering operation on the pixels of an image. They thereafter applied segmentation by dividing regions into sub parts of an image and examining each part of an image region. The tumor region was detected through segmentation, after which thresholding was applied and edge detection with clustering. Kalaiselvi et al. (2016) used MRI images for segmentation and detection of brain tumor. In a pre-processing step, a median filter is used to remove noise from the image. Thereafter, top-hat, bottom-hat, minima transformation and Otsu's segmentation are used to enhance the image. For the abnormality detection, the fuzzy symmetric measure is applied. Once the abnormal part of the brain is detected, Otsu's thresholding method is used to separate the abnormal part from the image from the background. The performances of all processes are also checked by false and true rate parameters.

Logeswari and Karnan (2010) presents a methodology for the detection of the brain tumor. An input image in JPEG format of 256×256 of grayscale is taken. The image is pre-processed, that is, noise is removed from the images using a weighted median filter. Artificial neural network self-organizing map (SOM) and hierarchical self-organizing map (HSOM) are used for image segmentation; SOM for the mapping and training and HSOM for the enhancement of the processes of the graphical map. Finally, the results of the tumor are calculated with pixels and tumor cells of the size $3 \times 3, 5 \times 5, 7 \times 7, 9 \times 9, 11 \times 11$ windows.

Handore and Patil (2015) conducted experiments on 42 MRI images of brain tumor. The images were converted into 256×256 gray scales. After removing the noise, the images were enhanced by a median filter and power law transformation. The regions were segmented so that the main region of the tumor was detected. For the segmentation threshold values, T1 minimum and T2 maximum were derived from the histogram. The tumor region of the image was cropped to calculate the features of the pixels. Mallikarjuna (2015) used MRI and CT scan images as input which were then enhanced using median filters. Results were drawn through K-means clustering, K-means segmentation and upgrading the maxima of morphological operators. In K-means algorithm, anew mechanism used in the image segmentation decreased the processing time and increased the accuracy of the problem-solving. In the case of Azhari et al. (2014), irregular noise was removed and different filters were applied for image enhancement and sharpening. For edge detection, the Canny edge detection algorithm was used to detect the boundary of the tumor. Gray level frequencies were obtained through histogram and clustering for the same neighbours. Finally, morphological operators were used for the addition and subtraction of the vector to enhance the maxima.

The work of Nisar et al. (2015) on segmentation and detection of brain tumor using MRI scans involves normalizing histograms. An MRI image is taken as input for the processing. Image enhancement facilitates the removal of noise and interruptions from the input image. Spatial filtering and median filters are used for the removal of noise. After converting the image into grayscale, a histogram for the image is obtained. The histogram produces two images: the original image and an image containing the brain tumor. The normalized histogram is built such that a decision sample is taken whether the tumor is detected or not.

The work of Işın et al. (2016) on segmentation and detection of brain tumor is multi-modular, including manual, semi-automatic and fully automatic segmentation methods. High performance deep learning is done through convolutional neural networks. The main objective is to enhance the results of positron emission tomography, magnetic resonance spectroscopy and diffusion tensor imaging. Nakhmani et al. (2014) also deployed semiautomatic segmentation of tumor and detection of necrosis on MRI scans. First the MRI images were converted into digital form. Then the Sobolev contour (snake) technique was applied using density estimation for adaptive segmentation. Finally, an expert system was used to decide using the resulting image.

Chandra and Rao (2016) propose a genetic algorithm for the segmentation. For the computation of maximum and the minimum chromosomes, fitness function is used and defined as f = 1/m. Achieved desired systemic vascular resistance values range from 20 to 44 and the accuracy of segmentation from 82 to 97%. The work of Abdel-Maksoud et al. (2015) on image segmentation algorithms is a comparative analysis based on time, performance and accuracy. A brain surface extractor algorithm is used for skull removal to decrease processing time. Hybrid clustering technique is used with the kernel-distance-based intuitionistic fuzzy c-means algorithm. The algorithms are implemented and desired results obtained in MATLAB (7.12.0).

The focus of Damodharan and Raghavan (2015) is on the complex structure of brain, the homogeneity of RF coil, brain tissue fragility and systematic conflicts arising from the segmentation process. The process begins with skull-striping techniques including binarization through thresholding, morphological operations with opening, closing, erosion and dilation and applying binary mask. The orthogonal polynomial transformation is applied with the cerebrospinal fluid, white matter and gray matter segmentation. Statistical features consist of mean, variance, entropy and wavelet-based energy function. Classification is done via a feed forward neural network (FNN). The network consists of 25 input layer neurons, a hidden layer with 5 neurons and 1 output layer neuron. FFN and multi-layer perceptron neural network models are used to drive the connections of neurons. K-nearest neighbour and Bayesian classification are used for comparison. The proposed nearest neighbour classifier produced better results than others.

Nabizadeh and Kubat (2015) proposed an algorithm of 2D single spectral anatomical MR images. Tumor detection is based on the comparison of two brain mutual-information histogram. Different features like Gabor, mean, average contrast, skewness, kurtosis, energy and entropy, gray level co-occurrence matrix features (homogeneity, absolute value, inverse difference and angular second momentum etc.) and gray level run length are extracted from a sliding window to obtain the tumor image. Histograms of oriented gradient and linear binary pattern features are applied. Performance is also calculated mathematically by aggregation. In the case of Subashini and Sahoo (2013), Otsu's method is used for binarization and then K-means, fuzzy-c-means and thresholding techniques applied for segmentation. The artificial neural network system architecture is proposed for classification. Havaei et al. (2017) analysed brain tumor segmentation using Brain Tumor Segmentation (BRATS) datasets. The convolution neural network is implemented as a classifier, and evaluation is through the online BRATS evaluation system.

Sugapriya (2017) proposed algorithms for different types of datasets for brain tumor segmentation. The first stage includes resizing and filtering, then the original image is converted into binary image, and crystal-based thresholding done using the Otsu method. A crystal oscillator-dependent PIC microcontroller is used.

Performance parameters and comparison are based on such mathematical formulations as DB, PBM and IFV unit.

El-Dahshan et al. (2014) reviews the computer-aided detection system of MRIs of the brain as well as hybrid machine learning for the detection of brain tumors through MRIs. The feature extraction is based on wavelet transform, minimizing the wavelet coefficients through principal analysis. For finding the normal and abnormal cells of the brain, feed forward back propagation neural network is applied. One hundred one images from brain tumor datasets are used for the experiment, achieving results up to 99% accuracy. The results by the authors reviewed adequately summarize the detailed discussions on brain tumor segmentation and classification.

T_2 relaxometry is a quantitative magnetic resonance tool that can be used to improve the sensitivity of identifying hippocampal abnormalities, being significantly better than mere visual assessment; and it can detect changes that would not be considered to be hippocampal sclerosis. Because it does not depend on side to-side comparison, it is particularly useful in detecting subtle abnormalities and, as in this chapter, to interrogate the hippocampus when it is not involved directly in the seizure origin. The main finding in this study is that there is a subtle prolongation of T_2 relaxation time in the hippocampi of patients with partial epilepsy when the hippocampus is not the primary seizure focus. This includes the hippocampus contra lateral to the seizure focus in patients with mesial temporal sclerosis (MTS), patients with lesion temporal lobe epilepsy where the lesion is the primary seizure focus, and patients with extra-temporal lobe epilepsy where the primary seizure focus is outside the temporal lobe. Significant prolongation of the T_2 relaxation time was identified in each group. Interestingly, the degree of T_2 prolongation was similar in all groups and less than is typically seen in sclerotic hippocampi. These abnormalities, which were not identified on visual assessment alone, are consistent with observations that the hippocampus can be abnormal secondary to seizures arising elsewhere. For example, there are histologically identified hippocampal changes in patients with extra hippocampal pathology that are presumed to be the seizure origin, and these pathological abnormalities are subtler, with less neuronal loss, than those seen in typical hippocampal sclerosis (Lieb and Babb, 1986; Levesque et al., 1991).

The data are most consistent with hippocampal injury in the hippocampi that are at a distance from the presumed seizure focus occurring as a consequence of ongoing seizures once partial epilepsy has been established. This is supported by the similarity in the degree of prolongation of T_2 relaxation time between the groups despite the wide variability in site of origin of partial epilepsy and in the number of different causes of the epilepsy. In addition, the number of patients with MTS and bilaterally abnormal T_2 relaxation times is less than the number of patients with bilateral abnormalities identified in adult studies (Jackson et al., 1993; Bernasconi et al., 2000; Mackay et al. 2000). The authors deduced that the data in the current study support the view that T_2 relaxation times are prolonged in hippocampi remote from a seizure focus in children and young adults with partial epilepsy compared with control subjects. They believe that these data are best interpreted as hippocampal injury occurring as a consequence of ongoing partial epilepsy irrespective of the cause of the epilepsy. This adds further evidence to the argument that uncontrolled seizures may cause progressive damage to the hippocampus brain organs attached to the brain, and that this can happen in childhood and young adulthood.

R is a programming language and environment commonly used in statistical computing, data analytics and scientific research. It is one of the most popular languages used by statisticians, data analysts, researchers and marketers to retrieve, clean, analyse, visualize and present data. Due to its expressive syntax and easy-to-use interface, it has grown in popularity in recent years. In this study we shall make use of R machine learning software, relaxometry data, for easy classification of the brain tumors classes and types. R is used for data analysis/science, statistical inference/computing and machine learning algorithm.

R has found a lot of use in predictive analytics and machine learning. It has various packages for common machine learning tasks such as linear and non-linear regression, decision trees, linear and non-linear classification and many more. Both machine learning enthusiasts and experts make use of R to implement machine learning algorithms in various research areas such as finance, genetics marketing, medical sciences and general health care research (Grolemund, 2014).

There are two basic types of learning used in data science. The first, supervised learning is a primary learning scheme that has been applied in recent radiomics studies. Supervised learning is conceptually divided into two phases. First, training samples with available class labels are used to build a classifier by finding a set of parameters to define a decision boundary among classes. Second, the learned classifier is used to predict class labels of unknown testing samples. Notably, the selection of classifiers depends on the desired properties of the classifier, including convergence and modelling assumptions. An example of this approach is a study that showed the selection of machine learning classifiers for supervised radiomics applications in detecting radiomics biomarkers. More examples include tumor subtype classification and survival time prediction, and these are discussed in clinical applications.

Secondly, deep learning is an area of machine learning that emerged from the intersection of neural networks, artificial intelligence, graphical modelling, optimization, pattern recognition and signal processing. "Deep learning networks are a revolutionary development of neural networks, and it has been suggested that they can be utilized to create even more powerful predictors." Faster computer processors, cheaper memory and the deluge of new forms of data allow businesses of all sizes to use deep learning to transform real-time data analysis (Hu et al., 2014).

Without knowing any prior labels of data, unsupervised learning algorithms group the data according to their similarity. For example, a knowledge-based unsupervised fuzzy clustering approach was proposed to automate brain tumor segmentation. It showed that tumor and healthy intracranial regions could be grouped using this clustering algorithm through a rule-based expert system. The growth of clinical imaging results in the onerous task of manually annotating tumors on imaging volumes. Therefore, there has been growing interest in exploring scalable algorithms for annotating large volumes of tumor imaging data and evaluating human inter-rater variability (Perkuhn et al., 2018; Roberts et al., 2013).

Many challenges can be addressed in the development of consistent, suitable, safe, flexible and real-time healthcare systems based on the Bloch NMR flow equation (Dada et al., 2017):

$$v^2 \frac{\partial^2 M_y}{\partial x^2} + 2v\frac{\partial^2 M_y}{\partial x \partial t} + \frac{\partial^2 M_y}{\partial t^2} + vT_0\frac{\partial M_y}{\partial x} + T_0\frac{\partial M_y}{\partial t} + \left(\gamma^2 B_1^2(x,t) + T_g\right)M_y$$
$$= \frac{M_o\gamma B_1(x,t)}{T_1} \tag{8.1}$$

γ is the gyromagnetic ratio, v is the fluid velocity, T_1 and T_2 are spin lattice and spin spin relaxation times respectively, $T_o = \frac{1}{T_1} + \frac{1}{T_2}$ and $T_g = \frac{1}{T_1 T_2}$ are the sum and product of the relaxation rates respectively. M_y is the NMR transverse magnetization. At any given time t, we can obtain information about the system, if appropriate boundary conditions are applied. The term $\frac{M_0\gamma B_1(x,t)}{T_1}$ is the forcing function. If this function is zero, a freely vibrating system results; else, the system is undergoing a forced vibration. The term $\gamma B_1(x,t)$ defines the radiofrequency identification system. Equation (8.1) is the generalized time-dependent non-homogenous second-order Bloch NMR flow differential equation derived from the Bloch NMR flow equations which have been modelled into basic and well-known equations such as Bessel equation, Diffusion equation, Wave equation, Schrödinger's equation, Legendre's equation, Euler's equation, etc. (Awojoyogbe et al., 2016a, b; Dada et al., 2015, 2017). In this contribution, R machine learning will be used for the classification of human brain tumors using magnetic resonance relaxometry data T_1, T_2, T_0 and T_g.

8.2   Dataset of NMR Relaxometry for Three Classes of Brain Tumor

The MRI Relaxometry data are obtained from 150 projections across the original object of three tumor types (glioblastoma multiforme, metastasis and lower grade glioma) with the relaxation rates and relaxation times. These original objects are T_1 and T_2 relaxation times from an experimental measurement in these tumour types (Badve et al., 2017). The upper and lower limits of each measurement were distributed over a sample space of 50 for each class. The dataset for training the R classifiers is then given in Table 8.1.

Table 8.1 Dataset of NMR relaxometry for three classes of brain tumor (glioblastoma multiforme, metastasis and lower grade glioma)

T_1 relaxation time	T_2 relaxation time	T_0 relaxation rate	T_g relaxation rate	Tumor type
1.886	0.16	6.780222694	3.313891835	Glioblastoma.Multiforme
1.875919	0.159102	6.818348169	3.350505066	Glioblastoma.Multiforme
1.865838	0.158204	6.856904904	3.387728567	Glioblastoma.Multiforme
1.855757	0.157306	6.895900259	3.42557598	Glioblastoma.Multiforme
1.845676	0.156408	6.935341762	3.464061329	Glioblastoma.Multiforme
1.835595	0.15551	6.975237115	3.503199035	Glioblastoma.Multiforme
1.825514	0.154612	7.015594199	3.543003929	Glioblastoma.Multiforme
1.815433	0.153714	7.056421077	3.583491266	Glioblastoma.Multiforme
1.805352	0.152816	7.097726002	3.62467674	Glioblastoma.Multiforme
1.795271	0.151918	7.139517421	3.666576496	Glioblastoma.Multiforme
1.78519	0.15102	7.181803982	3.709207153	Glioblastoma.Multiforme
1.775109	0.150122	7.224594538	3.752585814	Glioblastoma.Multiforme
1.765028	0.149224	7.267898156	3.796730084	Glioblastoma.Multiforme
1.754947	0.148326	7.311724119	3.841658091	Glioblastoma.Multiforme
1.744866	0.147428	7.356081939	3.887388503	Glioblastoma.Multiforme
1.734785	0.14653	7.400981358	3.933940546	Glioblastoma.Multiforme
1.724704	0.145632	7.446432357	3.981334026	Glioblastoma.Multiforme
1.714623	0.144734	7.492445166	4.029589351	Glioblastoma.Multiforme
1.704542	0.143836	7.539030268	4.078727548	Glioblastoma.Multiforme
1.694461	0.142938	7.58619841	4.128770294	Glioblastoma.Multiforme
1.68438	0.14204	7.633960606	4.179739932	Glioblastoma.Multiforme
1.674299	0.141142	7.682328155	4.2316595	Glioblastoma.Multiforme
1.664218	0.140244	7.731312639	4.284552758	Glioblastoma.Multiforme
1.654137	0.139346	7.780925942	4.338444213	Glioblastoma.Multiforme
1.644056	0.138448	7.831180251	4.393359146	Glioblastoma.Multiforme
1.633975	0.13755	7.882088074	4.449323647	Glioblastoma.Multiforme
1.623894	0.136652	7.933662243	4.506364641	Glioblastoma.Multiforme
1.613813	0.135754	7.985915931	4.564509922	Glioblastoma.Multiforme
1.603732	0.134856	8.038862659	4.623788188	Glioblastoma.Multiforme
1.593651	0.133958	8.09251631	4.684229076	Glioblastoma.Multiforme
1.58357	0.13306	8.146891141	4.745863197	Glioblastoma.Multiforme
1.573489	0.132162	8.202001792	4.808722178	Glioblastoma.Multiforme
1.563408	0.131264	8.257863307	4.872838701	Glioblastoma.Multiforme
1.553327	0.130366	8.314491138	4.938246543	Glioblastoma.Multiforme
1.543246	0.129468	8.371901167	5.004980628	Glioblastoma.Multiforme
1.533165	0.12857	8.430109717	5.073077065	Glioblastoma.Multiforme
1.523084	0.127672	8.489133569	5.142573202	Glioblastoma.Multiforme
1.513003	0.126774	8.548989977	5.213507677	Glioblastoma.Multiforme
1.502922	0.125876	8.609696687	5.285920468	Glioblastoma.Multiforme
1.492841	0.124978	8.671271951	5.359852957	Glioblastoma.Multiforme
1.48276	0.12408	8.733734547	5.435347979	Glioblastoma.Multiforme

(continued)

Table 8.1 (continued)

T_1 relaxation time	T_2 relaxation time	T_0 relaxation rate	T_g relaxation rate	Tumor type
1.472679	0.123182	8.797103801	5.512449895	Glioblastoma.Multiforme
1.462598	0.122284	8.861399603	5.591204647	Glioblastoma.Multiforme
1.452517	0.121386	8.926642428	5.671659834	Glioblastoma.Multiforme
1.442436	0.120488	8.992853361	5.753864782	Glioblastoma.Multiforme
1.432355	0.11959	9.060054116	5.837870618	Glioblastoma.Multiforme
1.422274	0.118692	9.128267064	5.923730351	Glioblastoma.Multiforme
1.412193	0.117794	9.197515251	6.011498955	Glioblastoma.Multiforme
1.402112	0.116896	9.267822434	6.101233459	Glioblastoma.Multiforme
1.392	0.116	9.33908046	6.193024178	Glioblastoma.Multiforme
1.597	0.132	8.201931652	4.743743003	Metastasis
1.585857	0.130898	8.270110006	4.817291929	Metastasis
1.574714	0.129796	8.33943362	4.892569489	Metastasis
1.563571	0.128694	8.409931644	4.969630433	Metastasis
1.552428	0.127592	8.481634231	5.048531702	Metastasis
1.541285	0.12649	8.554572574	5.129332538	Metastasis
1.530142	0.125388	8.628778956	5.21209459	Metastasis
1.518999	0.124286	8.704286796	5.296882035	Metastasis
1.507856	0.123184	8.781130697	5.383761708	Metastasis
1.496713	0.122082	8.859346504	5.47280323	Metastasis
1.48557	0.12098	8.938971358	5.56407915	Metastasis
1.474427	0.119878	9.020043754	5.657665098	Metastasis
1.463284	0.118776	9.102603606	5.753639941	Metastasis
1.452141	0.117674	9.186692308	5.852085951	Metastasis
1.440998	0.116572	9.272352811	5.953088985	Metastasis
1.429855	0.11547	9.35962969	6.056738673	Metastasis
1.418712	0.114368	9.448569227	6.163128621	Metastasis
1.407569	0.113266	9.539219488	6.272356625	Metastasis
1.396426	0.112164	9.631630416	6.384524898	Metastasis
1.385283	0.111062	9.725853918	6.499740312	Metastasis
1.37414	0.10996	9.821943967	6.61811466	Metastasis
1.362997	0.108858	9.919956703	6.739764925	Metastasis
1.351854	0.107756	10.01995055	6.864813578	Metastasis
1.340711	0.106654	10.12198631	6.993388891	Metastasis
1.329568	0.105552	10.22612733	7.125625267	Metastasis
1.318425	0.10445	10.3324396	7.2616636	Metastasis
1.307282	0.103348	10.44099188	7.401651656	Metastasis
1.296139	0.102246	10.5518559	7.54574448	Metastasis
1.284996	0.101144	10.66510647	7.694104833	Metastasis
1.273853	0.100042	10.7808217	7.846903656	Metastasis
1.26271	0.09894	10.89908312	8.004320579	Metastasis
1.251567	0.097838	11.01997591	8.166544447	Metastasis
1.240424	0.096736	11.14358912	8.333773908	Metastasis

(continued)

Table 8.1 (continued)

T_1 relaxation time	T_2 relaxation time	T_0 relaxation rate	T_g relaxation rate	Tumor type
1.229281	0.095634	11.27001585	8.506218021	Metastasis
1.218138	0.094532	11.39935352	8.684096932	Metastasis
1.206995	0.09343	11.5317041	8.867642581	Metastasis
1.195852	0.092328	11.6671744	9.057099477	Metastasis
1.184709	0.091226	11.80587635	9.25272553	Metastasis
1.173566	0.090124	11.94792729	9.454792941	Metastasis
1.162423	0.089022	12.09345035	9.663589171	Metastasis
1.15128	0.08792	12.24257477	9.879417989	Metastasis
1.140137	0.086818	12.39543631	10.10260059	Metastasis
1.128994	0.085716	12.55217764	10.33347683	Metastasis
1.117851	0.084614	12.71294882	10.57240653	Metastasis
1.106708	0.083512	12.87790775	10.81977093	Metastasis
1.095565	0.08241	13.04722074	11.07597423	Metastasis
1.084422	0.081308	13.22106301	11.34144528	Metastasis
1.073279	0.080206	13.39961935	11.61663945	Metastasis
1.062136	0.079104	13.58308478	11.90204057	Metastasis
1.051	0.078	13.77198761	12.19839469	Metastasis
1.797	0.225	5.000927472	2.473257899	Lower.Grade.Glioma
1.78896	0.222837	5.046569087	2.508488225	Lower.Grade.Glioma
1.78092	0.220674	5.093078994	2.544511521	Lower.Grade.Glioma
1.77288	0.218511	5.140482615	2.581352741	Lower.Grade.Glioma
1.76484	0.216348	5.188806383	2.619037862	Lower.Grade.Glioma
1.7568	0.214185	5.238077789	2.657593939	Lower.Grade.Glioma
1.74876	0.212022	5.288325441	2.697049158	Lower.Grade.Glioma
1.74072	0.209859	5.339579119	2.737432895	Lower.Grade.Glioma
1.73268	0.207696	5.391869834	2.778775781	Lower.Grade.Glioma
1.72464	0.205533	5.445229897	2.821109764	Lower.Grade.Glioma
1.7166	0.20337	5.499692983	2.864468186	Lower.Grade.Glioma
1.70856	0.201207	5.55529421	2.908885853	Lower.Grade.Glioma
1.70052	0.199044	5.612070209	2.95439912	Lower.Grade.Glioma
1.69248	0.196881	5.670059216	3.001045971	Lower.Grade.Glioma
1.68444	0.194718	5.729301155	3.048866117	Lower.Grade.Glioma
1.6764	0.192555	5.789837733	3.097901091	Lower.Grade.Glioma
1.66836	0.190392	5.851712545	3.14819435	Lower.Grade.Glioma
1.66032	0.188229	5.914971176	3.199791391	Lower.Grade.Glioma
1.65228	0.186066	5.979661324	3.252739867	Lower.Grade.Glioma
1.64424	0.183903	6.045832916	3.307089717	Lower.Grade.Glioma
1.6362	0.18174	6.113538246	3.3628933	Lower.Grade.Glioma
1.62816	0.179577	6.182832116	3.420205547	Lower.Grade.Glioma
1.62012	0.177414	6.253771986	3.479084115	Lower.Grade.Glioma
1.61208	0.175251	6.326418144	3.539589558	Lower.Grade.Glioma
1.60404	0.173088	6.400833873	3.601785506	Lower.Grade.Glioma

(continued)

Table 8.1 (continued)

T_1 relaxation time	T_2 relaxation time	T_0 relaxation rate	T_g relaxation rate	Tumor type
1.596	0.170925	6.47708565	3.665738868	Lower.Grade.Glioma
1.58796	0.168762	6.555243341	3.731520036	Lower.Grade.Glioma
1.57992	0.166599	6.63538043	3.799203118	Lower.Grade.Glioma
1.57188	0.164436	6.717574248	3.868866179	Lower.Grade.Glioma
1.56384	0.162273	6.801906235	3.940591511	Lower.Grade.Glioma
1.5558	0.16011	6.888462215	4.014465919	Lower.Grade.Glioma
1.54776	0.157947	6.977332695	4.090581029	Lower.Grade.Glioma
1.53972	0.155784	7.068613191	4.169033627	Lower.Grade.Glioma
1.53168	0.153621	7.162404578	4.249926024	Lower.Grade.Glioma
1.52364	0.151458	7.258813475	4.333366451	Lower.Grade.Glioma
1.5156	0.149295	7.35795266	4.419469492	Lower.Grade.Glioma
1.50756	0.147132	7.459941519	4.508356552	Lower.Grade.Glioma
1.49952	0.144969	7.564906543	4.600156366	Lower.Grade.Glioma
1.49148	0.142806	7.672981862	4.695005563	Lower.Grade.Glioma
1.48344	0.140643	7.784309834	4.793049268	Lower.Grade.Glioma
1.4754	0.13848	7.899041684	4.89444177	Lower.Grade.Glioma
1.46736	0.136317	8.017338211	4.999347257	Lower.Grade.Glioma
1.45932	0.134154	8.139370557	5.107940611	Lower.Grade.Glioma
1.45128	0.131991	8.26532105	5.220408288	Lower.Grade.Glioma
1.44324	0.129828	8.395384143	5.336949288	Lower.Grade.Glioma
1.4352	0.127665	8.529767432	5.45777622	Lower.Grade.Glioma
1.42716	0.125502	8.668692794	5.583116476	Lower.Grade.Glioma
1.41912	0.123339	8.812397635	5.713213534	Lower.Grade.Glioma
1.41108	0.121176	8.961136275	5.848328396	Lower.Grade.Glioma
1.403	0.119	9.116119719	5.989566176	Lower.Grade.Glioma

8.3 Statistical Summary of the MRI Relaxometry Dataset

In order to understand the nature of the dataset, it is important to check the behaviour of our data in statistical terms. Table 8.2 presents this summary and we can observe that each tumour class has the same number of instances and this occurs in 33% of the entire dataset. The mean, median, the minimum values, maximum values and some percentiles (25th, 50th and 75th) are also shown for each feature in the dataset of Table 8.1.

This table shows that there is no data imbalance in the dataset. This means that we do not need to compensate for data imbalance in the subsequent tumour classification algorithm.

Table 8.2 Statistical summary of NMR relaxometry dataset

T_1.Relaxation.Time	T_2.Relaxation.Time	T_0.Relaxation.Rate	T_g.Relaxation.Rate
Min.: 1.051	Min.: 0.0780	Min.: 5.001	Min.: 2.473
1st Qu.: 1.406	1st Qu.: 0.1159	1st Qu.: 7.037	1st Qu.: 3.799
Median: 1.523	Median: 0.1341	Median: 8.116	Median: 4.813
Mean: 1.514	Mean: 0.1375	Mean: 8.433	Mean: 5.417
3rd Qu.: 1.675	3rd Qu.: 0.1556	3rd Qu.: 9.344	3rd Qu.: 6.117
Max.: 1.876	Max.: 0.2250	Max.: 13.772	Max.: 12.198
Tumour.Type			
Glioblastoma.Multiforme: 40			
Lower.Grade.Glioma: 40			
Metastasis: 40			

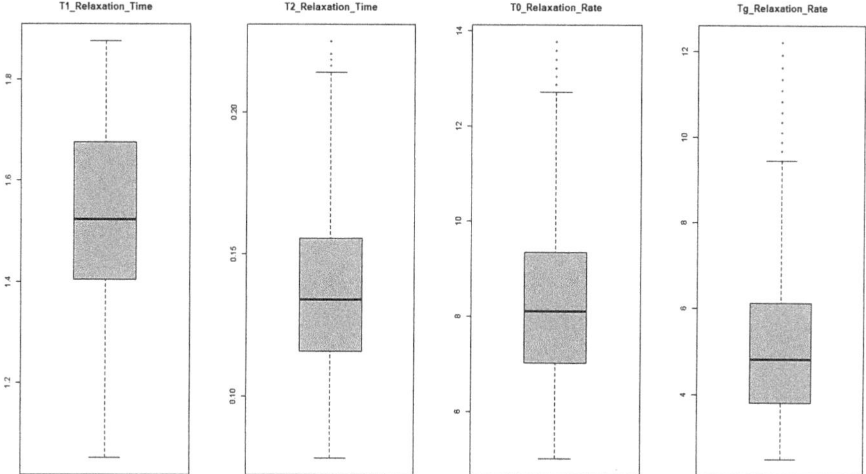

Fig. 8.1 Boxplot of each NMR relaxation features

8.4   Visualization of the MRI Relaxometry Dataset

Since we now have a basic understanding of the dataset, we may now visualize the components of the dataset to further elucidate the behaviour of the dataset. We will explore univariate plots to understand each NMR relaxation feature and multivariate plots to understand the relationship between the relaxation features. Figure 8.1 shows a univariate plots for the relaxation features in which each horizontal line starting from bottom will indicates the minimum, lower quartile, median, upper quartile and maximum value of T1.Relaxation.Time, T2.Relaxation.Time, T0. Relaxation.Rate and Tg.Relaxation.Rate.

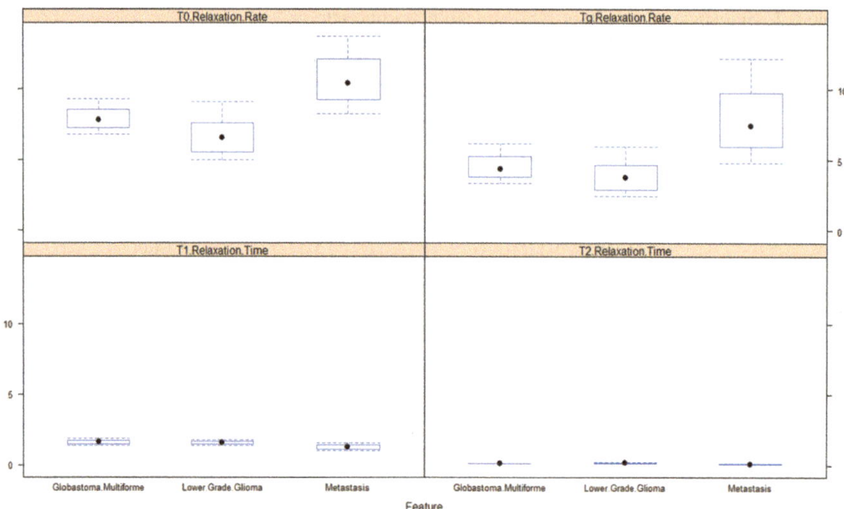

Fig. 8.2 Boxplot of each NMR relaxation features according to tumour type

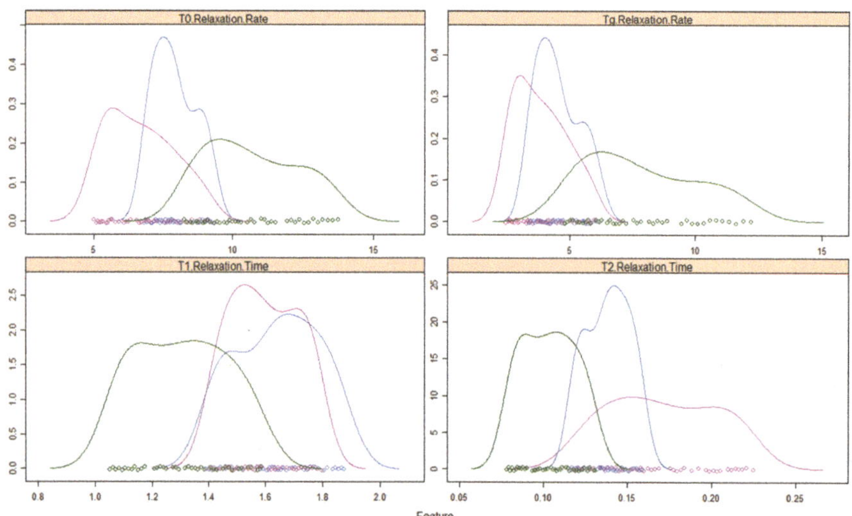

Fig. 8.3 Probability density plots of the NMR relaxation features

Figure 8.2 shows interactions between the NMR relaxation features for each of the tumour classes. These plots indicate some form of linear demarcations between the tumour classes.

Figure 8.3 shows the probability distribution of the features according to the tumour class. It will be observed that the distribution is Gaussian in nature. These profiles shows that the dataset can be conveniently shown as clusters for classification purposes and gives us an idea of the kind of model which is suitable for any

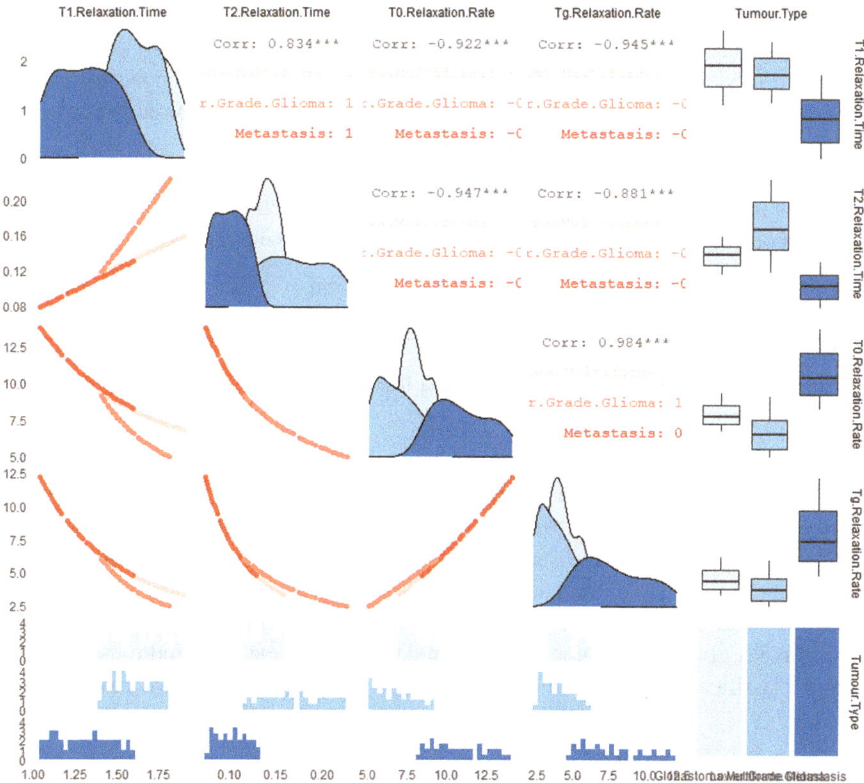

Fig. 8.4   Correlation matrix of the dataset

classification algorithm to be implemented on the dataset. Fortunately, ggally library provides a means of combining multivariate plots in a single visualization. Using the ggpairs() function, an informative scatterplot matrix is given in Fig. 8.4. This shows all the possible correlations that exist in the dataset and we can use this information to guide the selection of appropriate features that would be needed for model construction and neural network training for our classification algorithm.

8.5   Model Construction

We are now going to construct models based on R machine learning which understands the dataset so as to make predictions on entirely new set of data. We shall set-up the test harness (automated test framework) to use tenfold cross validation and build five different models which will be used to predict the type of tumour in our dataset and then determine the model which has the best performance. In this framework, the dataset of Table 8.1 will be split into ten parts in which nine parts were used for training while the remaining were used for testing. We then repeated the process three times for each algorithm with different data splits so as to obtain improved model performance. The metric of "Accuracy" was employed for evaluation of all the models. It is a measure of ratio of the number of correctly predicted instances to the total number of instances in the dataset.

Figures 8.2, 8.3, and 8.4 show that the tumour classes are partially and linearly separable in some dimensions. This suggests that the models should have fairly good results. For this analysis, we have made use of the following models:

1. Linear Discriminant Analysis (LDA) Classification
2. Regression Trees (CART)
3. k-Nearest Neighbors (kNN)
4. Support Vector Machines (SVM) with a linear kernel
5. Random Forest (RFt)

It is important to note that LDA is a simple linear model; CART and kNN are nonlinear models, while SVM and RFt are complex nonlinear models. The random number seed were reset prior to each run to ensure that each algorithm evaluation is performed using exactly the same data splits. This is aimed at ensuring that the results can be directly compared. The performances of these models are presented in Tables 8.3 and 8.4.

Table 8.3 gives the accuracy for the models and this metric is more useful on a binary classification than multi-class classification problems like we have in this

Table 8.3 Accuracy of the models

Summary.resamples (object = results)							
Models: LDA, CART, kNN, SVM, RFt							
Number of resamples: 10							
Accuracy							
	Min.	1st Qu.	Median	Mean	3rd Qu.	Max.	NA's
LDA	0.75	0.854167	0.916667	0.908333	1	1	0
CART	0.666667	0.833333	0.833333	0.816667	0.833333	0.916667	0
kNN	0.583333	0.833333	0.833333	0.841667	0.916667	1	0
SVM	0.666667	0.770833	0.875	0.85	0.916667	1	0
RFt	0.75	0.833333	0.875	0.866667	0.916667	1	0

Table 8.4 Cohen's Kappa obtained for the models

Kappa							
	Min.	1st Qu.	Median	Mean	3rd Qu.	Max.	NA's
LDA	0.625	0.78125	0.875	0.8625	1	1	0
CART	0.5	0.75	0.75	0.725	0.75	0.875	0
kNN	0.375	0.75	0.75	0.7625	0.875	1	0
SVM	0.5	0.65625	0.8125	0.775	0.875	1	0
RFt	0.625	0.75	0.8125	0.8	0.875	1	0

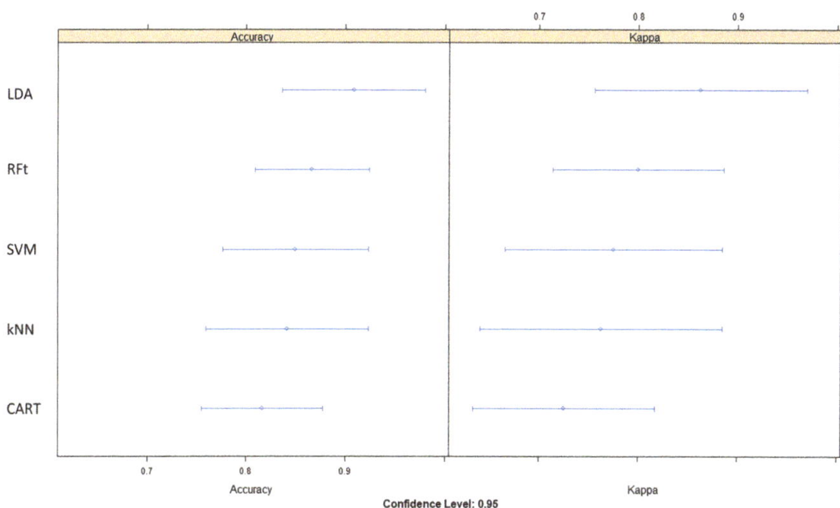

Fig. 8.5 Plots of the model evaluation results

Table 8.5 Summary of the results of LDA

120 samples
4 predictor
3 classes: 'Glioblastoma.Multiforme', 'Lower.Grade.Glioma', 'Metastasis'
No pre-processing
Resampling: Cross-validated (10 fold)
Summary of sample sizes: 108, 108, 108, 108, 108, 108, ...
Resampling results:

Accuracy	Kappa	Accuracy SD	Kappa SD
0.9083333	0.8625	0.09976825	0.1496524

chapter. The challenge in this situation is that accuracy in multi-class problem is not obvious enough when broken down across the classes. Table 8.4 presents Kappa or Cohen's Kappa which is like classification accuracy but normalized at the baseline of random chance on the dataset in Table 8.1. It is a more useful measure to use on problems in which the classes are not balanced.

A plot of the model evaluation results when compared to the spread and the mean accuracy of each model is given in Fig. 8.5. It is noteworthy that there is a population of accuracy measures for each model because the models were evaluated ten times (tenfold cross validation).

As shown in Fig. 8.5, LDA has the best performance. The summary of the results of LDA is given in Table 8.5.

This table provides information that was used in the model training and eventually produced an accuracy of 86.26 ± 14.9%.

8.6 Model Prediction

Since LDA was found to be the best model, we need to check its performance on the validation set. This allows for an independent and confirmatory check on the accuracy of LDA. After running the LDA model on the validation set, the result is shown in Table 8.6.

Table 8.6 shows that the ten samples of glioblastoma multiforme in the validation set were correctly predicted while ten samples of lower grade glioma were correctly predicted. However, only seven samples of Metastasis in the validation set were correctly classified while the remaining three samples were misclassified as glioblastoma multiforme. This shows that the LDA model performed impressively in glioblastoma multiforme and lower grade glioma but fairly good in metastasis.

The overall performance report of the LDA model is shown in Table 8.7 and this shows that the accuracy of the model is 90%, which is good for brain tumour classification using MR relaxation properties.

Finally, Table 8.8 presents the model performance for each tumour class. From this table, we see that LDA produced a mean accuracy of 92.5%, 100% and 85% for

Table 8.6 Confusion matrix of the LDA model

		Reference		
		Glioblastoma. Multiforme	Lower.Grade. Glioma	Metastasis
Prediction	Glioblastoma.Multiforme	10	0	3
	Lower.Grade.Glioma	0	10	0
	Metastasis	0	0	7

Table 8.7 Overall performance statistics of the LDA model

Accuracy	0.9
95% CI	(0.7347, 0.9789)
No information rate	0.3333
P-value [Acc> NIR]	1.67E−10
Kappa	0.85
Mcnemar's test P-value	NA

Table 8.8 Performance statistics of the LDA model according to tumour type

	Class: Glioblastoma. Multiforme	Class: Lower.Grade. Glioma	Class: Metastasis
Sensitivity	1	1	0.7
Specificity	0.85	1	1
Pos Pred value	0.7692	1	1
Neg Pred value	1	1	0.8696
Prevalence	0.3333	0.3333	0.3333
Detection rate	0.3333	0.3333	0.2333
Detection prevalence	0.4333	0.3333	0.2333
Balanced accuracy	0.925	1	0.85

glioblastoma multiforme, lower grade glioma and metastases respectively. Although the validation dataset is not large (20% of the entire dataset), the results we have obtained is in the region of the anticipated margin of $86.26 \pm 14.9\%$. This implies that the model is fairly reliable and accurate.

8.7 Discussion of Results

It is important for primary and metastatic malignant brain tumors to be detected early so that healthcare providers can ensure the employment of appropriate diagnostic and treatment procedures (Badve et al., 2017). Furthermore, proper and early delineation primarily vasogenic edema which are normally in the vicinity of metastatic lesions from edema with neoplastic cellular infiltration (around glioblastomas) can be of a great help in elucidating tumor margins towards effective treatment planning. Advanced magnetic resonance techniques such as diffusion tensor imaging, perfusion imaging and magnetic resonance spectroscopy have proved to be somehow useful in delineating brain metastases from glioblastomas while identifying regions of peritumoral infiltration (Kinoshita et al., 2010; Price et al., 2006). However, these techniques are not simple, rapid or quantitative enough, in probing tissue features for patients with brain tumors in terms of subtle microenvironment changes which are not appreciated by the human eye on standard qualitative clinical images.

These subtle microenvironment changes of tissues are usually impressed on the T_1 and T_2 relaxation times in MRI scans. This is the basis of magnetic resonance fingerprinting. In this study, we have made use of the experimental T_1 and T_2 relaxation time determination (MR relaxometry) for three different types of brain tumours and randomly distributed these measurements into 40 sample points each. T_0 and T_g relaxation rates were then computed for all the sample points to generate MR dataset given in Table 8.1. This dataset is prepared so that machine learning algorithms can easily look for trends in the dataset so that we can use the algorithm as a computer-based diagnostic tool. Table 8.2 and Figs. 8.1, 8.2, 8.3, and 8.4 provide a basic understanding of what the dataset looks like and the correlations between the features of the dataset. This knowledge is important towards the development of appropriate machine learning algorithm for a set of data. In this study, we have tested the machine learning models: LDA, CART, kNN, SVM and RFt in RStudio environment. According to Tables 8.3, 8.4 and Fig. 8.5, LDA has the best performance because it produced the best classification accuracy of 91.7% for normal metric and 86.25% for Cohen's Kappa metric. Details of LDA's performance are shown in Table 8.5. LDA was then implemented on the validation set (20% of the dataset of Table 8.1) and the confusion matrix from this task is presented in Table 8.6. All instances of glioblastoma multiforme and lower grade glioma were correctly predicted by the LDA model while only seven out of ten instances of metastasis were correctly predicted. Three cases were misclassified as glioblastoma multiforme. Coincidentally, this issue has been observed clinically because the differentiation between solitary metastasis and glioblastoma multiforme on conventional MRI has been challenging due to several overlapping imaging characteristics such that final diagnosis mostly requires surgical sampling. The advantage of the machine learning model that has been selected for brain tumour classification in this study is that we can easily separate clear cases of metastasis from cases with mixed features. This could help in improved delineation of inter-tumour and intra-tumour boundaries towards effective surgical intervention and treatment planning.

The overall model accuracy of LDA is 90% according to Table 8.7 while Table 8.8 gives the balanced accuracy of 92.5%, 100% and 85% for GBM, LGG and metastasis respectively. All tumour classes returned impressive specificity and a good sensitivity. This implies that the LDA model is a reliable model in computer-based diagnosis of brain tumours. Larger MR relaxation data may help in improving the accuracy of this model while reducing the mis-classification problems in metastasis cases.

8.8   Conclusion

This chapter has presented an algorithm to hierarchically classify human brain tumor into three classes using Linear Discriminant Analysis model. Although we have been able to achieve the aim of developing a simple and rapid method for brain tumour diagnosis, it is important to note that more advanced machine learning models can still be implemented on the dataset for improved detection accuracy. The advantage of this study is that it relies basically on the relaxometry data determined for the selected brain tumours. This means that problems normally encountered with MRI scans at low magnetic fields can be bypassed with the method we have developed in this chapter. Secondly, blood or tissue samples can be screened much faster so that increased number of patients will be diagnosed within a short period of time. Finally, this method could allow laboratories without sophisticated imaging facilities to conduct brain tumour diagnosis with just a relaxometer. However, the results obtained through this model needs to be mapped to ground truth before implementation in clinical diagnosis.

Chapter 9
Advanced Magnetic Resonance Image Processing and Quantitative Analysis in Avizo for Demonstrating Radiomic Contrast Between Radiation Necrosis and Tumor Progression

Abstract Radiation necrosis is the most common side effect of stereotactic radio-surgery after tumor treatment. The features of necrosis are similar to those of tumor progression on standard anatomical MRIs. Currently, it is clinically difficult to differentiate between tumor and necrosis; no imaging technique is able to provide a conclusive resolution of the problem. Furthermore, available imaging techniques require long follow-ups which are not only stressful for the patient but usually involve contrast agent administrators which could be toxic to the tissues. In order to address this problem by computational magnetic resonance imaging, this chapter will attempt to show how advanced image processing and quantitative analysis with Avizo software is used to extract radiomic features from the images to determine the values of T_1 and T_2 relaxation parameters. Efforts will be focused on the following: (1) Processing grayscale images of brain tumor and necrosis for binarization and separation (2) Applying the label analysis module in Avizo software to extract radiomic data (numerical features) from the processed images; and (3) Developing graphical profiles of the extracted radiomic data.

Keywords Magnetic resonance image processing · Radiation necrosis · Radiomics · Avizo software · Brain tumor

9.1 Potential Application of Radiomics

Numerous radiomic features—size- and shape-based features, descriptors of the image intensity histogram, descriptors of the relationships between image voxels (e.g. gray-level co-occurrence matrix (GLCM), run length matrix (RLM), size zone matrix, and neighbourhood gray tone difference matrix derived textures), texture extracted from wavelet and Laplacian of Gaussian filtered images, as well as fractal features—can be extracted from medical images (Haralick et al., 1973). Radiomic features do not only provide an objective and quantitative way to assess tumor phenotype, they also have a wide range of potential applications in oncology.

© The Author(s), under exclusive license to Springer Nature Switzerland AG 2021
M. O. Dada, B. O. Awojoyogbe, *Computational Molecular Magnetic Resonance Imaging for Neuro-oncology*, Biological and Medical Physics, Biomedical Engineering, https://doi.org/10.1007/978-3-030-76728-0_9

9.2 Factors Affecting Radiomic Feature Quantification

Despite the wide range of potential applications, radiomics features quantification may be sensitive to a number of technical factors. For example, Yip and Aerts (2016) assessed the variability of PET radiomic features due to different acquisition modes, matrix sizes, post-filtering widths, reconstruction algorithms and iteration numbers. Of these features, 40 were shown to have substantial variability with a relative difference of >30%. Only four features: intensity-histogram, derived entropy and energy, GLCM-maximal correlation co-efficiency and RLM-low gray level run emphasis, were found to have variability of <5%. The textures that are sensitive to acquisition modes and reconstruction parameters are thus not recommended for radiomic applications, such as malignant and benign tissue differentiation.

9.3 Machine Learning in Radiomics

Machine learning offers an approach for discovering predictive radiomic features. Here, the two investigator does not begin with any *a priori* biological hypothesis. Thus, the parameter space is searched for an imaging feature that is statistically associated with clinical outcomes. Before evaluating machine learning models, a specification for the medical diagnostic task is needed so that models can be appropriately trained. For example, supervised, unsupervised, and semi-supervised learning models are fundamental learning strategies used in accordance with the difference level of available clinical outcome labels. In supervised learning, the goal is to learn from a certain portion of trained samples with known class labels and to predict classes or numerical values for unknown patterns from large and noisy data sets. Conversely, unsupervised learning finds the natural structure from data without having any prior labels. As a hybrid setting, semi-supervised learning needs only a small portion of labelled trained data. The unlabelled data samples, instead of being discarded, are also used in the learning process. More recently, the rise of deep learning as a new frontier in machine learning has advanced large-scale medical image analysis. Zhou et al. (2018) describes these learning strategies and highlight specific clinical applications in the context of brain tumors. Supervised Learning and unsupervised learning have been discussed in the previous chapter.

9.4 Semi-Supervised Learning

Semi-supervised learning is designed specifically for tasks where it is difficult to obtain class labels for certain patients (e.g., the estimation of tumor progression). In other words, the advance of semi-supervised learning overcomes a limitation of conventional supervised learning that is incapable of making use of data with missing labels in training. Semi-supervised models have great potential to effectively enable predictive analysis with uncompleted clinical labels in training. For example, Zhou et al. (2018) predicted states of brain tumor prognosis using a semi-supervised learning model. With less than 26% of the available staging labels, a discriminative analysis was performed using the staging labels of patient with glioblastoma.

9.5 Deep Learning

Recently, deep learning has emerged as a powerful technique that defines a network architecture concatenating multiple neural-like processing layers with multiple levels of abstraction. Deep learning methods have achieved record-breaking performance in numerous computer vision applications when the number of available training samples is large enough. Convolutional neural networks (ConvNet), for example, are a deep learning model that incorporates concatenated convolutional layers and pooling layers, followed by fully connected layers to learn high-level representation of input data. A growing set of studies have shown superior results in the field of medical image analysis by applying ConvNet models. Zhou et al. (2018) introduced a ConvNet-based approach for brain tumor segmentation. With the design network architecture containing multiple small 3×3 kernels and deep layers, the model achieved strong segmentation performance (Dice score $= 0.88$) on the Brain Tumor Segmentation Challenge 2013 database.

9.6 Research Opportunities and Challenges

Substantial progress has been made from recent radiomic studies to deepen our understanding of imaging characteristic in cancer. The identified quantitative features can be used to alert radiologists to suspicious abnormalities, because radiomic models can capture even subtle variations in the tumor environment that are not easily perceived by human experts. Zhou et al. (2018) highlight the convergence of quantitative image feature extraction and machine learning techniques for supporting diagnosis, prognosis, and treatment predictions.

9.7 Procedures in Advanced Image Processing and Quantitative Analysis

Brain metastases are found in 20–40% of all cancer patients and stereotactic radiosurgery (SRS), in which high dose of radiation is delivered to the tumor has been an established method of treatment for patients with brain metastases (Mehrabian et al.,

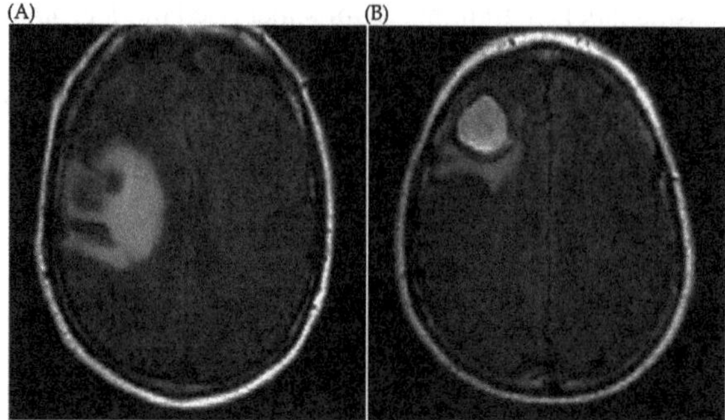

Fig. 9.1 T2-weighted FLAIR magnetic resonance imaging showing (**a**) progressive tumour, (**b**) radiation necrosis (Mehrabian et al., 2017)

2017). SRS tends to lead better local control as well as improved survival rate when compared to treatment with whole brain radiotherapy (WBRT). However, some patients have shown a higher rate of distant brain failure when treated with only SRS in comparison to treatment with WBRT. An additional disadvantage of SRS is a greater risk of radiation necrosis within the region of therapy. Radiation necrosis is the most common morphologic alteration after SRS which can lead to considerable neurologic morbidity depending on its location in the brain. However, the underlying process of radiation necrosis is currently unknown with varied definition. In fact, both tumor progression and radiation necrosis appear as an increased vasogenic edema on T2-weighted FLAIR MRI as shown in Fig. 9.1.

In this chapter, we shall explore Avizo software image analysis module to extract image features that are not immediately clear during visual observation. These features could provide a radiomic avenue for better understanding of radiation necrosis and provides data that could be used for building machine learning algorithms aimed at improved understanding of how radiation necrosis can be clearly differentiated from progressive tumours.

9.7.1 Processing Grayscale Images in Avizo

The data is loaded into Avizo. The data object appears in the project view. If *Auto Display* is disabled, an Ortho Slice module is attached to visualize the data by right-clicking on the green data icon. In the object popup, the category entry *Display* is selected, and then the entry Ortho Slice is double-clicked (alternatively, *Create* is clicked). Another possibility is to use the search field and type the first letter of the Ortho Slice. Selecting the module in the list will automatically create it in the project view. Two other Ortho Slice modules are then attached. In the properties panel of the second Ortho Slice, the xz orientation is selected and the yz orientation is selected in the properties panel of the third Ortho Slice. Improving image quality is often necessary to obtain the best results with image analysis.

The next step illustrates how to process images with an image filter commonly used for smoothing or noise reduction.

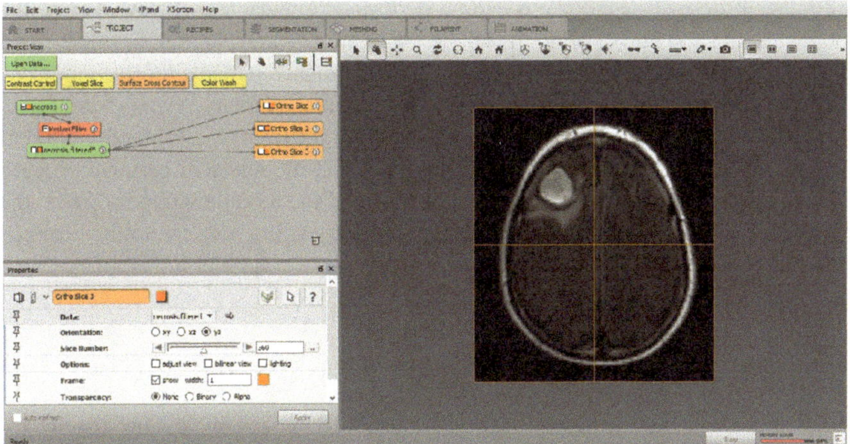

Fig. 9.2 Removal of noise from the necrosis image using median filter

1. A Median Filter module is attached to the data (this can be done by using the search field and the first letters of the median filter, then, selecting the module in the list).
2. 3D is selected in the interpretation port of Median Filter (Fig. 9.2).
3. Then the *Apply* button is clicked.

Once computed, the result is stored in a new image object *necrosis.filtered* and *tumor.filtered*.

The median filter smoothing modules use low-pass filters to reduce the contrast and soften the edges of objects in an image. A low-pass filter lets low frequencies go through but attenuates high frequencies and noise. It reduces contrast but tends also to defocus the image (it is worthy of note that the filtered image may be normalized by dividing the output gray levels by the sum of absolute values of the kernel coefficients. If not, overflow may occur). This module uses morphological operators to set the pixel value to the median for the defined neighbourhood. The Median Filter usually works well when images contain non-Gaussian noise and/or very small artifacts. It does not cause blurring to the same extent as the Box Filter; however, it takes a considerably longer period to execute. The gray levels of all pixels in the neighbourhood are sorted from the smallest value to the largest one. The central pixel in the sort is then the median value, i.e., the value for which there are as many lower gray levels as higher ones. The process may be iterated.

9.7.2   Interpretation as 2D Image or 3D Stack

Sometimes, it is useful to interpret the input data of an image processing algorithm as a 3D volume or as a sequence of 2D planes. For instance, a number of image filters and image processing algorithms can be performed either on each XY-slice or the volume using a 2D kernel or on the whole volume using a 3D kernel. In some cases, a 2D algorithm may be preferred, either for performance or for a more appropriate effect depending on the data and the desired outcome. In many Avizo modules, an interpretation port shows the state of the current module (that is, XY planes or 3D). If the state of the port is XY planes, it means that the module will perform on each XY-slice. If the state of the port is 3D; it means that the module will perform on the entire three-dimensional image at once. In some cases, the port cannot be changed (it is grayed), for instance, when processing can only be applied in XY planes.

9.7.3 *Binarization of Grayscale Images*

Binarization means transforming a grayscale image into a binary image, i.e., a label image with only interior and exterior materials. Threshold binarization is used when the relevant information in the grayscale image corresponds to a specific gray level interval. Thresholding is a simple segmentation method; more sophisticated automatic, semi-automatic or manual segmentation tools are also available in Avizo. Threshold binarization can be done with the interactive thresholding module which prompts you to set the levels with a visual feedback.

1. An interactive thresholding module is attached to the filtered data (Fig. 9.3).
2. The threshold values can be interactively modified with immediate 2D or 3D visual feedback. The selected pixels appear in blue in the displayed image.
3. In the properties panel of the module, the intensity range port can be set to the range 0–30.

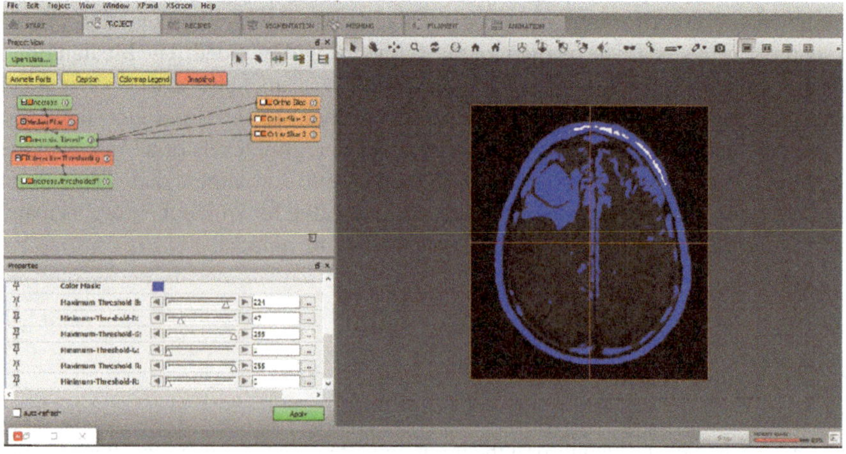

Fig. 9.3 Interactive thresholding on the necrosis image

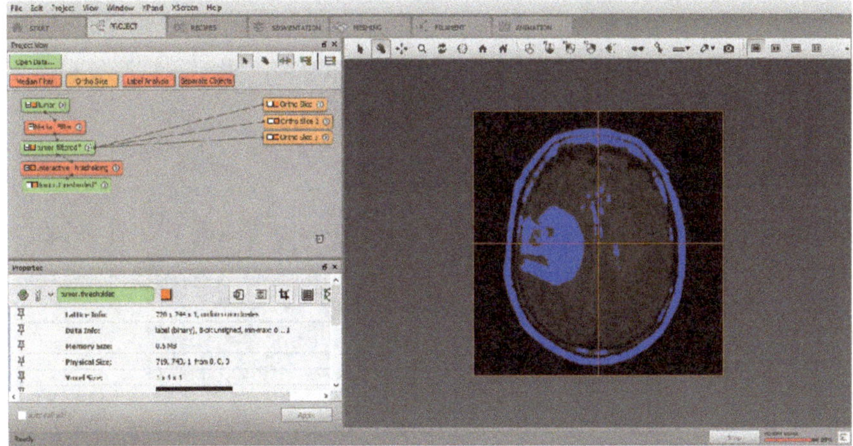

Fig. 9.4 Interactive thresholding on the tumor image

4. The 3D option in the preview type port can be checked and unchecked to get a2D only or 3D preview (see Fig. 9.4).
5. The apply button is clicked to start the module.
6. The output binary image named *necrosis.thresholded* and *tumor.thresholding* is generated in the Project view.

Binarization transforms a gray level image into a binary image. This method is used when the relevant information in the gray level image corresponds to specific gray level interval. This tool allows thresholds to be selected interactively. The current selection is displayed in blue by default over an OrthoSlice representation with gray level data as background. This preview can also be displayed in 3D with volume rendering.

The *Apply* button is clicked to create the binary output image. A new field is created, that is, 1 for each value within the threshold interval and 0 for all other field values. The interactive thresholding preview is hidden by switching of the orange visibility button in the module icon in the Project View or next to the module name in the Properties panel. Then, the first Ortho Slice is connected to the thresholded data.

9.7.4   Image Separation

In the dataset shown in Figs. 9.5 and 9.6, some of the pores in the necrosis and tumor appear to be touching, but ideally, they should be separated for proper analysis. With thresholding, this kind of output cannot be avoided when the acquisition is too coarse or noisy, because the gray levels of the considered objects are not uniform enough across the volume, or because the resolution is too low to distinguish some objects' boundaries. In such cases, the Separate Objects module can be used to separate connected particles.

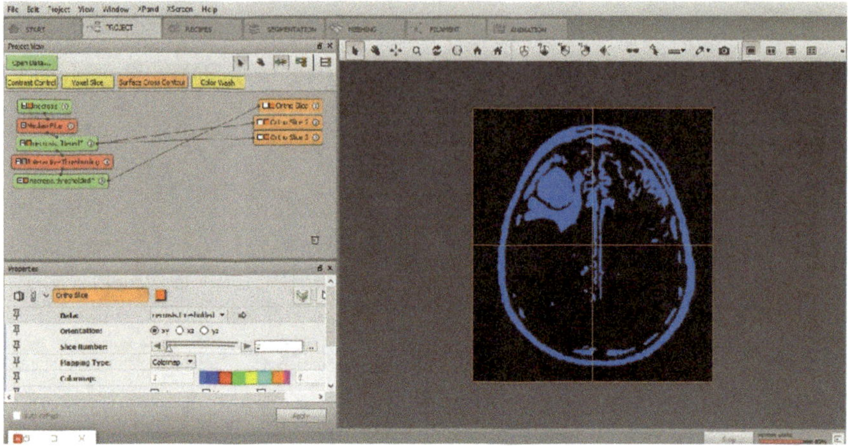

Fig. 9.5 Binary image of necrosis displayed with OrthoSlice

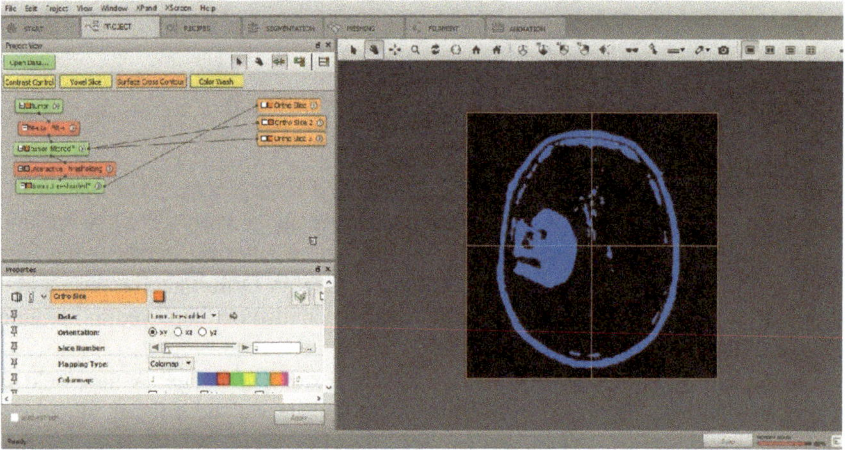

Fig. 9.6 Binary image of tumor displayed with an OrthoSlice

1. First, a Separate Objects module is attached to the thresholded data.
2. Then the Marker Extent port value is set to 1 instead of the default value 4. This is a contrast factor that controls the size of seeds causing the objects to be separated. Increasing this value can merge some markers and therefore decrease the number of separated objects.
3. When the Apply button is clicked, *necrosis.Separate* and *tumor.Separate* data are generated in the project view.
4. Then the first Ortho Slice is attached to the new data.

The principle of the Separate Objects module is to compute watershed lines on a distance map. The Separate Objects module is a high-level combination of the

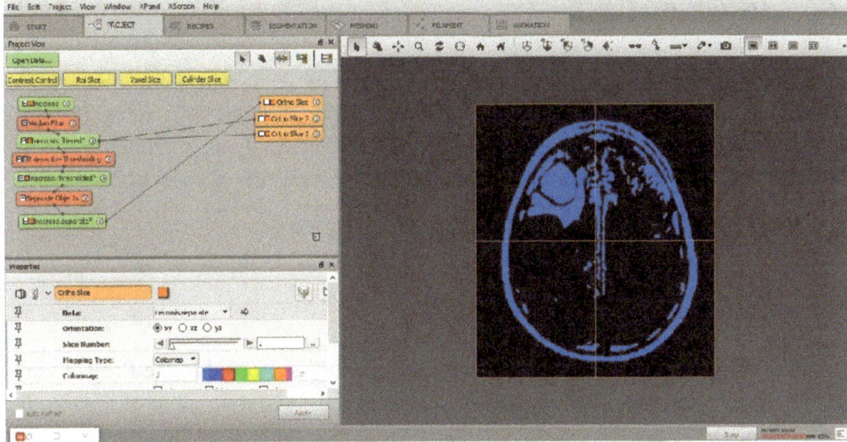

Fig. 9.7 Necrosis image component separation

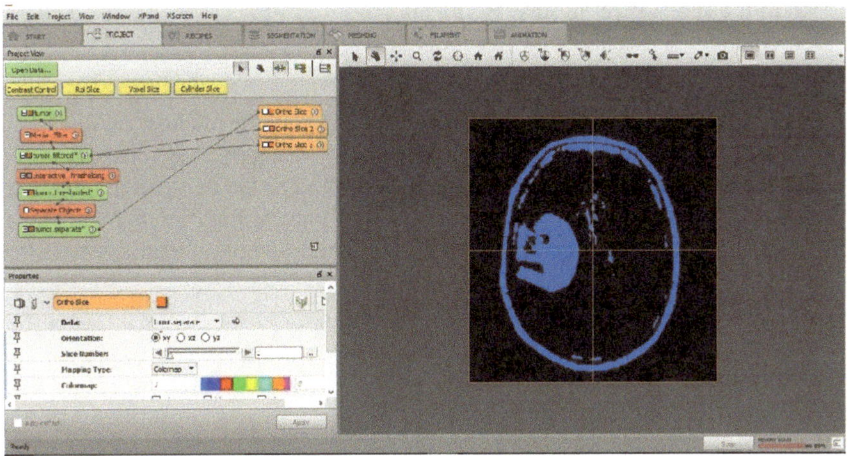

Fig. 9.8 Tumor image component separation

watershed, distance map and H-Maxima. It can be used as a simple and straight forward separation tool, which is satisfactory in many cases. Some separation may be missing or unwanted, in particular with non-convex shapes (considering also 3D) (Figs. 9.7 and 9.8).

The Separate Objects module computes the watershed lines of a binary image. The contrast level which is used to reduce the number of markers can be adjusted for the watershed process. This module is a high-level combination of watershed, distance transform and numerical reconstruction algorithms. There is a limitation to the separating ability: if some particles overlap too much, they will not be separated. This module can be used on the gradient modulus to compute best-fit contours.

9.7.5 *Image Analysis*

An Analyze module can then be used to get the volume, surface area, mean value, number of voxels, etc. individually for each separate particle. This analysis on the stack of images is undertaken by using the Label Analysis module to extract statistical and numerical information, including the measure of objects.

1. A Label Analysis module is attached to the separate data.
2. *necrosis.am* and *tumor.am* is set as intensity image in the dedicated port of the module.
3. The Apply button is clicked.

A new data object *necrosis.Label* and *tumor.Label* is created in the Project view, and the Tables panel is displayed, showing a spreadsheet-style table of results: the analysis *necrosis-analysis* and *tumor-analysis,* is also created in the Project View.

The toolbar offers the possibility to:

1. Copy parts of the table
2. Export the spreadsheet in several formats
3. Sort columns in ascending or descending order
4. Label seek execution.

The basic measures (as selected in the Measures port of the module) are displayed in the tables' panel. The area column in the lower spreadsheet is selected. In the Measures port, basic is a group of measures that contains the most commonly used measures: Volume3d, Area3d, BaryCentreX, BaryCentreY, BaryCentreZ and Mean. It is possible to define new groups of measures, composed of pre-defined measures as well as user-defined measures.

9.7.6 *Extracted Numerical Data*

After carrying out the analysis of radiomic contrast between tumor progression and radiation necrosis images (shown in Fig. 9.1), the results obtained are discussed as follows:

A Label analysis module was used to obtain the spreadsheet and the bar chart. The Label Analysis command computes a group of measures on each cell or connected component of the input images and generates a spreadsheet with values. If the input image provides a coordinate unit, result values are expressed according to that coordinate unit. The analysis results are read-only spreadsheets and yet, as with any spreadsheet, it can be displayed in the Tables panel, its histogram computed, plotted, queried from Tcl scripts, and imported into an external spreadsheet program like Microsoft Excel. In order to edit cell values, rows or columns could be added or removed and the converter Analysis to Spreadsheet can be used. In the Label Analysis, there is a Label Analysis panel with the spreadsheet. Above the spreadsheet, there is a toolbar that offers the possibilities to copy parts of the table, export the spreadsheet in several formats, sort columns in ascending or descending order and do label seek.

A spreadsheet-style table of the data containing statistical and numerical information was obtained. A two spreadsheet appears in the same table. The spreadsheet contains measures such as Area, BaryCenterX, BaryCenterY, Mean and Index. It is possible to define new groups of measures, composed of pre-defined measures as well as user-defined measures.

9.7.7   Visualizations of Extracted Data

The histogram port is located above the spreadsheet in the Label Analysis panel. It gives a graphical representation of the data with a bar chart. Avizo does the labelling of the axis. For clarity, the bar chart can be coloured. i.e., each bar has a color corresponding to each particle. This can be done with the editor, Colour Map. Clicking on the Colour Map displays a list. Clicking on Option in the list will make another list to pop up. Clicking on Edit Colour Map will produce another background showing the coloured bar chart. Furthermore, we have made use of other applications to make 2D and 3D visualizations of Tables 9.1 and 9.2.

As shown in Figs. 9.9 and 9.10, Area is the most prominent measure extracted from the images. Image feature (index) 11 has the largest area in the case radiation necrosis and feature 25 has the largest area in the case of tumour progression. Figures 9.11, 9.12, and 9.13 show how all the measures vary with the image elements for both cases under investigation. Figure 9.13 shows that BaryCenterY is the poorest measure to show contrasts between radiation necrosis and tumour progression. Fortunately, BaryCenterX and Area show better contrast at unique image indices and these points may prove to be regions of emphasis for effective clinical diagnosis. Figure 9.14 and 9.15 show 3D distribution of the area extracted from both images as functions of BaryCenterX and BaryCenterY. Obviously, Fig. 9.14 demonstrate that the higher magnitudes of area extracted from the tumour progression image are mostly concentrated around the tumour while extracted areas from radiation necrosis image have unique distribution in some directions.

9.7.8   Interactive Selection of the Data

Avizo allows the images in the 3D viewer to be linked to their corresponding rows in the analysis panel in order to locate individual objects with corresponding measures.

1. When the label seek button of the analysis panel toolbar is clicked, anew Ortho Slice is automatically attached to the separated image and displayed in the 3D viewer, as well as a point dragger.

Table 9.1 Numerical data extracted from necrosis image

Area	BaryCenterX	BaryCenterY	Mean	Index
3814	430.821	75.6411	1	1
5644	271.276	92.2624	1	2
746	374.426	97.4209	1	3
1913	519.742	125.045	1	4
184	493.69	140.19	1	5
4227	586.914	230.148	1	6
1864	160.805	194.899	1	7
29	253.172	153.103	1	8
244	214.529	170.906	1	9
10	364.3	162.1	1	10
9110	252.47	235.973	1	11
86	212.709	186.57	1	12
424	530.41	195.71	1	13
807	319.037	215.674	1	14
31	199.645	196.194	1	15
37	384.324	198.919	1	16
450	545.236	217.784	1	17
91	185.308	212.538	1	18
16	366.125	212	1	19
261	174.529	232.521	1	20
1	357	220	1	21
919	568.577	253.786	1	22
79	373.544	235.013	1	23
27	390.185	235.815	1	24
222	161.266	253.689	1	25
26	482	245.269	1	26
995	121.42	279.289	1	27
4061	284.046	311.773	1	28
447	183.11	278.134	1	29
2060	198.195	306.656	1	30
213	389.93	298.16	1	31
101	370.653	323.426	1	32
3289	90.9428	419.8	1	33
187	137.781	325.187	1	34
616	602.333	348.482	1	35
4440	646.468	446.164	1	36
7	371	417.714	1	37
887	95.3224	544.459	1	38
163	623.54	541.589	1	39
2	122.5	546	1	40
2413	622.825	608.45	1	41
2466	142.371	635.063	1	42

(continued)

Table 9.1   (continued)

Area	BaryCenterX	BaryCenterY	Mean	Index
164	147.049	601.469	1	43
34	601.588	599.324	1	44
79	587.38	621.595	1	45
1207	565.935	685.035	1	46
21	537.524	677.095	1	47
443	213.503	705.163	1	48
14	339.786	703.357	1	49
1019	509.164	725.474	1	50
356	240.635	719.149	1	51
549	343.967	717.233	1	52
1	479	711	1	53
457	426.109	719.271	1	54
38	459.053	715.763	1	55
488	270.264	732.902	1	56
2350	385.926	747.14	1	57

2. When the cell of the analysis lower table is selected, the point dragger will move
   to the corresponding object location in the 3D view. If the histogram of one of the
   analyses measures is displayed, a vertical line appears that displays the position
   and value of the selected row.
3. In the interviewer window, the point dragger can be moved using the rectangular
   handles. Upon button release, the analysis table will highlight the corresponding
   object row. In order to move the dragger, the viewer must be put into interaction
   mode and the mouse moved over one of the dragger's crosshairs and the left
   mouse button pressed. The colour of the picked crosshair changes. The movement of the dragger is restricted to the corresponding plain.
4. It is also possible to click with the middle mouse button over a pickable object in
   the scene displayed in the 3D viewer, for instance, over a particular pore on a
   displayed slice. The dragger will move to the picked point and the corresponding
   spreadsheet row will also be highlighted. The entire process is demonstrated in
   Figs. 9.16 and 9.17.

9.8   Image-Based Filtering

Image elements displayed in the 3D viewer can be filtered. For example, we may
choose to visualize only particles in which Area3d belongs to a specified range.

1. An Analysis Filter is attached to *necrosis.Label-Analysis* and *tumor.Label-
   Analysis.*
2. The Image port is connected to *necrosis.Label* and *tumor.Label.*

Table 9.2 Numerical data extracted from tumor image

Area	BaryCenterX	BaryCenterY	Mean	Index
2380	355.097	61.4382	1	1
517	244.277	74.5957	1	2
732	464.128	82.1694	1	3
917	303.418	81.2181	1	4
1639	373.82	83.4002	1	5
323	428.43	91.5635	1	6
3663	148.301	197.615	1	7
48	260.208	87.6458	1	8
746	229.151	112.783	1	9
966	466.161	110.257	1	10
1437	524.452	137.26	1	11
7	496.286	136	1	12
113	196.903	153.593	1	13
105	182.257	174.419	1	14
56	523.196	176.393	1	15
1749	576.09	240.869	1	16
737	543.193	217.794	1	17
28	168.607	205.321	1	18
266	154.556	229.583	1	19
24	371.875	218.667	1	20
23	368.13	223.739	1	21
31	356.258	233.452	1	22
74	360.932	252.608	1	23
376	352.021	290.809	1	24
22820	244.845	384.84	1	25
73	368.671	281.384	1	26
160	136.125	308.831	1	27
2512	599.95	377.458	1	28
1598	160.589	328.503	1	29
50	366.56	331.86	1	30
127	131.709	338.016	1	31
61	352.836	335.443	1	32
47	569.617	333.66	1	33
2069	100.868	420.854	1	34
78	571.359	352.423	1	35
6	369	350.5	1	36
961	153.528	381.009	1	37
68	411.309	396.5	1	38
2257	580.652	533.691	1	39
703	111.226	514.858	1	40
40	570.775	503.75	1	41
3713	180.377	620.208	1	42

(continued)

Table 9.2   (continued)

Area	BaryCenterX	BaryCenterY	Mean	Index
27	549.37	559.778	1	43
118	540.288	578.11	1	44
2041	507.972	644.532	1	45
81	306.617	671.185	1	46
2397	396.423	696.987	1	47
85	323.353	675.094	1	48
226	370.429	678.164	1	49
1116	295.574	695.012	1	50

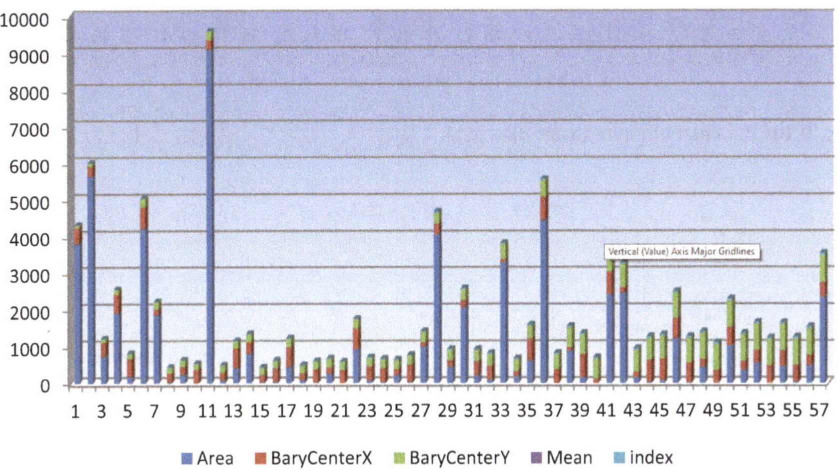

Fig. 9.9   Bar chart of necrosis image data

3. A new filter is created by entering Area3d>=300 in the filter port. Area3d can be inserted in the formula by typing it in or by double-clicking on it in the list displayed below the formula field.
4. When the Apply button is clicked, a new analysis with fewer objects is created.
5. To verify that fewer objects have been created, the Ortho Slicer can be connected to the new label image *necrosis.Label-filtering* and *tumor.Label-filtering*.

The Analysis Filter removes from the input label analysis labels whose measure does not fulfil a filter formula. Additionally, a label image can be processed in the same way. This process can drastically reduce time for image processing during MRI-based diagnosis. Figures 9.18 and 9.19 showed the images that have been reconstructed using the filtered data. They also showed a segmented image with which several brain tissues can be easily identified. For example, the segmentation showed that the brain linings which are mostly made up of similar tissues have non-uniform intensity values indicated by different colours in Figs. 9.18 and 9.19. This may help in delineating the disease states of these tissues since the varying

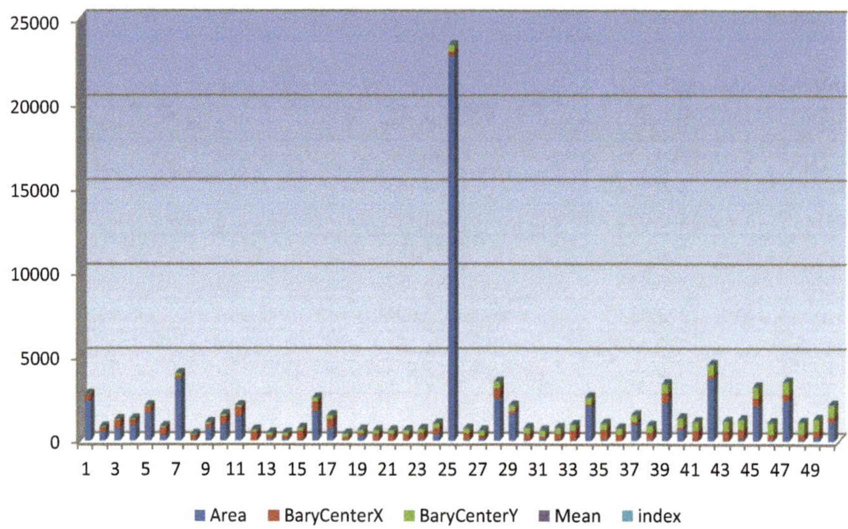

Fig. 9.10 Bar chart of tumor image data

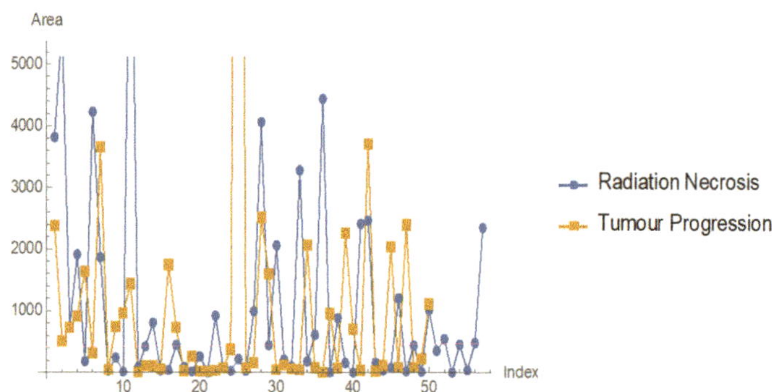

Fig. 9.11 Plot of the area against image elements (index) for radiation necrosis and tumour progression

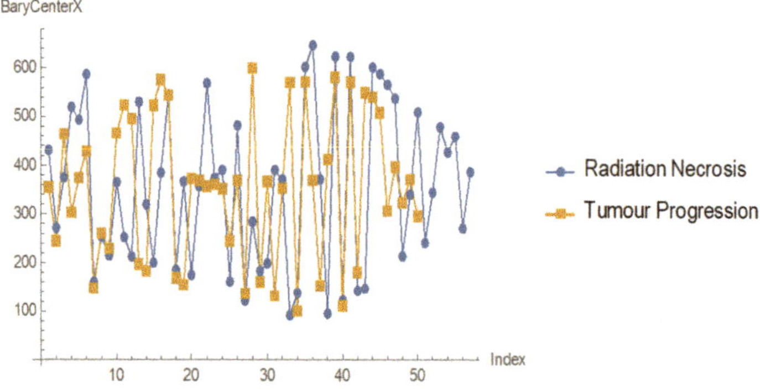

Fig. 9.12 Plot of the BaryCenterX against image elements (index) for radiation necrosis and tumour progression

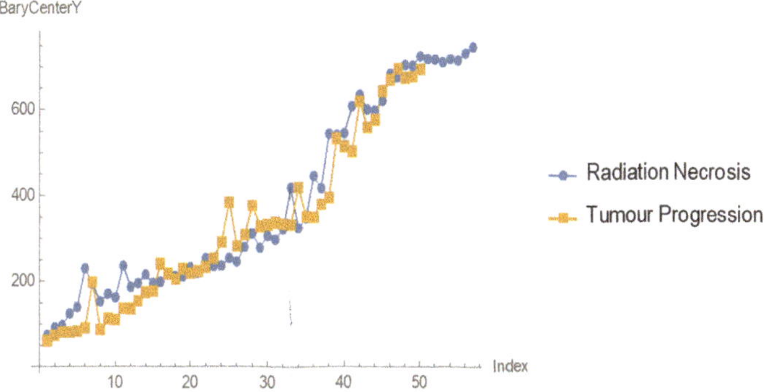

Fig. 9.13 Plot of the BaryCenterY against image elements (index) for radiation necrosis and tumour progression

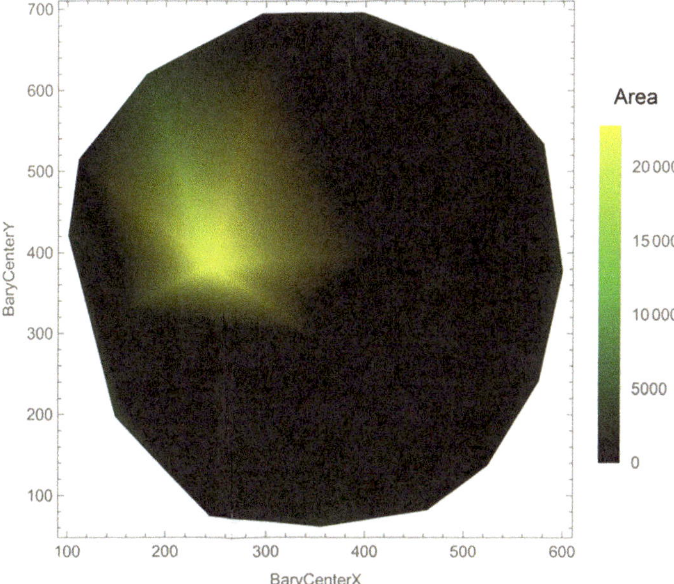

Fig. 9.14   3D plot of area against BaryCenterX and BaryCenterY for tumour progression image

intensity are mostly due to the unique microstructure profiles of the brain tissues. Generalized visualizations of the filtered data are shown in Figs. 9.20 and 9.21 in which tissue area is the most prominent characteristics. Figures 9.22 and 9.23 bring out the unique behaviour of the filtered area with respect to BaryCenterX and BaryCenterY for radiation necrosis and tumour progression. The 3D plot of area for tumour progression showed a very large range of area when compared to case of radiation necrosis even if this observation is mostly due to the image element with index $= 25$. This is a confirmation of the observation that was made in Figs. 9.14 and 9.15. It is important to show the variations of the various filtered measures in Tables 9.3 and 9.4 according to the indexed elements.

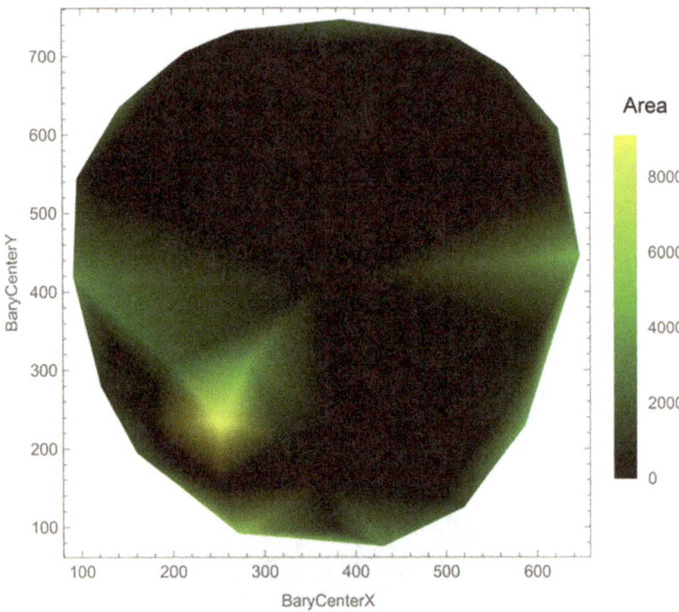

Fig. 9.15 3D plot of area against BaryCenterX and BaryCenterY for radiation necrosis image

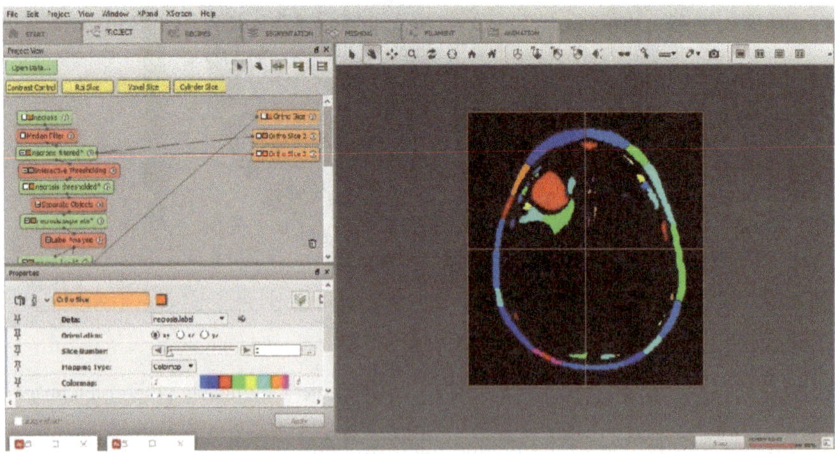

Fig. 9.16 Interactive selection of necrosis image

Figures 9.24, 9.25, and 9.26 show contrasts between the two conditions under investigation with BaryCenterY showing the list difference. However, in terms of BaryCenter, tumour progression appears to show distinct behaviour around index 25. These radiomic properties may contribute to improved follow-up after tumour treatment.

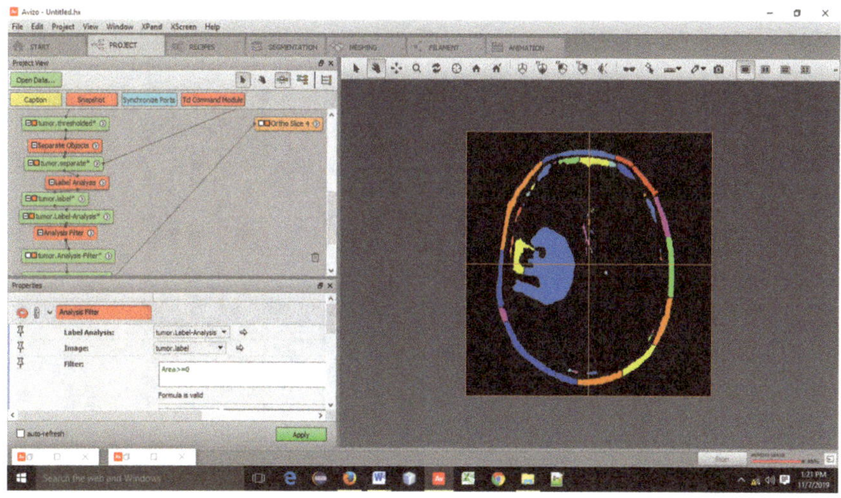

Fig. 9.17 Interactive selection of tumor image

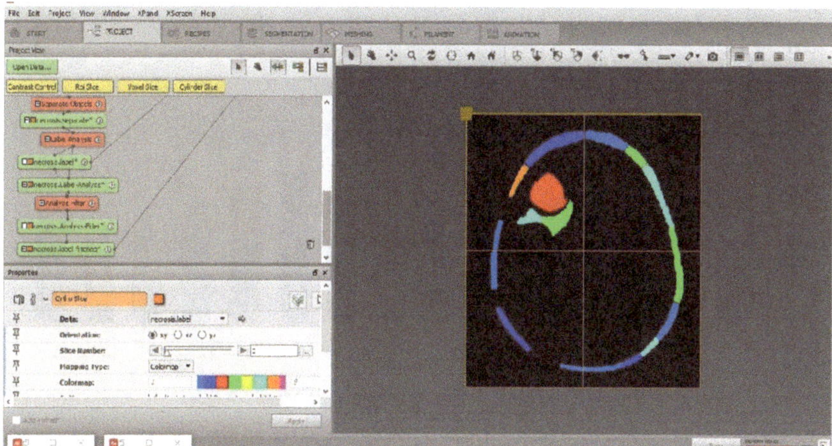

Fig. 9.18 Reconstruction of necrosis image from filtered data

9.9 Discussion

A designed system useful for processing and analysing MR images have been demonstrated for brain scans during radiation necrosis and tumour progression. Avizo was used to show the radiomic contrast between these conditions by employing the images in Fig. 9.1. The Median filter module (which filters the images) and the interactive thresholding module (which prompts the setting of the levels with visual feedback) have been employed for image processing. The Label

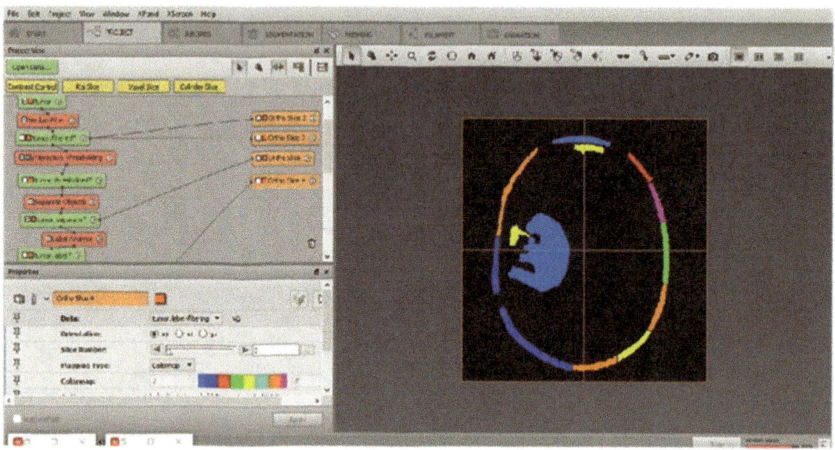

Fig. 9.19 Reconstruction of tumor image from filtered data

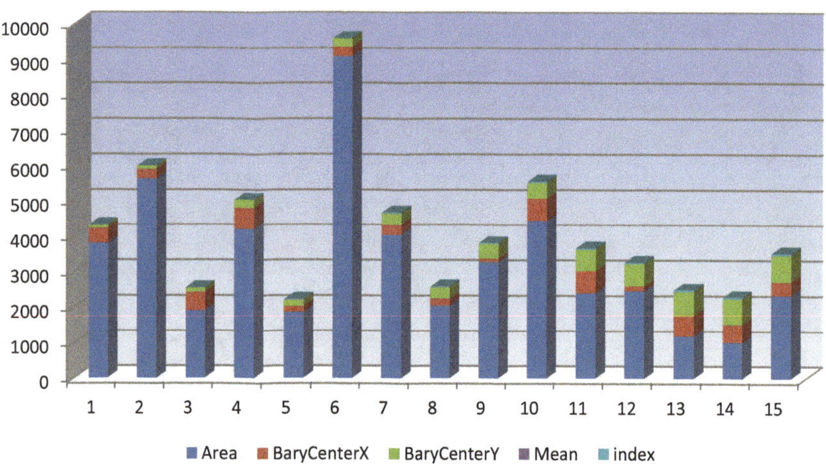

Fig. 9.20 Filtered extracted data bar chart for radiation necrosis

Analysis module was used to acquire numerical and statistical data from the images as shown in Tables 9.1, 9.2, 9.3, and 9.4. Using Label Analysis, these tables were obtained from a comprehensive spreadsheet that provides information on the area, volume, barycenterX, barycenterY, and Mean of the image data. Figures 9.9 and 9.10 show a generalized representation of the extracted data in Tables 9.1 and 9.2. Figures 9.11, 9.12, and 9.13 show graphs of each extracted measure with respect to the image elements index. These figures showed some unique points where differences in radiation necrosis and tumour progression can be easily identified. Since the area of image elements is the most prominent measure, Figs. 9.14 and 9.15 demonstrate how the variations occur with respect to the barycenter measures. Tumour

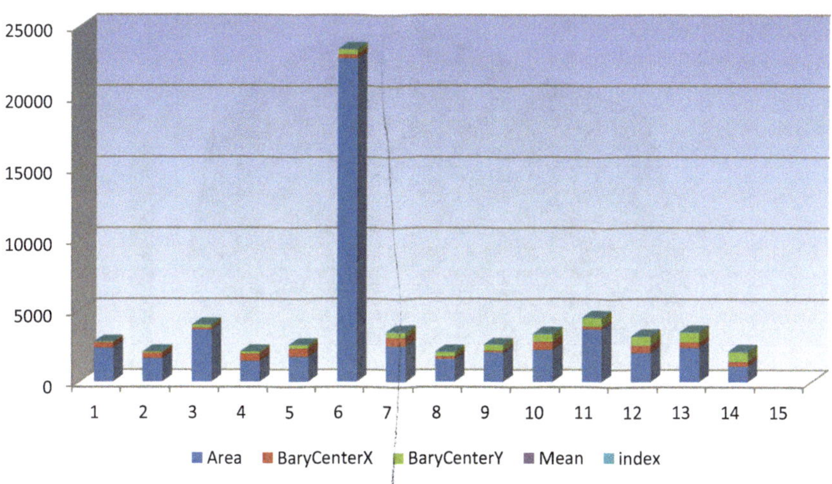

Fig. 9.21 Filtered extracted data bar chart for tumour progression

progression image has the largest areas which are concentrated at a specific point of BaryCenterX and BaryCenterY. Detailed segmented processed images of the conditions under investigation are given in Figs. 9.16 and 9.17. Each element of the images has been shown in unique colours according to their intensity on MRI scanner. By visual inspection, these images can also be used to differentiate between necrosis and tumour progression while the radiomic data could serve as confirmation as well as giving information on hidden image features. A filter was used to exclude features with Area3d values with values of 300 or more. We notice that this filter removed most image elements leaving the areas of confusion (voxels where we are not sure if we have necrosis or tumour progression). The filtered data are shown in Tables 9.3 and 9.4 while the generalized visualizations are shown in Figs. 9.20 and 9.21. Area, being the most prominent feature, was represented as 3D plot in Figs. 9.22 and 9.23. These figures show the magnitude of extracted filtered area in unique version as well as the distribution. The filtered image data were used to reconstruct the images and shown in Figs. 9.18 and 9.19. A confirmation of the unique features noticed in the filtered data has been presented in Figs. 9.24, 9.25, and 9.26 where the points to be focused on for delineation have been demonstrated.

9.10   Conclusion

It is important to note that we have made use of JPEG version of the MR images which are probably a bit low in quality. We believe that using a DICOM version would produce improved results. One other advantage of this study is that the

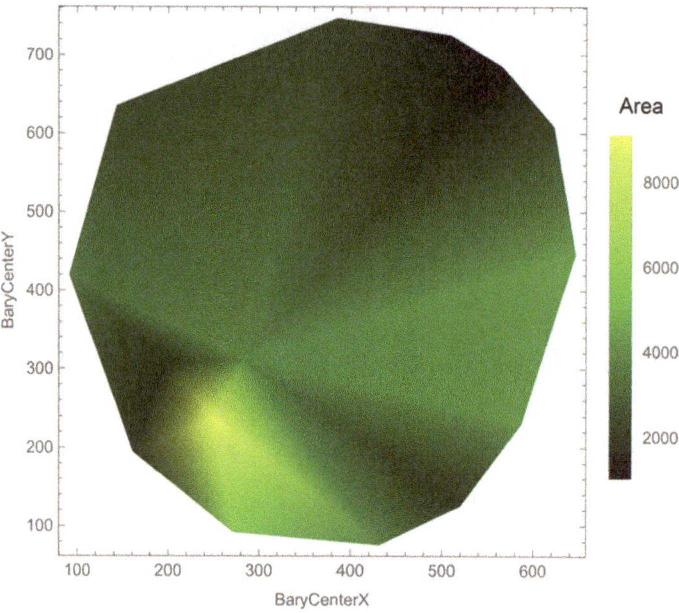

Fig. 9.22 3D plot of area against BaryCenterX and BaryCenterY for radiation necrosis image using filtered data

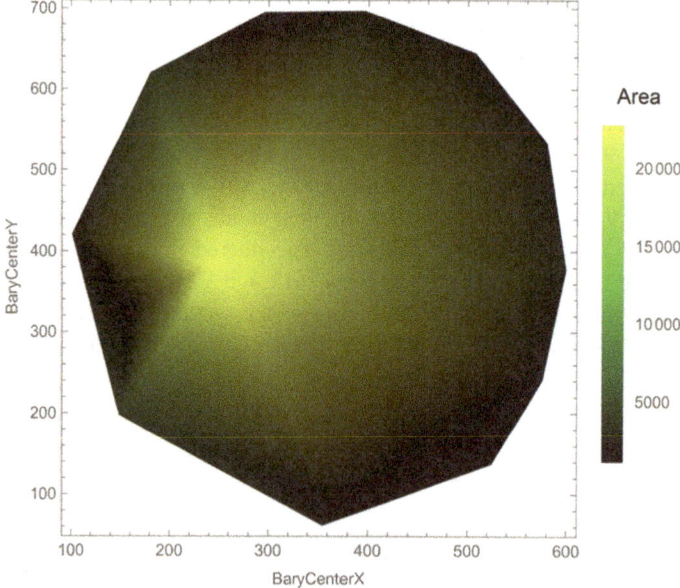

Fig. 9.23 3D plot of area against BaryCenterX and BaryCenterY for tumour progression image using filtered data

Table 9.3 Filtered extracted numerical data from the necrosis image

Area	BaryCenterX	BaryCenterY	Mean	Index
3814	430.821	75.6411	1	1
5644	271.276	92.2624	1	2
1913	519.742	125.045	1	4
4227	586.914	230.148	1	6
1864	160.805	194.899	1	7
9110	252.47	235.973	1	11
4061	284.046	311.773	1	28
2060	198.195	306.656	1	30
3289	90.9428	419.8	1	33
4440	646.468	446.164	1	36
2413	622.825	608.45	1	41
2466	142.371	635.063	1	42
1207	565.935	685.035	1	46
1019	509.164	725.474	1	50
2350	385.926	747.14	1	57

Table 9.4 Filtered extracted numerical data from the tumor image

Area	BaryCenterX	BaryCenterY	Mean	Index
2380	355.097	61.4382	1	1
1639	373.82	83.4002	1	5
3663	148.301	197.615	1	7
1437	524.452	137.26	1	11
1749	576.09	240.869	1	16
22820	244.845	384.84	1	25
2512	599.95	377.458	1	28
1598	160.589	328.503	1	29
2069	100.868	420.854	1	34
2257	580.652	533.691	1	39
3713	180.377	620.208	1	42
2041	507.972	644.532	1	45
2397	396.423	696.987	1	47
1116	295.574	695.012	1	50

extracted data are interesting sources of data for machine learning and artificial intelligence for clinical diagnosis. Any machine learning-based models developed with these data may be integrated with MRI scanners for easy delineation of radiation necrosis and tumour progression. The indices are indicative of several brain features as observed by the image processing module. The features they represent may not be the same for both images. Hence, it may be interesting if future

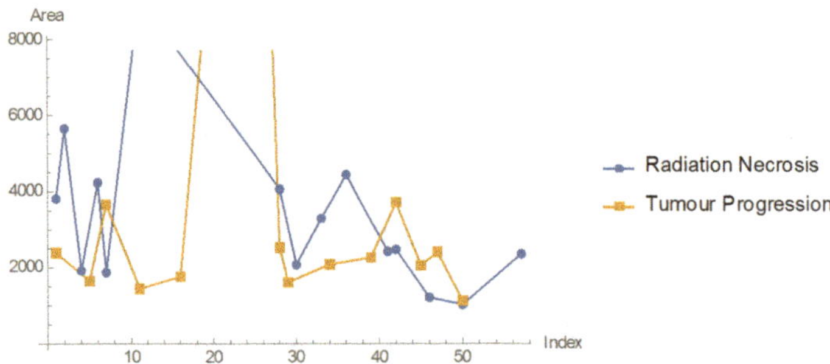

Fig. 9.24 Plot of the filtered area against image elements (index) for radiation necrosis and tumour progression

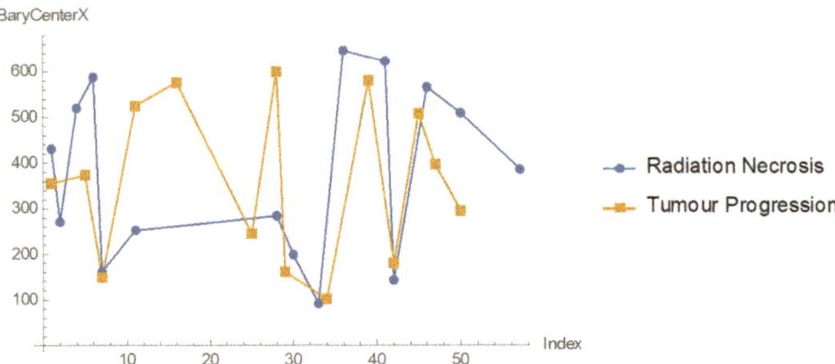

Fig. 9.25 Plot of the filtered BaryCenterX against image elements (index) for radiation necrosis and tumour progression

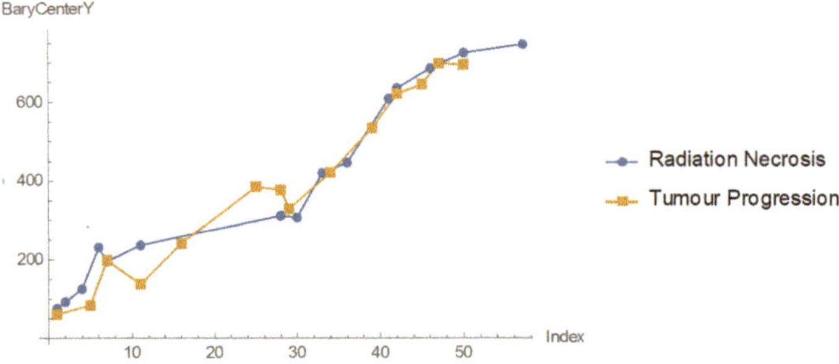

Fig. 9.26 Plot of the filtered BaryCenterY against image elements (index) for radiation necrosis and tumour progression

studies can harmonize the indices on both images so as to be able to identify the specific tissues they represent on the scans. Once this is achieved, the particular tissue where the most changes are noticed for both conditions can now be the object of focus for future clinical examinations. This will lead to understanding the processes that occur at the molecular level.

Chapter 10
Computational Analysis of Magnetic Resonance Imaging Contrast Agents and Their Physico-Chemical Variables

Abstract This chapter shows how time-independent NMR Bloch flow equation can be used for comparative studies of contrast agent's NMR response via the transverse magnetization. The solution provides a means of studying the transverse magnetization from the selected contrast agents without making any assumption to make the problem tractable. We applied various computational approaches to the assessment of MRI contrast agents and demonstrated an example of linking some of the approaches to imaging experiments. We show how computational methods can be used to make improvement to MRI scans and extract quantitative variables to aid effective diagnosis. A simplified form of MRI signal equation in FLASH sequences was used to obtain an expression for the concentration of contrast agents during scanning. This expression was then applied to compute the concentration across the post contrast images for patients. The resulting data was used together with a wolfram Mathematica program to reconstruct concentration maps from the extracted data. These maps imply that we can actually extract concentration dynamics of contrast media from MRI scans without too much complicated designs of the imaging media.

10.1 Analytical Method for Monitoring the Dynamics of Responsive Contrast Agents

In Sect. 2.2 of Chap. 2, we discussed MR contrast agents and their physicochemical basis. Responsive contrast agents are MR diagnostic agents whose contrasting features are sensitive to a given physicochemical variable that characterizes the micro-environment in which the probe is found after administration (Aime et al., 2005). The crucial diagnostic parameters are pH, temperature, enzyme activity, redox potential and concentration of specific ions, and low weight metabolites.

However, the clinical use of responsive contrast agents is merely a future possibility because the accurate measurement of any of the parameters above with a given probe requires a precise knowledge of the local concentration of the contrast medium in the region of interest. If this can be achieved, we can then safely attribute observed changes in relaxation rate of water protons to changes in the parameters to be measured.

Although novel classes of MRI paramagnetic contrast media which address this problem have been successfully developed, the challenges above may actually be a blessing in disguise. That is, changes in the local concentration of the contrast medium in the region of interest may serve as additional source of tissue contrast provided the local concentration changes can be inferred directly from measurements of MRI signals.

10.2 Bloch NMR Flow Model for Contrast Agents Moving in 1 D Tissue Spaces

In order to address this problem, a computational model which relates the measured MRI signal in terms of the transverse magnetisation to the tissue temperature and concentration of the contrast agents is considered. Such a model may be tested *in-vitro* for setting reference points for parameter relationships and then used to predict contrast media dynamics *in-vivo*. Since the contrast agents are meant to improve MR signals at different points within a physiological system, it is important to measure the spatial transverse magnetisation in relation to the expected variations in the relaxation times. In such a case, the constant velocity time-independent Bloch NMR flow equation given in Eq. (2.274) shall be employed. Within the small RF limit, $\gamma^2 B_1{}^2 T_1 T_2 < \, < 1$, Eq. (2.274) becomes:

$$v^2 \frac{d^2 M_y}{dx^2} + vT_0 \frac{dM_y}{dx} + T_g M_y = \frac{M_0}{T_1} \gamma B_1(x) \tag{10.1}$$

where T_0, T_g and all other terms retain their usual meanings. For this formulation, it is assumed that the resonance condition exists at Larmor frequency as well. For paramagnetic complexes, Eqs. (2.10) and (2.22) may be combined such that the Curie law becomes:

$$M_0 = \rho \frac{I(I+1)\gamma^2 \hbar^2 B_0}{3k_B T} \tag{10.2}$$

where ρ is the spin density (which cannot be altered in living beings (Brix et al., 2008)), k_B is the Boltzmann's constant, T is the measured body temperature, $h = 2\pi\hbar$ is the Planck constant, B_0 is the polarizing field and I is the spin quantum number of the paramagnetic complex to be administered.

If the applied RF field is related to the magnetic gradient field G_x and given as follows

$$\gamma B_1(x) = \gamma G_x(t)x \tag{10.3}$$

$$v^2 \frac{d^2 M_y}{dx^2} + vT_o \frac{dM_y}{dx} + T_g M_y = \frac{\rho}{T_1} \frac{I(I+1)\gamma^3 \hbar^2 B_0}{3k_B T} G_x(t)x \tag{10.4}$$

the complementary function, which is a solution to the homogeneous part of Eq. (10.4), is given as:

$$M_{yc} = C_1 e^{-\frac{T_o}{2v}(1-\alpha)x} + C_2 e^{-\frac{T_o}{2v}(1+\alpha)x} \tag{10.5}$$

where $\alpha = \sqrt{1 - \frac{4T_g}{T_0^2}}$ is a dimensionless relaxation weighting parameter. Due to the nature of the right-hand side of Eq. (10.4), the following solution may be assumed:

$$M_{yp} = K_1 x + K_0 \tag{10.6}$$

Substituting for Eq. (10.6) in Eq. (10.4):

$$vT_o K_1 + T_g K_1 x + T_g K_0 = \frac{\rho}{T_1} \frac{I(I+1)\gamma^3 \hbar^2 B_0}{3k_B T} G_x(t)x. \tag{10.7}$$

Equating the coefficients of like powers of x, the following expressions are obtained:

$$K_0 = 0, K_1 = \frac{\rho}{T_1 T_g} \frac{I(I+1)\gamma^3 \hbar^2 B_0}{3k_B T} G_x(t) = \rho T_2 \frac{I(I+1)\gamma^3 \hbar^2 B_0}{3k_B T} G_x(t). \tag{10.8}$$

Therefore, the general solution ($M_{yc} + M_{yp}$) is given as:

$$M_y(x) = Ce^{-\frac{T_o}{2v}(1-\alpha)x} + C_2 e^{-\frac{T_o}{2v}(1+\alpha)x} + \rho T_2 \frac{I(I+1)\gamma^3 \hbar^2 B_0}{3k_B T} G_x(t)x, \tag{10.9}$$

$$M_y(x) - \rho T_2 \frac{I(I+1)\gamma^3 \hbar^2 B_0}{3k_B T} G_x(t)x = e^{-\frac{T_0}{2v}x}(C_1 e^{\alpha x} + C_2 e^{-\alpha x}), \tag{10.10}$$

$$T_0 = -\frac{2v}{x} \ln \left[\frac{M_y(x) - \rho T_2 \frac{I(I+1)\gamma^3 \hbar^2 B_0}{3k_B T} G_x(t)x}{(C_1 e^{\alpha x} + C_2 e^{-\alpha x})} \right]. \tag{10.11}$$

10.3   Application of Model to Cardiovascular Disease

In contrast-enhanced magnetic resonance angiography, contrast agents are administered to monitor blood flow in human blood vessels. Cardiovascular diseases occur when blood vessels are blocked by fatty deposits. When a contrast agent is injected, the T_1 of blood is shortened from $T_{1blood} = 1.2$ s (at $B_0 = 1.5$ T) to less than 100 ms during the first bolus passage (Brix et al., 2008). The relaxation rate as a function of the local contrast agent concentration (C) is given by Eq. (2.9). For blood, this equation becomes:

$$R_1 = \frac{1}{T_1} = \frac{1}{T_{1blood}} + r_1 C \tag{10.12}$$

If the contrast agent does not have any significant influence on the transverse relaxation time T_2, Eq. (10.11) becomes:

$$r_1 C = -\frac{2v}{x} \ln \left(\frac{M_y(x) - \rho T_2 \frac{I(I+1)\gamma^3 \hbar^2 B_0}{3 k_B T} G_x(t) x}{(C_1 e^{\alpha x} + C_2 e^{-\alpha x})} \right) - \left(\frac{1}{T_{1blood}} + \frac{1}{T_2} \right) \tag{10.13}$$

10.4   Computational Monitoring the Dynamics of Contrast Agents

Using the experimental values of relaxation parameters (Brix et al., 2008): $T_{1blood} = 1.2$ s, $T_2 = 0.15$ s, $r_1 = 10$/mmol/s, $v = 1.0$ m/s (in the aortic arch), $\rho = 6.022 \times 10^{26}$ M^{-1}, $T = 37$ °C, $G_x = 0.2$ T m^{-1}. A Wolfram Mathematica computer code based on the expressions of Eq. (10.13) gives the following visualizations:

Figure 10.1 shows the distribution of the concentration of the contrast agent according to Eq. (10.13). These profiles demonstrate that the concentration increases with increase in the transverse magnetization at higher values of the spatial parameter x. The value of the concentration is significantly lower when x is not far from zero. However, it appears that this behaviour is local because as the voxel size becomes smaller such as in the case of Fig. 10.1c, d, the concentration increases. This is probably due to the fact that there are no larger spaces over which the contrast agents could diffuse. Since the transverse magnetization is directly proportional to the MR signal, this model could offer us an idea of how to extract the in-vivo behaviour of contrast agents by observing the dynamic behaviour of the MR signal in accordance with the spatial ranges over which the observation is made.

We used the idea of the profiles of Fig. 10.1 to investigate what the concentration distribution will be in different types of blood vessels. Figure 10.2 has been

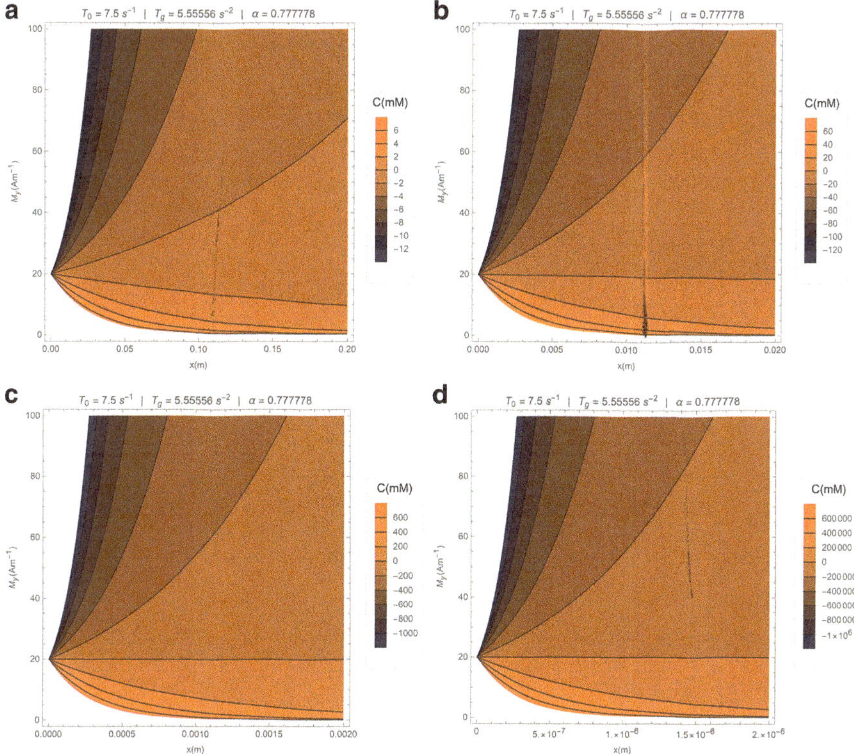

Fig. 10.1 (**a–d**) Contour profiles of contrast agent concentration as a function of the transverse magnetisation and x using relaxation parameters of human blood at 1.5 T for various spatial resolutions

developed using hemodynamic properties of human blood vessels (Glaser, 2012; Tortora & Derrickson, 2012; Marieb & Hoehn, 2013) and NMR relaxation features of human blood at 1.5 T [$T_{1blood} = 1.2$ s and $T_2 = 0.15$ s] (Moratal et al., 2006). Figure 10.2 has a similar pattern with Fig. 10.1 with larger blood vessels showing lower concentration values in comparison to smaller blood vessels. However, Fig. 10.2b, d have almost identical concentration distribution with slightly different pattern. This implies that voxel size is not the only determinant factor in the distribution of contrast agent's concentration within blood vessels but rather, a combination of the voxel size and the speed of blood flow. This result may prove be useful not only to monitoring contrast agents after ingestion but also monitor the speed at which the blood moves within the vessels under investigation. For example, if the transverse magnetization does not match the expected concentration values for a particular blood vessel, vessel blockage could be suspected.

As promising as this model could be, there is still a need for a standardized means of extracting quantitative MR signal values or transverse magnetization from MR scans. In addition to this, a constant fluid velocity has been assumed in this model

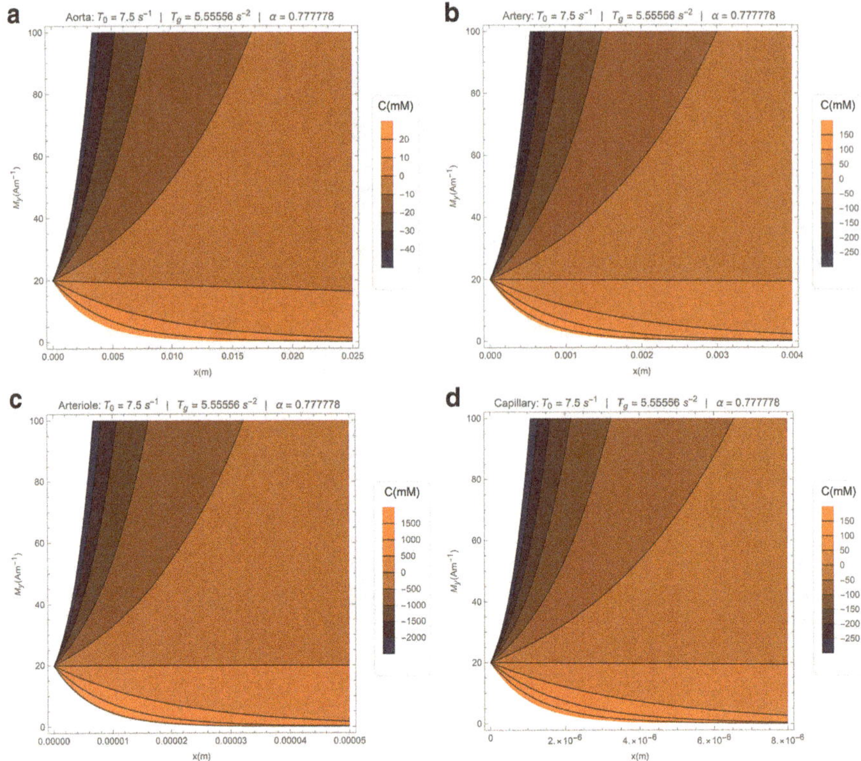

Fig. 10.2 Contour profiles of concentration as a function of transverse magnetisation and x using relaxation properties of human blood at 1.5 T and hemodynamic properties of (**a**) aorta (**b**) artery (**c**) arteriole (**d**) capillary

and this could make it difficult to apply this model to physiological situations in which the fluid velocity shows spatial variations. In subsequent sections, we shall consider cases with variable fluid velocity.

10.5   Computational Model for Comparative Analysis of MRI Contrast Agents

The efficacy of magnetic resonance imaging contrast agents is not only determined by their pharmacokinetic properties but also by their magnetic properties as given by their T_1-and T_2-relaxivities (Brix et al., 2008; Rohrer et al., 2005). It is very important to conduct comparative investigation on a large number of magnetic resonance imaging contrast agents because their properties are dependent on different physiological environments and binding to macromolecules in the blood (Rohrer et al., 2005). Hence, different contrast agents are designed for different imaging objectives. However, the individual linearities of relaxation rates as functions of concentration is difficult to quantify using the experimental measurements in a wider concentration range due to the large number of samples investigated at several magnetic field strengths and in different solvents. To address this problem, a model and computer program for comparative analysis of NMR profiles of contrast agents in different physiological environments have been developed in this section.

We consider spin dynamics of contrast agents and surrounding tissues based on time-independent Bloch NMR flow equation as given by Eq. (2.325). Within a rotating frame, Larmor condition exists such according to Eq. (10.14).

$$f_0 = \gamma B - \omega = 0 \tag{10.14}$$

In the instance in which the molecules of the contrast agents are in motion at a variable velocity $v(x)$ and the relaxation times of the solvent without the contrast agent is T_{i0} ($i = 1, 2$), Eq. (2.325) is written as:

$$v^2(x)\frac{d^2M_y}{dx^2} + v(x)\left(T_0 + \frac{dv}{dx}\right)\frac{dM_y}{dx} + \left(\gamma^2 B_1^2(x) + T_g\right)M_y = \frac{M_0}{T_1}\gamma B_1(x) \tag{10.15}$$

where

$$T_0 = \frac{1}{T_{10}} + \frac{1}{T_{20}} + r_1 C + r_2 C, \quad T_g = \left(\frac{1}{T_{10}} + r_1 C\right)\left(\frac{1}{T_{20}} + r_2 C\right) \tag{10.16}$$

Note that C is the concentration of the contrast agent while r_1 and r_2 are the relaxivities. Assuming that fluid velocity and the applied field have the following forms:

$$v(x) = \frac{x}{\delta} \tag{10.17}$$

$$B_1(x) = \frac{T_1}{T_2}Gx \tag{10.18}$$

Then, Eq. (10.15) becomes:

$$\frac{x^2}{\delta^2}\frac{d^2M_y}{dx^2} + \frac{x}{\delta}\left(T_0 + \frac{1}{\delta}\right)\frac{dM_y}{dx} + \left(\left(\gamma\frac{T_1}{T_2}Gx\right)^2 + T_g\right)M_y = \frac{M_0}{T_1}\frac{T_1}{T_2}\gamma Gx \tag{10.19}$$

$$x^2\frac{d^2M_y}{dx^2} + x(1 + \delta T_0)\frac{dM_y}{dx} + \left(\left(\gamma\frac{T_1}{T_2}G\delta x\right)^2 + \delta^2 T_g\right)M_y = \frac{M_0\delta^2}{T_2}\gamma Gx \tag{10.20}$$

Setting

$$\xi = \frac{T_1}{T_2}\gamma G \delta \tag{10.21}$$

Equation (10.20) becomes:

$$x^2 \frac{d^2 M_y}{dx^2} + x(1 + \delta T_0)\frac{dM_y}{dx} + \left(\xi^2 x^2 + \delta^2 T_g\right)M_y = \frac{M_0}{T_1}\xi x \delta \tag{10.22}$$

Equation (10.22) is not a standard differential equation but it is similar to the Lommel differential equation (Zwillinger, 1997). Because of the difficulties in obtaining analytical solution to Eq. (10.22), we employed Wolfram Mathematica software to solve this equation. The solution is given as:

$$
\begin{aligned}
M_y(x) &= \frac{\pi M_0}{T_1}\xi x \delta \left(2^{-2-\alpha}\right)(\xi x)^{-\alpha}\left\{\Gamma\left(\frac{\delta T_0}{4}+\frac{\alpha}{2}+\frac{1}{2}\right)(\xi x)^{2\alpha}Y_\alpha(\xi x)\,{}_1\tilde{F}_2\left(\left(\frac{\delta T_0+2\alpha+2}{4}\right);1+\alpha,\left(\frac{\delta T_0+2\alpha+6}{4}\right);-\frac{(\xi x)^2}{4}\right)\right\} \\
&\quad -\frac{\pi M_0}{T_1}\xi x \delta\left(2^{-2-\alpha}\right)(\xi x)^{-\alpha}\left\{\cot(\alpha\pi)\Gamma\left(\frac{\delta T_0}{4}+\frac{\alpha}{2}+\frac{1}{2}\right)(\xi x)^{2\alpha}J_\alpha(\xi x)\,{}_1\tilde{F}_2\left(\left(\frac{\delta T_0+2\alpha+2}{4}\right);1+\alpha,\left(\frac{\delta T_0+2\alpha+6}{4}\right);-\frac{(\xi x)^2}{4}\right)\right\} \\
&\quad +\frac{\pi M_0}{T_1}\xi x \delta\left(2^{-2-\alpha}\right)(\xi x)^{-\alpha}\left\{(2)^{2\alpha}\csc(\alpha\pi)\Gamma\left(\frac{\delta T_0}{4}-\frac{\alpha}{2}+\frac{1}{2}\right)J_\alpha(\xi x)\,{}_1\tilde{F}_2\left(\left(\frac{\delta T_0-2\alpha+2}{4}\right);1-\alpha,\left(\frac{\delta T_0-2\alpha+6}{4}\right);-\frac{(\xi x)^2}{4}\right)\right\} \\
&\quad +(x)^{-\frac{\delta T_0}{2}}\left\{C_1 J_\alpha(\xi x)+C_2 Y_\alpha(\xi x)\right\}
\end{aligned}
\tag{10.23}
$$

where $\alpha = \frac{\delta}{2}\sqrt{T_0^2 - 4T_g}$ and ${}_1\tilde{F}_2$ is the regularized generalized hypergeometric function. Note that M_0 is the equilibrium magnetization, csc is the Cosecant function, Γ is the Gamma function, J_α is the Bessel function of first kind, Y_α is the Bessel function of the second kind, while C_1 and C_2 are both constants. In order to demonstrate how Eq. (10.23) can be used in the comparative analysis of agents, four contrast agents have been selected. Their experimentally determined NMR relaxometric properties are given in Table 10.1

A Wolfram Mathematica computer code has been developed according to the expression of Eq. (10.23). The data in Table 10.1 were taken at a temperature of 37 °C and uniform magnetic field, $B_0 = 1.5$ T. Other parameters used in the

Table 10.1 NMR relaxometric properties of four contrast agents (Rohrer et al., 2005) in three physiological environments (PE)

Unit of r_1 and r_2: (L/mmol s)	Blood ($1/T_{10} = 0.9$, $1/T_{20} = 4.6$) $T_{10} = 1.11$, $T_{20} = 0.22$		Plasma ($1/T_{10} = 0.69$, $1/T_{20} = 2.6$) $T_{10} = 1.45$, $T_{20} = 0.38$		Water ($1/T_{10} = 0.27$, $1/T_{20} = 0.9$) $T_{10} = 3.7$, $T_{20} = 1.11$	
	r_1	r_2	r_1	r_2	r_1	r_2
Magnevist	4.3	4.4	4.1	4.6	3.3	3.9
Gadovist	5.3	5.4	5.2	6.1	3.3	3.9
Prohance	4.4	5.5	4.1	5	2.9	3.2
Multihance	6.7	8.9	6.3	8.7	4	4.3

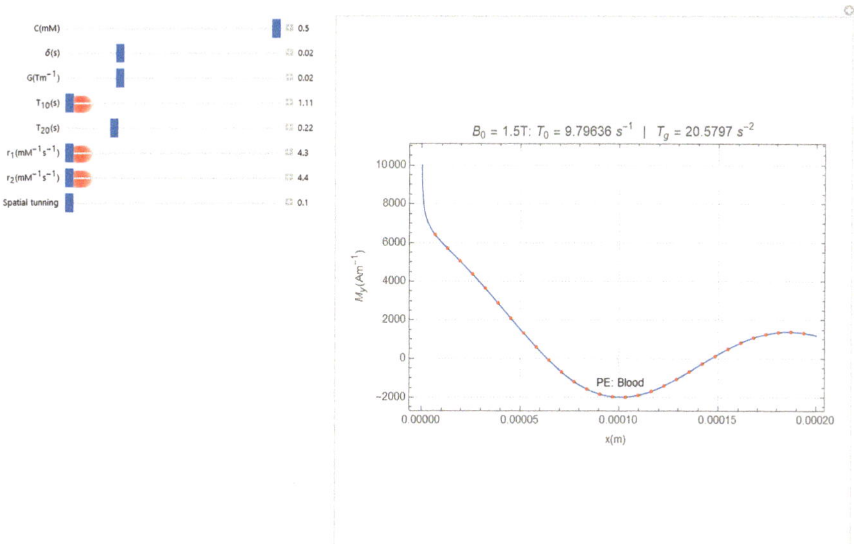

Fig. 10.3 Graphic user interface of the simulation computer program

simulations are: $C = 0.5$ mmol/L, $C_1 = 2000$, $C_2 = 100$, $G = 0.2$ T/m, $\delta = 0.02$ s and $M_0 = 100$ A/m.

Figure 10.3 shows a graphic user interface of the simulation of the MR transverse magnetization according to Eq. (10.23) for the selected contrast agents in Table 10.1. Once the precontrast-T_1 and -T_2 relaxations times for the ET are available and the relaxivities as well as other imaging parameters are known, this computer program can be used study the profiles of the transverse magnetization obtained after the contrast agents have been injected into the physiological environments. Figures 10.4, 10.5, and 10.6 give the profiles of the spatial variations of transverse magnetization for due to the selected contrast agents in blood, plasma and water. Analyses of contrast agent's relaxation behaviour in these physiological environments can give information of their response when injected into living tissues.

In Figs. 10.4, 10.5, and 10.6, we observe that T_0 and T_g relaxation rates increases with increase in r_1 and r_2 as expected. Although the patterns of the transverse magnetization appear similar in these figures, the values are actually of M_y are quite different at different spatial points. For example, at the point, $x = 200$ μm, $M_y = 1700$ A/m, 600 A/m, 1900 A/m and 2500 A/m for Fig. 10.4a–d respectively. The values at the same point are 800 A/m, 1000 A/m, 200 A/m and 1100 A/m for Fig. 10.5a–d respectively. However, the values of transverse magnetization at this point are all negative at $x = 200$ μm when the physiological environment is water (Fig. 10.6). This unique behaviour is observable at several other spatial points in the various physiological environments. In addition to this, we observed that at higher spatial resolution, the crests and troughs of the transverse magnetization are much more and extremely compressed such that it becomes very difficult to pinpoint the

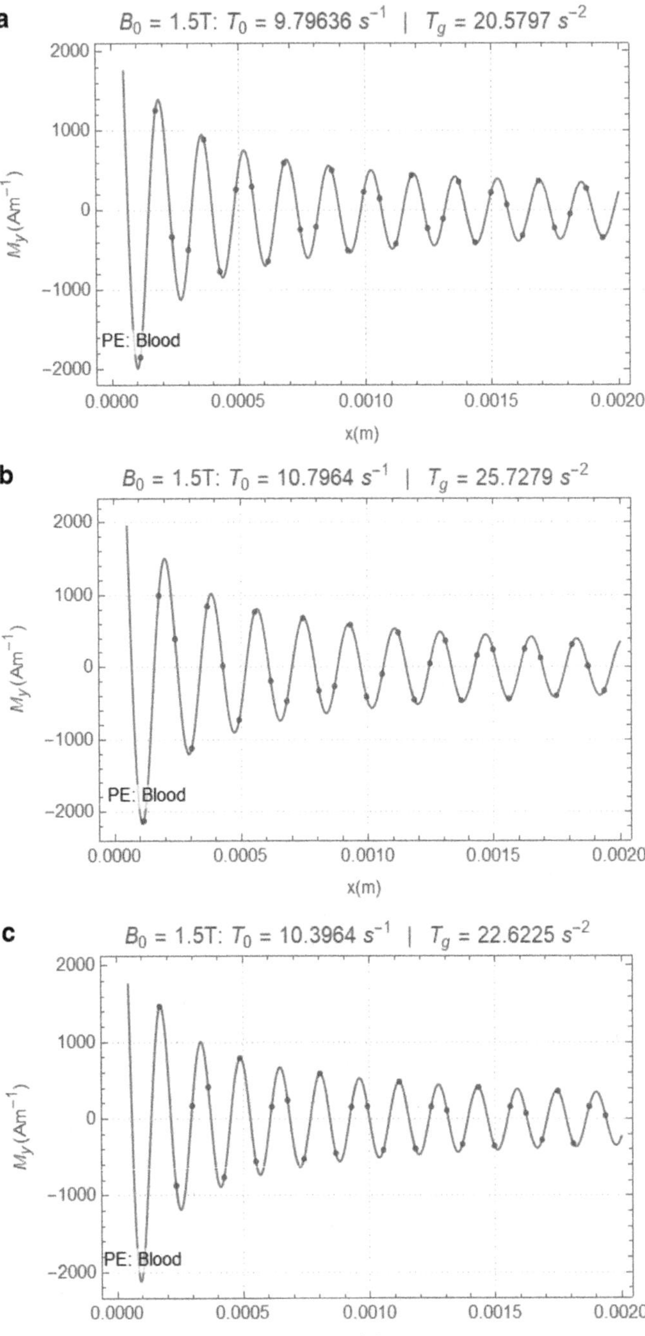

Fig. 10.4 Profiles of spatial variations of M_y determined from relaxivity values measured in the blood for (**a**) Magnevist (**b**) Gadovist (**c**) Prohance (**d**) Multihance

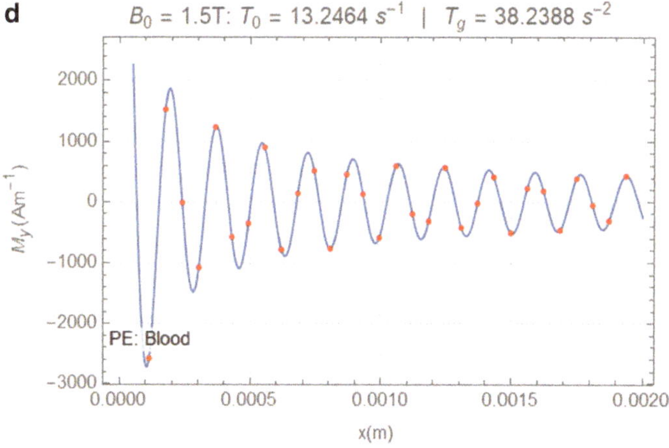

Fig. 10.4 (continued)

variations in M_y. It follows that this simulation is most useful when the spatial resolution is small and this points to a potential application of this model in molecular imaging. It is important to note that despite the significant closeness of the r_1 and r_2 values for some of the contrast agents, the simulation was still able to show unique contrasts through the transverse magnetization except the case of Fig. 10.6a, b in which r_1 and r_2 are essentially the same.

One other importance of this model is that large number of contrast agents may be compared within a very short time using this simulation. The simulation program could make life much easier for NMR scientists in terms of managing contrast agents for different physiological interests and they can run as many simulations as possible before injecting the CAs into living systems. Meanwhile, is quite easy to modify the program for time-dependent response and nonlinear behaviour of the relaxation rates.

10.6   Development of Machine Learning Classification Algorithm for Screening MRI Contrast Agents

Machine learning and deep learning algorithms are becoming increasingly useful in improving workflows in medical imaging and help scientists in task automation (Willemink et al., 2020). These algorithms have found impressive applications in lesion detection, triaging screening mammograms, elimination of gadolinium-based contrast agents for MRI and reduction of radiation doses during CT imaging. In the fields of drug discovery and development, machine learning algorithms have proved to be useful in the development of novel drug candidates (Patel et al., 2020). The methods for drug targets design and novel drug discovery are currently having

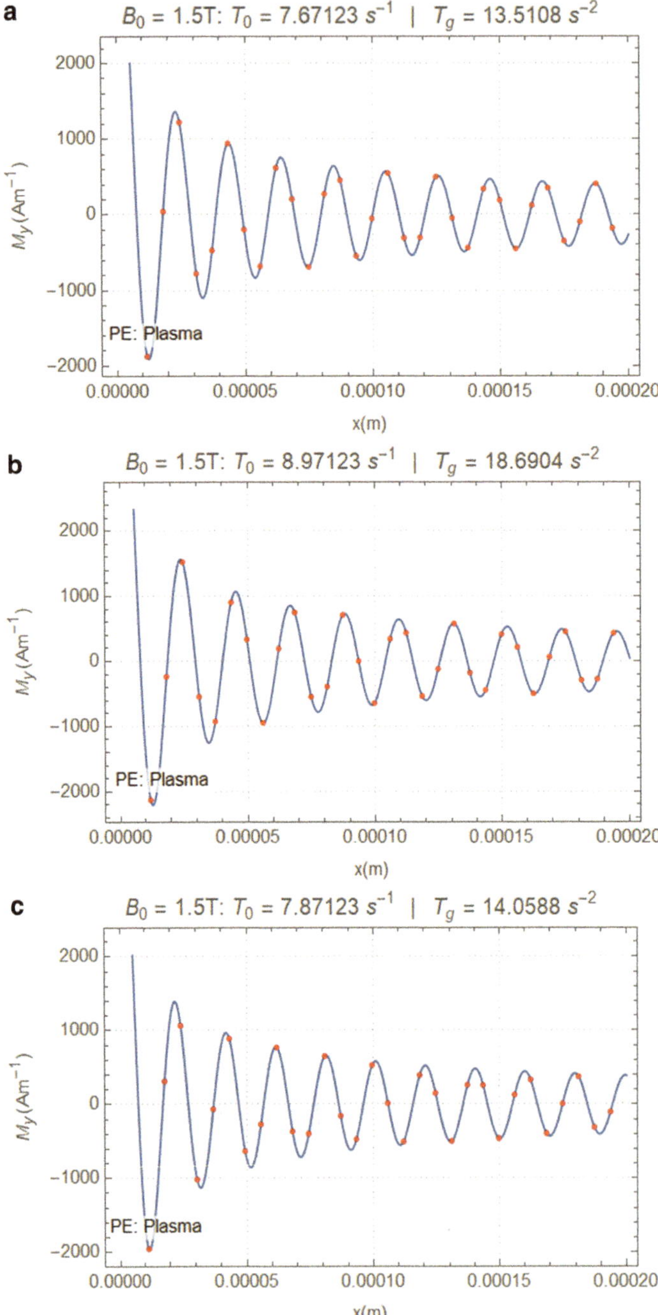

Fig. 10.5 Profiles of spatial variations of M_y determined from relaxivity values measured in plasma for (**a**) Magnevist (**b**) Gadovist (**c**) Prohance (**d**) Multihance

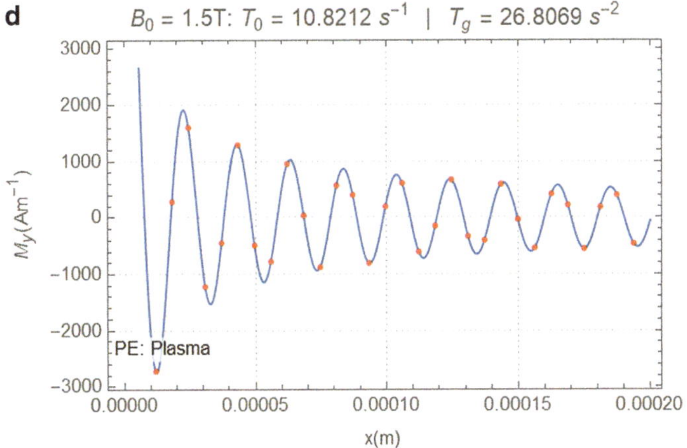

Fig. 10.5 (continued)

machine learning and deep learning algorithms incorporated to improve efficacy, efficiency and quality of final. The process generates numerous data and high-throughput screening and computational analysis (making use of machine learning) of databases of big data towards lead and target discovery. In a similar manner to drug discovery, the merit of a comparative investigation on a large number of MRI contrast agents with an experimental approach is complemented by the limitation that, because of the large amount of samples investigated at several magnetic field strengths (B_0) and in different solvents, the individual linearities of relaxation rates as functions of concentration could not be verified by further relaxation rate measurements in a wider concentration range (Rohrer et al., 2005).

To demonstrate how machine learning could be used to extract statistical information from experimental relaxivity measurements and help in improved screen of contrast agents with respect to different solvents, we shall make use of a Python-based classification algorithm which can be used to assess the behaviour of contrast agents within the solvents. This algorithm has the potential of predicting the in-vivo behaviour of contrast agents and thus helps in improving their responsiveness in disease diagnosis.

10.6.1   Data Preparation

The r_1 and r_2 relaxivities have been experimentally determined at a field of 1.5 T and temperature of 37 °C (Rohrer et al., 2005). These measurements are presented in Table 10.2.

The unique ranges of these measurements present an opportunity for the generation of synthetic dataset from these measurements. We employed a wolfram

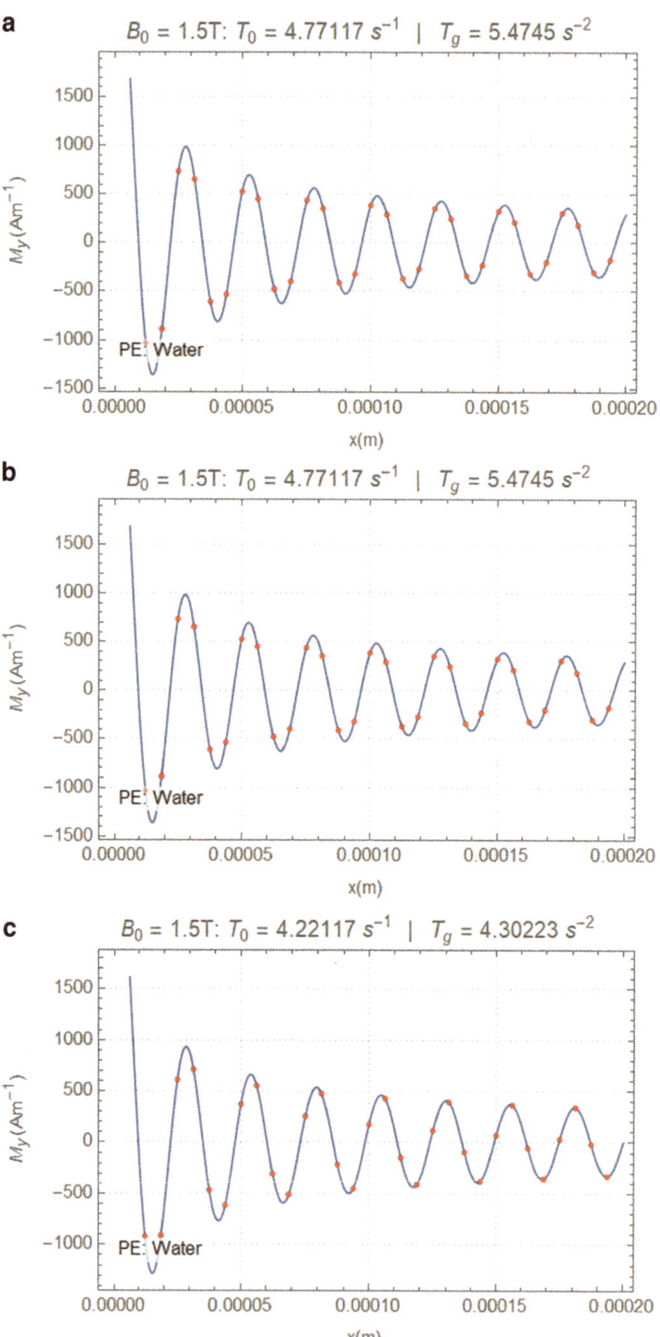

Fig. 10.6 Profiles of spatial variations of M_y determined from relaxivity values measured in water for (**a**) Magnevist (**b**) Gadovist (**c**) Prohance (**d**) Multihance

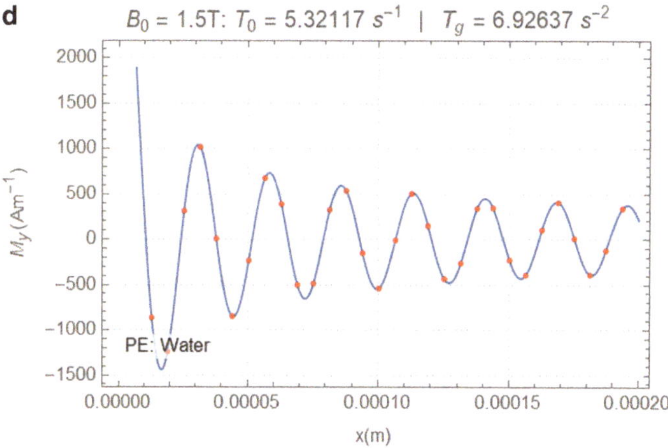

Fig. 10.6 (continued)

mathematical program to randomly generate sample points between the lower and higher limits for each of these measurements. The program generated 200 data points each for the respective solvents, making a total of 600 data points within the generated dataset. These data points imply that we are now investigating 200 samples of various contrast agents in water, plasma and blood. To see what the dataset looks like, we have selected five samples and presented them in Table 10.3.

Statistical summaries of the datasets represented in Tables 10.3 and 10.4 are given in Tables 10.5 and 10.6.

Tables 10.5 and 10.6 show the mean, median, maximum values and the quartiles of the generated r_1 and r_2 in the entire dataset.

10.6.2 Data Visualization

In order to implement machine learning algorithms on the generated dataset, it is import to understand how the r_1 and r_2 relaxivities vary from one contrast agents to the other as well as their variations across the solvents. A way of getting this done is the use of graphical representations of the synthesized data. These visualizations are shown in Figs. 10.7, 10.8, and 10.9. Twelve contrast agents have been selected for the visualization in Figs. 10.7 and 10.8 while the entire contrast agents investigated have been used in the visualization of Fig. 10.9.

Figures 10.7 and 10.8 show the distributions of the relaxivities of all the investigated contrast agents in different solvents. Resovist and Gadomer demonstrate overlapping values of r_1 across the solvents and this could indicate similar behaviour when injected into living tissues. That is, in-vivo r_1 response may not be significantly different from in-vitro response. Magnevist and Optimark also show similar

Table 10.2 NMR relaxivities of investigated contrast three solvents at $B_0 = 1.5$ T

Trade name	Water		Plasma		Blood	
	r_1	r_2	r_1	r_2	r_1	r_2
Magnevist	3.1–3.5	2.8–5.0	3.9–4.3	3.8–5.4	4.0–4.6	3.6–5.2
Gadovist	3.1–3.5	3.1–4.7	4.9–5.5	5.2–7.0	5.0–5.6	4.6–6.2
Prohance	2.7–3.1	2.5–3.9	3.9–4.3	4.2–5.8	4.1–4.7	5.0–6.0
Multihance	3.8–4.2	3.8–4.8	6.0–6.6	7.8–9.6	6.3–7.1	7.9–9.9
Dotarem	2.7–3.1	2.5–3.9	3.4–3.8	3.4–5.2	3.9–4.5	6.0–7.4
Omniscan	3.1–3.5	3.0–4.2	4.0–4.6	4.2–6.2	4.3–4.9	5.5–8.3
Teslascan	1.5–1.7	1.4–2.8	3.4–3.8	5.7–8.5	4.9–5.5	8.2–9.6
Optimark	3.6–4.0	3.5–4.9	4.4–5.0	4.3–6.1	4.9–5.5	5.4–6.6
Resovist	8.2–9.2	54–68	7.0–7.8	86–104	7.5–8.5	71–83
Feridex	4.4–5.0	39–43	4.2–4.8	31–35	6.6–7.4	61–71
Gadomer	16.4–18.2	21–23	15–17	18–20	16–18	21–23
MS-325	4.9–5.5	5.3–6.5	18–20	32–36	18–20	35–39
Primovist	4.5–4.9	4.5–5.7	6.5–7.3	7.8–9.6	6.9–7.7	8.2–10.0
SH U 555 C	12.5–13.9	41–47	10.1–11.3	36–40	13–15	82–98

Unit of r_1 and r_2: L mmol^{-1} s^{-1}

overlapping of r_2 values across the solvents. Other contrast agents show overlapping of r_1 and r_2 values in two solvents only while some demonstrate unique values across all solvents. In-vivo responses of relaxivities for the contrast agents without overlapping of values may be difficult to predict because they are expected to carry over their unique response into living tissues. Figure 10.9 shows the correlation matrices of r_1 and r_2 values for all the investigated contrast agents. Contrast agents whose relaxivity values are most correlated are close to 1 while those with least relationship are close to -1. This visualization could help in screening large number of contrast agents so as to have knowledge of those contrast agents that can be used for similar purposes or diagnostic capabilities.

10.6.3   XGBoost Model for Classification

For machine learning model development, XGBoost will be employed because of the impressive manner it handles multiclass classification (which is consistent with the dataset). XGBoost implements gradient boosted decision trees, makes use of XGBClassifier and is designed for speed and performance that is impressive in machine learning. Before model training is performed, the entire dataset of Tables 10.3 and 10.4 were divided to two parts such that 70% were used for model training while the remaining 30% were used for model testing. The classes were also re-labelled as follows: Plasma-"0", Blood-"1" and Water "2". The performance of this model after training is presented in Table 10.7.

Table 10.3 A cross view of the r_1 relaxivity dataset

Magnevist	Gadovist	Prohance	Multihance	Dotarem	Omniscan	Teslascan	Optimark	Resovist	Feridex	Gadomer	MS_325	Primovist	SH_U_555_C	Solvent
4.183977	5.320111	4.054651	6.021254	3.581298	4.017195	3.639867	4.433588	7.559365	4.790149	16.17368	18.27208	6.70939	10.15344	Plasma
4.288152	5.084797	4.174562	6.129345	3.542068	4.124192	3.581931	4.890428	7.103967	4.312816	15.38277	19.82747	6.559944	10.95	Plasma
4.078259	5.461538	4.06782	6.364655	3.64567	4.390743	3.568992	4.633015	7.724847	4.517058	16.81496	18.93943	6.91257	10.34042	Plasma
4.230766	5.462046	4.281536	6.176826	3.499193	4.292881	3.512299	4.724966	7.220783	4.768826	16.24946	18.7034	6.97353	10.62455	Plasma
3.904467	5.075417	4.106338	6.500629	3.460292	4.027267	3.6119	4.954624	7.070182	4.404169	15.06749	18.10764	6.502861	10.78195	Plasma
4.445033	5.301947	4.596748	6.479361	4.429533	4.314254	5.347198	5.27458	8.288252	6.9419	17.14284	19.16149	7.532964	14.27743	Blood
4.366562	5.587928	4.350226	6.870281	4.442669	4.567189	4.953289	4.961482	8.261848	6.906266	16.30656	18.02419	6.956654	14.79297	Blood
4.128819	5.117684	4.462479	6.47984	4.426864	4.4607	5.480318	5.289617	8.194023	6.820625	17.91358	18.05623	7.198531	13.59655	Blood
4.080217	5.203437	4.292919	6.91831	3.996649	4.37737	5.271233	5.291404	7.78597	7.041767	17.90736	18.20129	7.37367	13.5172	Blood
4.52167	5.250683	4.16263	6.842767	4.024746	4.552114	5.005643	5.366806	8.160071	6.628473	17.75434	18.91403	7.258086	14.79248	Blood
3.244572	3.320893	2.814092	3.834581	3.034005	3.218902	1.614204	3.840861	9.060702	4.433578	17.81052	5.140061	4.572094	12.78344	Water
3.105803	3.49562	2.768097	4.17841	2.802492	3.252208	1.673514	3.801356	9.081361	4.934797	18.11824	5.425514	4.654931	13.0331	Water
3.218608	3.25506	2.965164	3.918066	2.787508	3.492503	1.523565	3.626771	8.524813	4.99866	16.6436	4.970785	4.548383	12.87051	Water
3.261829	3.377474	2.707106	4.193683	2.84909	3.339626	1.605744	3.705725	8.660552	4.833866	16.4332	5.221445	4.707339	12.68391	Water
3.155755	3.119837	2.869638	4.038726	2.931766	3.371733	1.601753	3.924431	9.033182	4.727876	17.52364	5.320162	4.584184	13.22365	Water

Table 10.4 A cross view of the r_2 relaxivity dataset

Magnevist	Gadovist	Prohance	Multihance	Dotarem	Omniscan	Teslascan	Optimark	Resovist	Feridex	Gadomer	MS_325	Primovist	SH_U_555_C	Outcome
4.810409	6.563675	4.561137	8.559423	4.280777	4.378407	8.253789	4.433636	103.4613	32.33991	19.53303	34.01918	8.063709	38.1428	Plasma
4.393785	5.463803	4.259487	7.917686	5.033305	5.93667	7.559069	5.737722	97.18514	32.13304	19.28872	32.15285	7.927836	37.09955	Plasma
4.732124	6.149301	4.387776	8.78354	4.530559	4.847706	5.841845	5.805874	98.5682	34.36437	19.566	32.50815	9.03069	37.91161	Plasma
4.168443	6.352781	5.668554	8.89684	4.94879	4.852749	7.52688	5.689501	96.58311	31.33351	18.28416	34.13254	9.591314	36.95925	Plasma
4.856175	6.147937	5.714862	9.339996	4.735418	5.03403	7.103576	4.610192	92.34854	33.55408	19.36376	32.89611	8.941388	37.9948	Plasma
4.908656	4.878625	5.406435	9.286289	7.379055	5.581476	8.570995	5.537285	78.19584	63.71601	21.7141	37.43673	9.28957	94.89383	Blood
3.666954	4.819269	5.357414	9.385978	7.385954	6.986634	8.238838	5.912424	80.45162	68.61801	22.07692	36.71081	9.457932	93.6963	Blood
4.967279	6.091459	5.988076	8.02357	6.041266	7.563644	9.062144	5.693916	80.38073	65.4223	22.52636	38.2558	9.238472	86.08224	Blood
5.192266	5.475204	5.043077	8.818476	6.103333	5.633788	8.8091	6.089413	81.12466	64.78554	21.52138	37.19124	9.301729	85.18828	Blood
4.298235	4.680619	5.463877	8.722458	7.045449	6.763239	9.548754	6.431428	72.40934	65.37094	22.66534	37.94837	9.994438	88.37534	Blood
4.06363	3.194898	2.70248	4.218562	3.029377	3.252605	2.576023	4.499172	57.21716	41.34941	22.46509	5.375585	4.984821	44.70757	Water
4.987682	4.159274	3.689154	4.146177	3.147652	3.89303	2.583722	4.805092	55.79715	40.66634	21.57844	6.121671	5.517384	43.04123	Water
4.563426	3.546662	3.763217	4.205425	3.522599	3.349342	1.73498	4.42155	65.38935	39.60388	21.58266	5.864732	5.631411	45.62803	Water
3.142713	3.587571	3.106847	4.596256	3.211739	3.485094	1.594288	4.080881	66.13695	41.72183	21.67959	5.300138	5.531874	46.00382	Water
3.236735	4.24512	3.283051	3.865407	3.279568	3.017957	2.610584	3.518058	67.98623	40.78133	21.69882	5.689483	5.393077	41.49749	Water

Table 10.5 Statistical summary of the r_1 relaxivity dataset

	Magnevist	Gadovist	Prohance	Multihance	Dotarem	Omniscan	Teslascan	Optimark	Resovist	Feridex	Gadomer	MS_325	Primovist	SH_U_555_C
count	600	600	600	600	600	600	600	600	600	600	600	600	600	600
mean	3.888845	4.60022	3.807061	5.682428	3.568841	4.069624	3.461437	4.559374	8.030065	5.404589	16.73591	14.4013	6.295976	12.59137
std	0.455847	0.934329	0.663799	1.211173	0.545997	0.590378	1.473284	0.608272	0.59105	1.147868	0.787304	6.538802	1.160766	1.460334
min	3.100653	3.101144	2.700071	3.800763	2.702134	3.102855	1.500225	3.600872	7.011455	4.200954	15.00214	4.908217	4.502005	10.10384
25%	3.4024	3.409364	3.010715	4.104529	3.003108	3.383866	1.652787	3.884255	7.553283	4.530447	16.20572	5.324938	4.808364	10.96426
50%	4.049325	5.098336	4.10803	6.322609	3.592485	4.322569	3.580843	4.677179	8.009729	4.763706	16.7627	18.53787	6.892404	13.05879
75%	4.226312	5.322923	4.282605	6.57093	4.041311	4.526163	5.032127	5.056532	8.451555	6.808875	17.32394	19.23955	7.221973	13.6516
max	4.593352	5.59506	4.697983	7.094195	4.49631	4.896438	5.498249	5.499544	9.199919	7.398257	18.19826	19.99692	7.690572	14.99608

Table 10.6 Statistical summary of the r_2 relaxivity dataset

	Magnevist	Gadovist	Prohance	Multihance	Dotarem	Omniscan	Teslascan	Optimark	Resovist	Feridex	Gadomer	MS_325	Primovist	SH_U_555_C
count	600	600	600	600	600	600	600	600	600	600	600	600	600	600
mean	4.305797	5.131735	4.57443	7.305747	4.730596	5.208886	5.989145	5.137798	77.6933	46.75514	20.97659	25.59366	7.617645	57.42735
std	0.59039	1.035197	1.084392	2.173229	1.53529	1.477031	2.923864	0.831732	14.59041	14.26775	1.53125	14.04112	1.877842	23.51497
min	2.810127	3.104216	2.515884	3.801391	2.501098	3.010411	1.401177	3.501056	54.01006	31.05421	18.00159	5.300138	4.511247	36.02046
25%	3.908906	4.26858	3.487743	4.559168	3.481558	3.799232	2.485531	4.439515	65.12173	33.96686	19.50703	6.170528	5.386978	38.93239
50%	4.34607	5.255798	5.010905	8.360252	4.275304	5.126684	6.952857	5.264936	77.31478	41.18262	21.57189	34.01099	8.416497	43.95168
75%	4.778055	5.95099	5.490533	8.99761	6.350389	6.197782	8.520063	5.806562	90.17105	63.45978	22.1712	35.93695	9.103022	86.23936
max	5.380807	6.989858	5.999688	9.892741	7.385954	8.229957	9.597876	6.595629	103.9983	70.99838	22.98391	38.97438	9.999602	97.92428

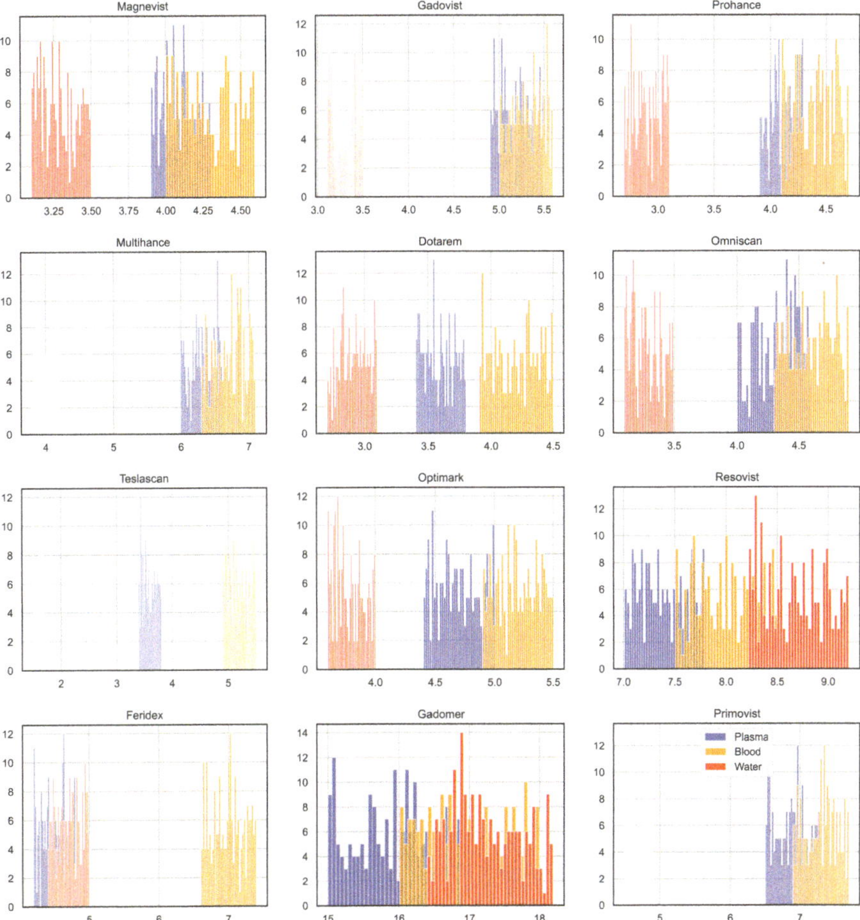

Fig. 10.7 r_1 relaxivity distribution for selected contrast agents

Table 10.7 shows that the performance of the classification algorithm which is based on XGBoost is perfect. It is worthy of note that in practical situations, the performance metrics are not expected to be 100%. The confusion matrices for r1 and r2 relaxivities are presented in Fig. 10.10.

We see that the confusion matrices produced by the classification algorithm is the same for both r_1 and r_2 relaxivities. 59, 69 and 52 samples were correctly classified as belonging to plasma, blood and water environments respectively. In both cases, there were no misclassifications. The structure of the data of original measurement and the closeness of some of the inputs are responsible to the perfect performance of this model. It is expected that if the data points are all from real measurements, the accuracy of this model may be lower but still will be good enough for making classifications based on the values of measured relaxivities. In addition, for large

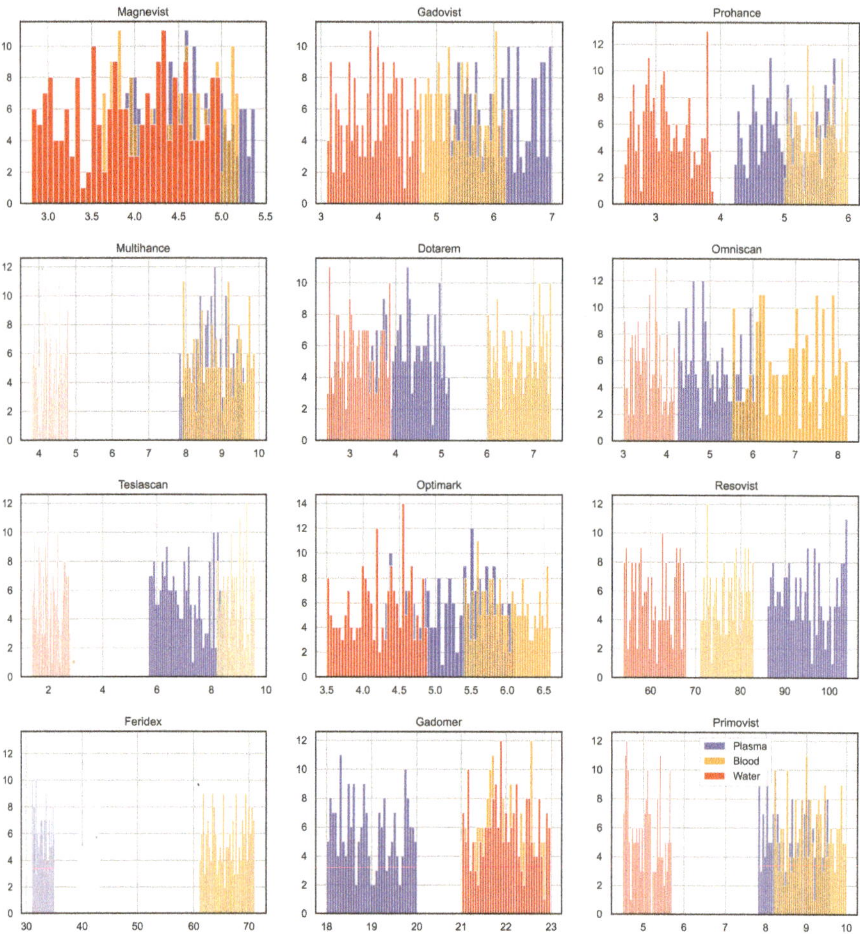

Fig. 10.8 r_2 relaxivity distribution for selected contrast agents

scale applications, extensive data are requirement for model training so as to improve the ability of the model to make better predictions. The advantage of this model is that the solvents used in this study are found within living tissues and this implies that this model can potentially be used to predict variations to contrast agent's relaxivities within living tissues and thus help in tracing them immediately after injection. This may not be possible without additional features such as T_1 and T_2 relaxation times as well as the MRI signal. Hence, this model can be improved upon by introducing additional magnetic resonance relaxation and tissue physiological features into the dataset for a more robust machine learning model.

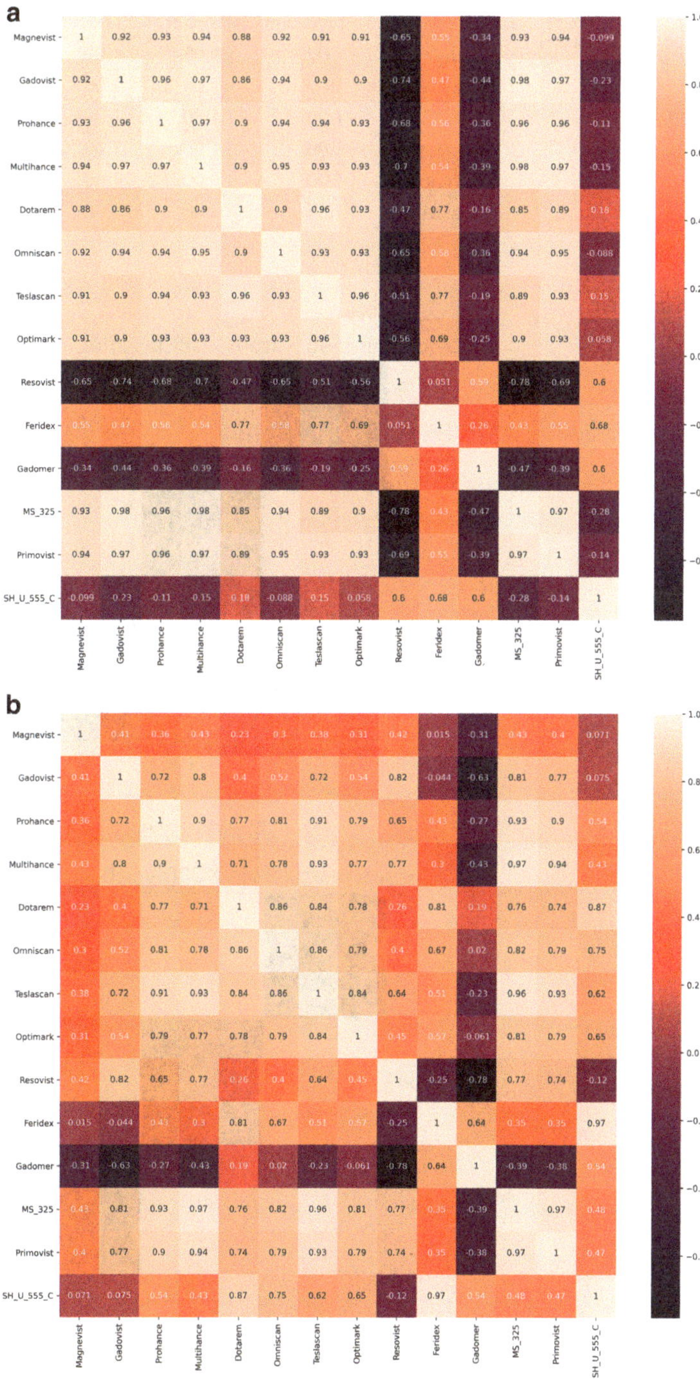

Fig. 10.9 Correlation matrix of the dataset of (**a**) r_1 (**b**) r_2 relaxivities

Table 10.7 Performance of XGBClassifier on the dataset

	Precision	Recall	f1-score	Support
Plasma (0)	1.00	1.00	1.00	59
Blood (1)	1.00	1.00	1.00	69
Water (2)	1.00	1.00	1.00	52
Accuracy	1.00	1.00	1.00	180
Macro avg	1.00	1.00	1.00	180
Weightrfavg	1.00	1.00	1.00	180

Fig. 10.10 Confusion matrix from XGBClassifier for (**a**) r_1 relaxivity (**b**) r_2 relaxivity

(A)

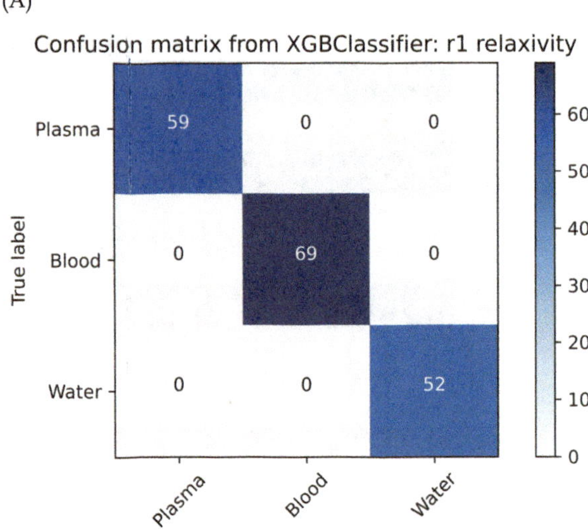

(B)

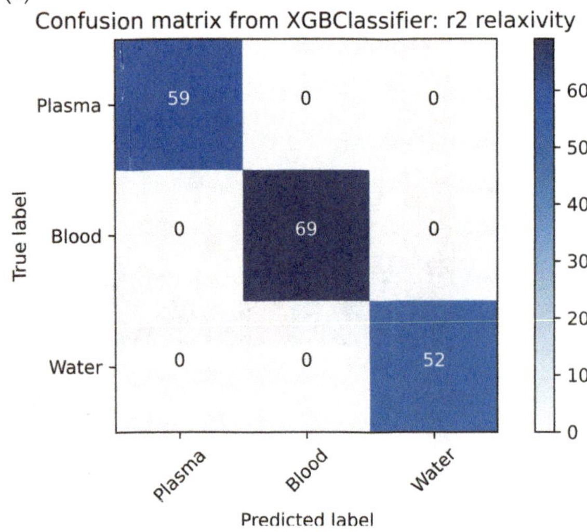

10.7 Bridging the Gap Between Computational Models and Experimental Imaging

Earlier sections have presented several computational methods for assessment of MRI contrast agents. Although some of these models can be used for clinical analysis, it is still imperative to find ways in which experimental imaging can easily address the challenge of having responsive probes with less expensive designs. This section presents a method with which probe concentration dynamics can be extracted from MRI scans using various computational methods.

We have use scans obtained with a 1.5 T unit (Magnetom, Siemens). T_1-weighted fast low angle shot (FLASH) sequences with repetition time of 0.153 s, echo time of 0.005 s, flip angle of 70°, effective section thickness of 8 mm, matrix size of 128×256, field of view of 350 mm FOV, one acquisition and acquisition time of 22 s, was used (Kim et al., 1997). Gadopentetate dimeglumine (Magnevist, Schering) dose of 0.1 mmol/kg was injected as a rapid bolus injection. Postcontrast T_1-weighted FLASH scan was performed immediately at 1, 2, and 3 min after injection of gadolinium in all patients and thereafter. Postcontrast images were obtained after 5 min and 10 min in 22 and 47 hemangiomas respectively.

The FLASH images obtained from Kim et al. (1997) were then processed with ImageJ software (National Institutes of Health, Bethesda, MD, USA). The MRI signals were then extracted as functions pixel values across the whole image using this software and saved as .csv files. For quantitative assessment of the MR scans, we used a wolfram Mathematica computer program to reconstruct the images from the data extracted with ImageJ. The reconstructed images are Figs. 10.11, 10.12, and 10.13.

The MRI signal equation for FLASH sequence is given as (Buckley & Parker, 2005):

$$S = g\rho \frac{\sin\alpha \left(1 - \exp\left(-\frac{TR}{T_1}\right)\right)}{\left(1 - \exp\left(-\frac{TR}{T_1}\right)\cos\alpha\right)} \exp\left(-\frac{TE}{T_2^*}\right) \tag{10.24}$$

This equation is difficult to resolve at a flip angle is 70°, hence, we shall set $\alpha = 90°$ so that we can demonstrate how we hope to resolve this complicated problem. Equation (10.24) then becomes:

$$S = g\rho \left(1 - \exp\left(-\frac{TR}{T_1}\right)\right) \exp\left(-\frac{TE}{T_2^*}\right) \tag{10.25}$$

Assuming the ions of the contrast agents have no effect on ρ and that the TE is so short as to be able to neglect (Buckley & Parker, 2005) the influence of T_2^* (or changes in T_2^* during the time series), Eq. (10.25) becomes:

Fig. 10.11 Reconstructed (**a**) precontrast image (**b**) postcontrast image at 1 min (**c**) postcontrast image at 10 min, of small hemangioma in a 53 years old man

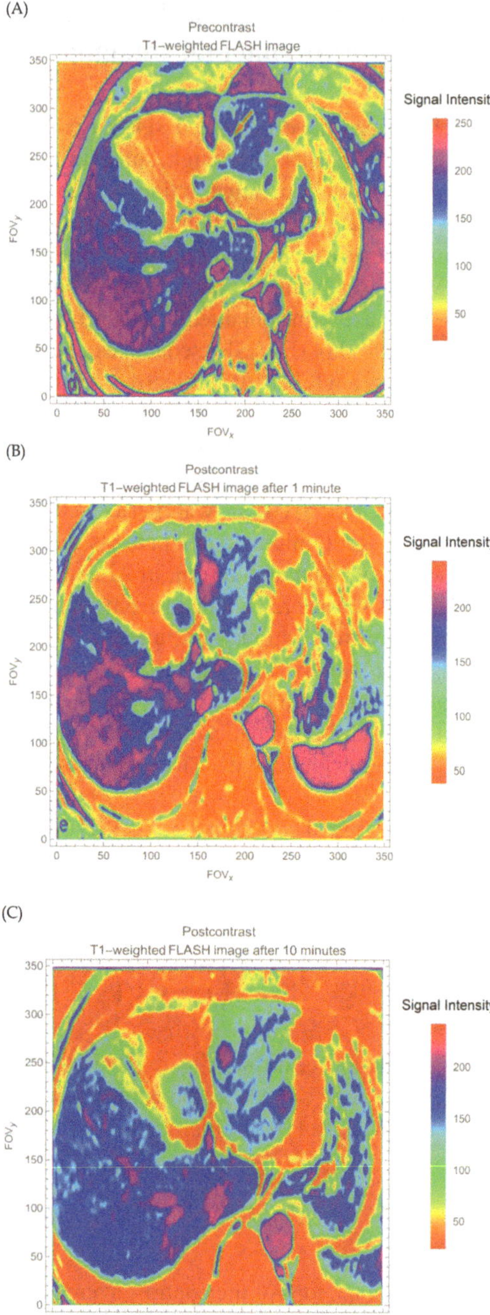

Fig. 10.12 Reconstructed (**a**) precontrast image (**b**) postcontrast image at 1 min (**c**) postcontrast image at 10 min, of medium-sized hemangioma in a 43 years old man

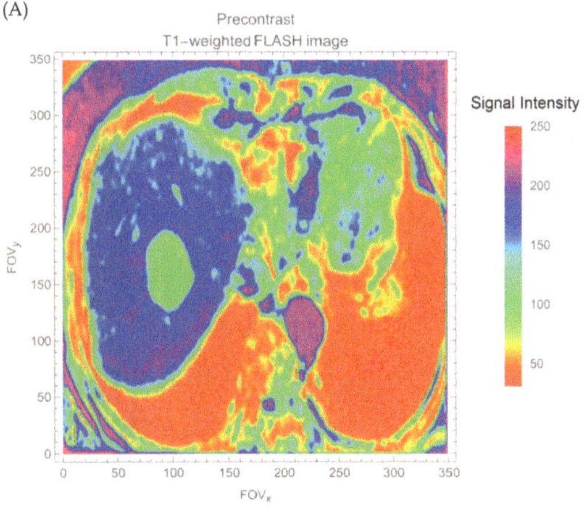

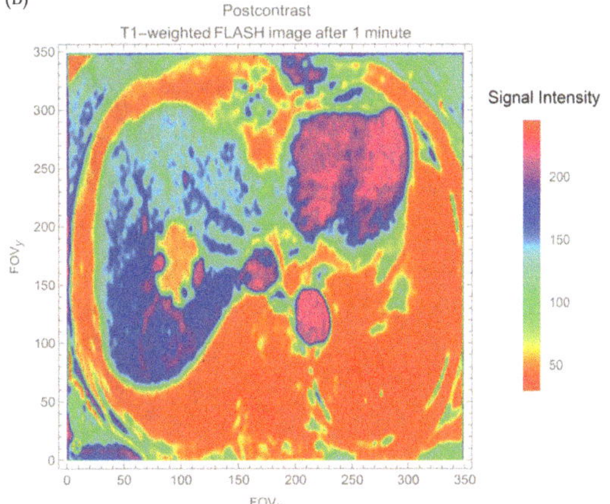

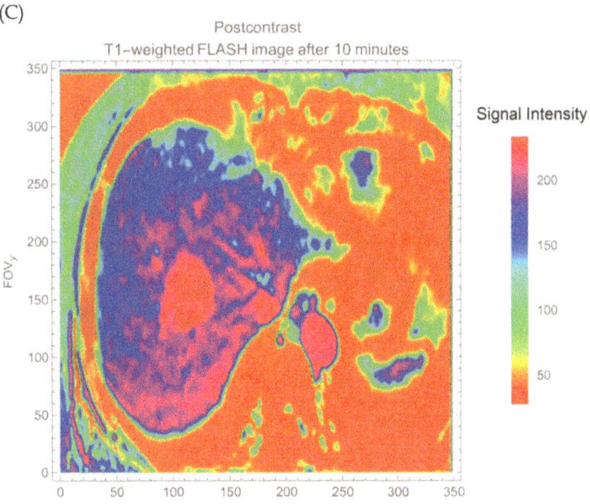

Fig. 10.13 Reconstructed (**a**) precontrast image (**b**) postcontrast image at 1 min (**c**) postcontrast image at 10 min, of large hemangioma in a 51 years old woman

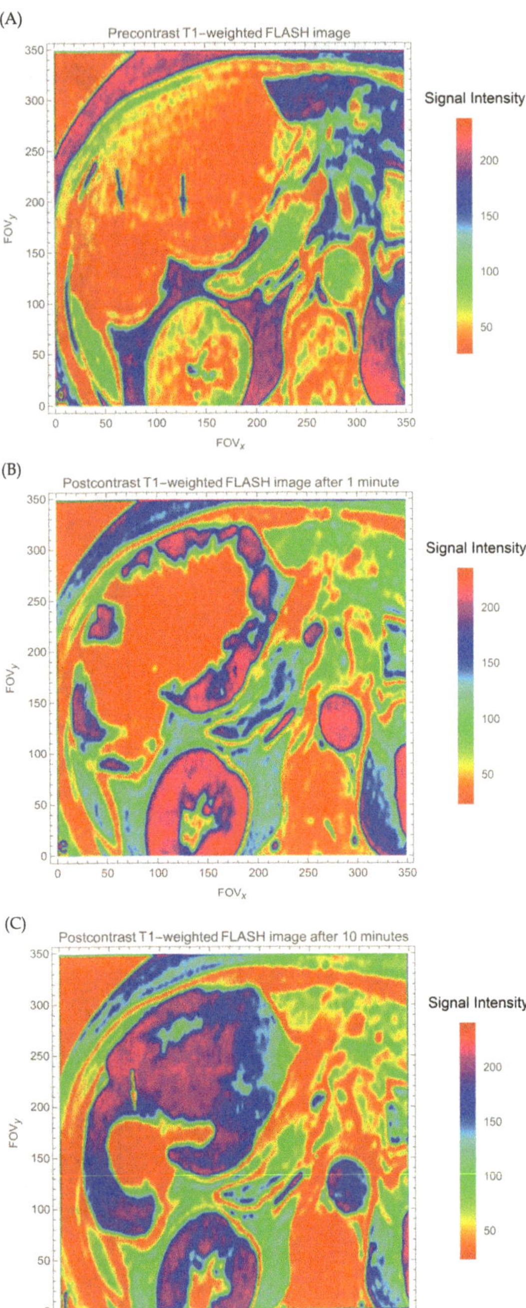

$$S = g\rho \left(1 - \exp\left(-\frac{TR}{T_1} \right) \right) \tag{10.26}$$

The pre-contrast signal may be expressed as:

$$S_0 = g\rho \left(1 - \exp\left(-\frac{TR}{T_{10}} \right) \right) \tag{10.27}$$

$$\frac{S}{S_0} = \frac{\left(1 - \exp\left(-\frac{TR}{T_1} \right) \right)}{\left(1 - \exp\left(-\frac{TR}{T_{10}} \right) \right)} = \frac{\left(1 - \exp\left(-\frac{TR}{T_{10}} - r_1(TR)C \right) \right)}{\left(1 - \exp\left(-\frac{TR}{T_{10}} \right) \right)} \tag{10.28}$$

$$\exp\left(-\frac{TR}{T_{10}} - r_1(TR)C \right) = 1 - \left(1 - \exp\left(-\frac{TR}{T_{10}} \right) \right) \frac{S}{S_0} \tag{10.29}$$

$$-\frac{TR}{T_{10}} - r_1(TR)C = \ln\left[1 - \left(\frac{S}{S_0} - \frac{S}{S_0} \exp\left(-\frac{TR}{T_{10}} \right) \right) \right] \tag{10.30}$$

$$C = -\frac{\frac{TR}{T_{10}} + \ln\left[1 - \frac{S}{S_0} + \frac{S}{S_0} \exp\left(-\frac{TR}{T_{10}} \right) \right]}{r_1(TR)} \tag{10.31}$$

The signals of Figs. 10.11a, 10.12a, and 10.13a can be represented by the precontrast signal S_0 while those of Figs. 10.11b, c, 10.12b, c, and 10.13b, c can be appropriately represented as S. Using FLASH sequence TR, r_1 in Table 10.1 and T_{10} from a related study (Farraher et al., 2006), the concentration distribution was computed for each of the postcontrast images. The computed concentration values of magnevist for the postcontrast images were also prepared in .csv format. Using a similar wolfram mathematica program, concentration maps were reconstructed from the extracted data and presented in Figs. 10.14, 10.15 and 10.16.

Figures 10.11, 10.12, and 10.13 show the importance of computational methods in improving quantitative imaging techniques. These images not only show the variations of the MR signal across the field of view but also the amount of MR signals at different locations in the liver. Figures 10.4, 10.5, and 10.6 are interesting because they demonstrate the use of computational approaches in making simple MRI contrast agents to be more responsive than they used to be without special agent's design. This implies that contrast agents usually leave trails of their dynamics on MRI signal but we just need to find ways of getting such information out of our MRI scans.

Fig. 10.14 Reconstructed (**a**) postcontrast concentration image at 1 min (**b**) postcontrast concentration image at 10 min image, of small hemangioma in a 53 years old man

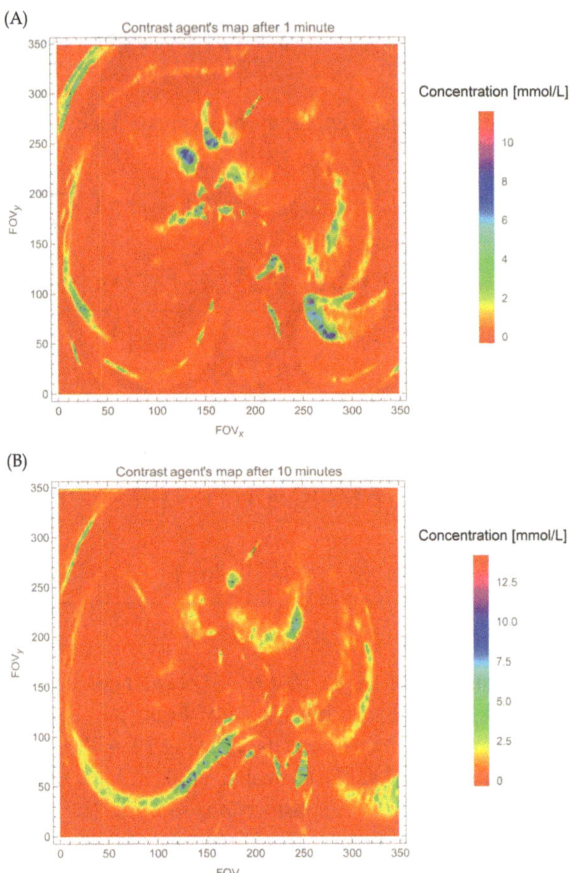

10.8   Discussion

In this section, we have made use of various computational approaches in the assessment of MRI contrast agents and demonstrated an example of linking some of the approaches to imaging experiments.

Figures 10.1 and 10.2 demonstrate concentration profiles according to the model Eq. (10.13) which is useful for assessment of blood pool agents. The concentration profile showed dependence on size of blood vessels as well as the speed of blood flow in these vessels. Concentration was found to increase with decrease in the vessel diameter provided that the blood flow speed is not too small. These profiles are quite useful in in-vitro laboratory studies of contrast agents towards understanding the influence physiological parameters on concentration variations. However, it is important to note that this model is based on the linear Eq. (10.12) which is the simplest case. It will be interesting to see how this model works in non-linear situations.

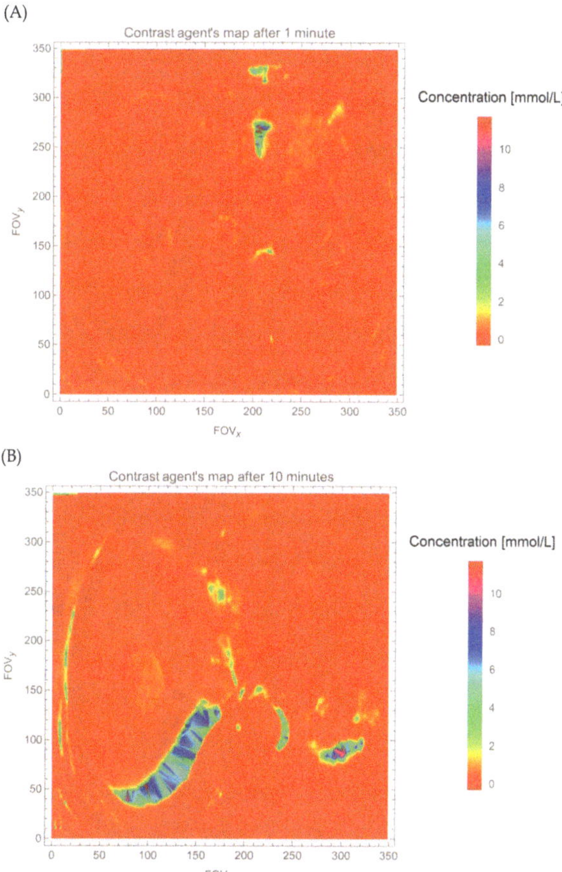

Fig. 10.15 Reconstructed (**a**) postcontrast concentration image at 1 min (**b**) postcontrast concentration image at 10 min image, of medium-sized hemangioma in a 43 years old man

Section 10.5 shows how time-independent NMR Bloch flow equation can be used comparative studies of contrast agent's NMR response via the transverse magnetization. The solution as given by the complicated Eq. (10.23) provided a means of studying the transverse magnetization from the selected contrast agents without making any assumption to make the problem tractable. However, wolfram Mathematica was able to resolve this equation and a simulation was built from this solution. The graphic user interface from this simulation program is given in Fig. 10.3. Using the data of Table 10.1, we obtained the profiles in Figs. 10.4, 10.5, and 10.6 and despite the closeness of some of the data in Table 10.1, unique transverse magnetization profiles were obtained. These profiles can potentially be used for laboratory assessment of contrast agent's NMR response if spatially dependent transverse magnetizations database is made available through extensive application of this simulation method on many contrast agents.

Section 10.6 presents a machine learning technique for screening large amount of contrast agents. Machine learning algorithms are well suited for analysing large data and the enormous amount of contrast agents that may need to be screened during

Fig. 10.16 Reconstructed (**a**) postcontrast concentration image at 1 min (**b**) postcontrast concentration image at 10 min image, of large hemangioma in a 53 years old woman

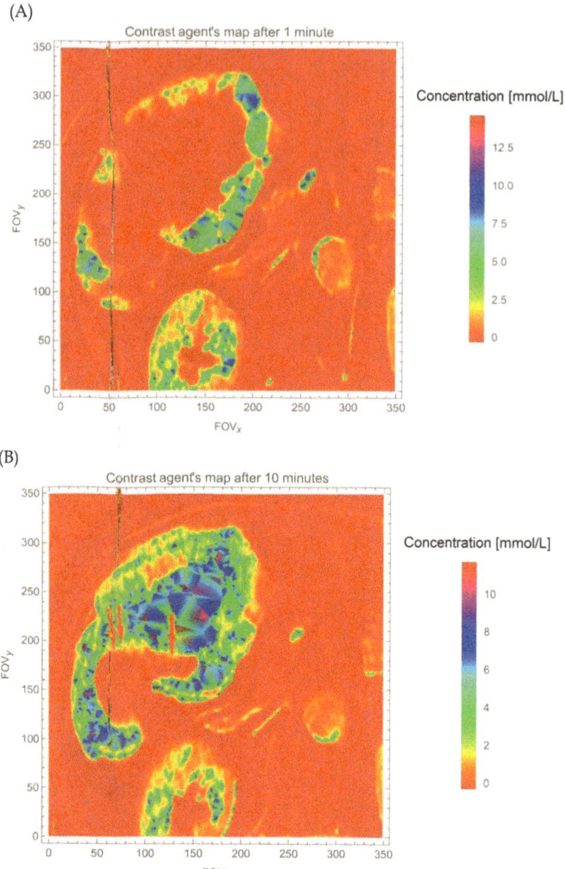

laboratory trials may benefit incredibly from machine learning methods. In place of a large experimental dataset, another algorithm was developed to generate 200 samples of synthetic data for each of the solvents under investigation. Figures 10.7 and 10.8 give the graphical representation of the dataset while Fig. 10.9 presents correlation matrices of r_1 and r_2 values of all the contrast agents; thus, showing which relaxivities are closely related to each other. XGBoost model was then used to train the dataset and accuracy of 100% was returned. Details of the performances of XGBClassifier are presented in Table 10.7 and Fig. 10.10. Despite the performances of the XGBoost model, it will be quite interesting to see how other machine learning models fare with the dataset generated in this section.

Figures 10.11, 10.12, and 10.13 shows MR images reconstructed from data extracted from grayscale images of three patients with different sizes of hepatic hemangioma. These figures show how computational methods can be used to make improvement to MRI scans and bring out quantitative variables to aid effective diagnosis. We used a simplified form of MRI signal equation in FLASH sequences to get an expression for the concentration of the contrast agents during scanning.

This expression was then used to compute the concentration across the postcontrast images for the three patients. The resulting data was used together with a wolfram Mathematica program to reconstruct concentration maps from the extracted data. These maps are presented in Figs. 10.14, 10.15, and 10.16. These figures imply that we can actually extract concentration dynamics of contrast media from MRI scans without too much complicated designs of these imaging media. However, we employed a simplified MRI signal equation via the FLASH imaging sequence. It will be interesting to investigate how these computational method works with sequences with less simple signal equation.

10.9 Conclusion

One of the challenges faced in this study is availability of suitable data for computational model testing and machine learning. Data science skills will play prominent roles in the future where dataset suitable for machine learning are properly curated from laboratory measurements. Secondly, we tried solving the signal equation from several other sequences to compute the concentration but the problem could not be solved even with advanced computational software. It will be in the interest of research of MRI contrast agents to see that MRI sequence gives the best condition for computing postcontrast concentration and the specific imaging parameters (especially, recovery time, echo time and flip angle) required for such scenario.

Chapter 11
General Conclusion

In this book, we have discussed practical ways that computational MRI can provide innovative data-driven solutions for the prediction, simplification and characterization of brain tumor as well as delineating their intrinsic mechanisms based on Bloch NMR flow equation.

Brain tissue diagnosis can be performed based on the availability of relaxometric data and simple computer program proficiency. These programs are easy to use, highly interactive and the data processing is fast and unambiguous. The results can motivate the use of data to produce data-driven predictions that may lead to the applications of artificial intelligence (AI) and machine learning in drug delivery and development. The applications may raise opportunities for the development of new diagnostics and treatments tools for tissue management specifically cancer care. Many challenges can be addressed in the development of consistent, suitable, safe, flexible and real-time healthcare systems based on the information presented in this book.

At any given time t, we can obtain detailed information about any tissue, if appropriate boundary conditions are applied. Radiofrequency (RF) controlled transverse magnetizations and signals can be obtained from the Bloch NMR flow equation that may form a significant component of the Internet of Things (IoT) concept as shown in Fig. 11.1.

Some of the computational results based on the Bloch NMR flow equation have already been applied for detailed analysis of brain tissues and neuro-oncology (Awojoyogbe & Dada, 2018; Dada et al., 2017; Awojoyogbe et al., 2016a, b).

Figure 11.1 presents workflow of radiomics in neuro-oncology and their relationship to the Bloch NMR flow equation and RF excitation: (I) After pre-processing steps and multimodal MR images are segmented (II) feature extraction with use of a variety of different techniques. (III) Machine learning methods are then trained on the features to generate models of underlying molecular markers and predict survival (Rudie et al., 2019). Deep learning models can be used for performing each of the described steps individually or in a more comprehensive fashion.

© The Author(s), under exclusive license to Springer Nature Switzerland AG 2021   347
M. O. Dada, B. O. Awojoyogbe, *Computational Molecular Magnetic Resonance Imaging for Neuro-oncology*, Biological and Medical Physics, Biomedical Engineering, https://doi.org/10.1007/978-3-030-76728-0_11

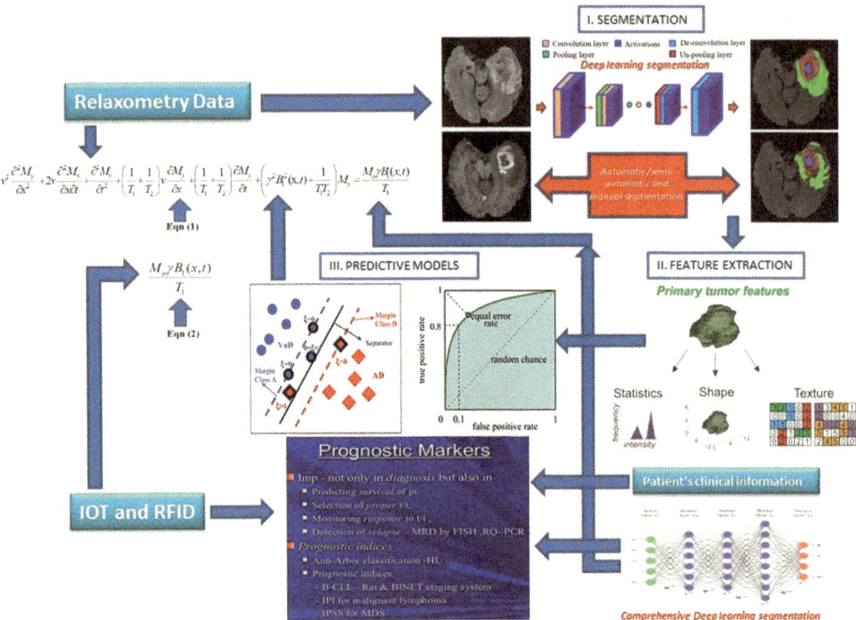

Fig. 11.1 Workflow of radiomics in neuro-oncology and their relationship to the Bloch NMR flow equation and RF excitation

An example of the applications of the book currently under investigation is the development of artificial intelligence algorithm for computer aided diagnosis of brain tumor (CADBRAT) using tensor flow, keras library and magnetic resonance images where two set of brain MR images are used (Confidence et al., 2021). In this study, the first dataset containing 253 images was used and were classified into normal brain tissue and abnormal (tumor) tissue while the second dataset contains five classes: Astrocytoma, Glioblastoma (Glioblastoma Multiforme), Oligodendroglioma, healthy tissue and unidentified tumor. Two less complex convolution neural network (CNN) networks (both containing 11 layers each) that can be trained comfortably without a GPU support were developed; AWSNeT1 and AWSNet2. AWSNeT1 was compiled using Keras while AWSNet2 was compiled using TFlearn python library. In contrast to other CNN networks, this network has max pooling layer after each convolution layer. The images of the first dataset were pre-processed (cropped, thresholded, eroded, brightened and dilated) to reduce the region of interest and noise. The pre-processed data was used to train AWSNeT1 and facilitate the brain tumor classification and prediction into two different classes of brain MR images. The training accuracy, validation accuracy and validation loss were calculated. The training accuracy was found to be 96.37% while validation accuracy of 96.00% and validation loss of 0.13395 were obtained. A Python library Flask was used to set up AWSNeT1 saved model functions which can be invoked via the web, making classification of tumor and non-tumor MR image diagnosis much easier.

The second set of data were serialized and fed into the AWSNet2 network for learning. The performance of the AWSNet2 was tested and compared with an intelligent method that is available in the literature. Classification accuracy of 99.95% and F1-score of 99.48% were obtained with the proposed AWSNet2 method. AWSNet2 structure is found to be an important tool that can be used in computerized brain tumor diagnosis.

An IOT based computer-aided brain tumor classifier (CADBRAT) system is being developed as an API application on the internet using IOT technologies that will make AWSNet1 and AWSNet2 available for use on the internet for personalized diagnosis.

The goal is to develop training software that can simulate MRI experiments and provide visual training tools to help understand MRI technology. The software would be a good way of starting to develop expertise and training that might provide support for the development, maintenance and operation of appropriate MRI devices. The software would be tailored to the users' educational experience without the need to understand complex mathematics, complex code, complex hardware or complex physics.

Bibliography

Abdel-Maksoud, E., Elmogy, M., & Al-Awadi, R. (2015). Brain tumor segmentation based on a hybrid clustering technique. *Egyptian Informatics Journal, 16*(1), 71–81.

Abragam, A. (1961). *The principles of nuclear magnetism.* Clarendon Press.

Abragam, A. (1996). Nuclear ferromagnetism and antiferromagnetism. In D. M. Grant & R. K. Harris (Eds.), *The encyclopaedia of nuclear magnetic resonance.* Wiley.

Abramowitz, M., & Stegun, I. A. (Eds.). (1964). *Handbook of mathematical functions with formulas, graphs, and mathematical tables* (Vol. 55). US Government Printing Office.

Acosta-Martínez, J., López-Haldón, J. E., Gutiérrez-Carretero, E., Díaz-Carrasco, I., Hajam, T. S., & Ordóñez Fernández, A. (2013). Radial and circumferential strain as markers of fibrosis in an experimental model of myocardial infarction. *Revista Española de Cardiología, 66*(6), 508–509.

Aggarwal, R., & Kaur, A. (2014). Efficient analysis of brain tumor detection and identification using different algorithms. *International Journal of Engineering Sciences & Research Technology, 3*(65), 606–610.

Ai, L., & Vafai, K. (2006). A coupling model for macromolecule transport in a stenosed arterial wall. *International Journal of Heat and Mass Transfer, 49*(9), 1568–1591.

Aime, S., Baranyai, Z., Gianolio, E., & Terreno, E. (2007). Paramagnetic contrast agents. In M. J. M. Modo & J. W. M. Bulte (Eds.), *Molecular and cellular MR imaging* (pp. 1–9). CRC Press - Taylor & Francis Group.

Aime, S., Botta, M., & Terreno, E. (2005). Gd (III) – Based contrast agents for MRI. *Advances in Inorganic Chemistry, 57*, 173–227.

Aime, S., Barge, A., Delli Castelli, D., Fedeli, F., Mortillaro, A., Nielsen, F. U., & Terreno, E. (2002). Paramagnetic lanthanide (III) complexes as pH-sensitive chemical exchange saturation transfer (CEST) contrast agents for MRI applications. *Magnetic Resonance in Medicine, 47*(4), 639–648.

Aime, S., Botta, M., Fasano, M., Crich, S. G., & Terreno, E. (1996). Gd (III) complexes as contrast agents for magnetic resonance imaging: a proton relaxation enhancement study of the interaction with human serum albumin. *JBIC Journal of Biological Inorganic Chemistry, 1*(4), 312–319.

Akram, M. U., & Usman, A. (2011, July). Computer aided system for brain tumor detection and segmentation. In *International Conference on Computer Networks and Information Technology* (pp. 299–302). IEEE.

© The Author(s), under exclusive license to Springer Nature Switzerland AG 2021   351

M. O. Dada, B. O. Awojoyogbe, *Computational Molecular Magnetic Resonance Imaging for Neuro-oncology*, Biological and Medical Physics, Biomedical Engineering, https://doi.org/10.1007/978-3-030-76728-0

Allen, T. M., & Cullis, P. R. (2004). Drug delivery systems: Entering the mainstream. *Science, 303* (5665), 1818–1822.

Alsop, D. C., & Detre, J. A. (1998). Multisection cerebral blood flow MR imaging with continuous arterial spin labeling. *Radiology, 208*(2), 410–416.

Ambily, P. K., James, S. P., & Mohan, R. R. (2015). Brain tumor detection using image fusion and neural network. *International Journal of Engineering Research and General Science, 3*(2), 1383–1388.

Anderson, A. W., & Gore, J. C. (1994). Analysis and correction of motion artifacts in diffusion weighted imaging. *Magnetic Resonance in Medicine, 32*(3), 379–387.

Andrews, D. W., Scott, C. B., Sperduto, P. W., Flanders, A. E., Gaspar, L. E., Schell, M. C., Werner-Wasik, M., Demas, W., Ryu, J., Bahary, J. P., & Souhami, L. (2004). Whole brain radiation therapy with or without stereotactic radiosurgery boost for patients with one to three brain metastases: Phase III results of the RTOG 9508 randomised trial. *The Lancet, 363*(9422), 1665–1672.

Andrews, L. C. (1992). *Special functions of mathematics for engineers* (2nd ed.). McGraw-Hill.

Angoth, V., Dwith, C. Y. N., & Singh, A. (2013). A novel wavelet based image fusion for brain tumor detection. *International Journal of Computer Vision and Signal Processing, 2*(1), 1–7.

Anitha, V., & Murugavalli, S. (2016). Brain tumour classification using two-tier classifier with adaptive segmentation technique. *IET Computer Vision, 10*(1), 9–17.

Antoine, X., Bao, W., & Besse, C. (2013). Computational methods for the dynamics of the nonlinear Schrödinger/Gross–Pitaevskii equations. *Computer Physics Communications, 184* (12), 2621–2633.

Aoki, I., Naruse, S., & Tanaka, C. (2004). Manganese-enhanced magnetic resonance imaging (MEMRI) of brain activity and applications to early detection of brain ischemia. *NMR in Biomedicine, 17*(8), 569–580.

Archana, S., & Elangovan, K. (2014). Survey of classification techniques in data mining. *International Journal of Computer Science and Mobile Applications, 2*(2), 65–71.

Artemov, D. (2003). Molecular magnetic resonance imaging with targeted contrast agents. *Journal of Cellular Biochemistry, 90*, 518–524.

Aruldhas, G. (2009). *Quantum mechanics*. PHI learning private limited.

Atalar, E., & McVeigh, E. R. (1994). Optimization of tag thickness for measuring position with magnetic resonance imaging. *IEEE Transactions on Medical Imaging, 13*(1), 152–160.

Auchter, R. M., Lamond, J. P., Alexander, E., III, Buatti, J. M., Chappell, R., Friedman, W. A., Kinsella, T. J., Levin, A. B., Noyes, W. R., Schultz, C. J., & Loeffler, J. S. (1996). A multiinstitutional outcome and prognostic factor analysis of radiosurgery for resectable single brain metastasis. *International Journal of Radiation Oncology, Biology, Physics, 35*(1), 27–35.

Awojoyogbe, B. O., & Dada, M. O. (2018). Computational design of an RF controlled theranostic model for evaluation of tissue biothermal response. *Journal of Medical and Biological Engineering, 38*(6), 993–1013.

Awojoyogbe, B. O., Dada, M. O., Onwu, S. O., Ige, T. A., & Akinwande, N. I. (2016a). Computational diffusion magnetic resonance imaging based on time-dependent Bloch NMR flow equation and Bessel functions. *Journal of Medical Systems, 40*(4), 106.

Awojoyogbe, O., & Dada, M. (2011a). Basis for the application of analytical models of the Bloch NMR flow equations for functional magnetic resonance imaging (fMRI): A review. *Recent Patents on Medical Imaging, 1*(1), 33–67.

Awojoyogbe, O. B. (1997). *Studies on the application of nuclear magnetic resonance (NMR, MRI) in the estimation of human blood flow rates*. Unpublished doctoral dissertation. Federal University of Technology, Minna, Nigeria.

Awojoyogbe, O. B. (2002). A mathematical model of the Bloch NMR equations for quantitative analysis of blood flow in blood vessels with changing cross-section—I. *Physica A: Statistical Mechanics and its Applications, 303*(1–2), 163–175.

Awojoyogbe, O. B. (2003). A mathematical model of Bloch NMR equations for quantitative analysis of blood flow in blood vessels of changing cross-section—PART II. *Physica A: Statistical Mechanics and its Applications, 323*, 534–550.

Awojoyogbe, O. B. (2004). Analytical solution of the time-dependent Bloch NMR flow equations: A translational mechanical analysis. *Physica A: Statistical Mechanics and its Applications, 339* (3–4), 437–460.

Awojoyogbe, O. B. (2007). A quantum mechanical model of the Bloch NMR flow equations for electron dynamics in fluids at the molecular level. *Physica Scripta, 75*(6), 788–794.

Awojoyogbe, O. B., & Boubaker, K. (2009). A solution to Bloch NMR flow equations for the analysis of hemodynamic functions of blood flow system using m-Boubaker polynomials. *Current Applied Physics, 9*(1), 278–283.

Awojoyogbe, O. B., & Dada, M. (2011b). The dynamics of NMR - Diffusion equation for the analysis of hemodynamic and metabolic changes in biological tissue. In E. T. Berg (Ed.), *Fluid transport: Theory, dynamics and applications* (pp. 183–217). Nova Science Publication.

Awojoyogbe, O. B., Dada, M., Faromika, O. P., Moses, O. F., & Fuwape, I. A. (2009). Polynomial solutions of Bloch NMR flow equations for classical and quantum mechanical analysis of fluid flow in porous media. *Open Magnetic Resonance Journal, 2*, 46–56.

Awojoyogbe, O. B., Dada, M., & Folorunsho, O. M. (2008). Mathematical analysis of adiabatic model of Bloch NMR flow equations for functional magnetic resonance imaging. In *Proceedings of the first International Seminar on Theoretical Physics & National Development* (pp. 177–201).

Awojoyogbe, O. B., & Dada, O. M. (2013). Mathematical design of a magnetic resonance imaging sequence based on Bloch NMR flow equations and Bessel functions. *Chinese Journal of Magnetic Resonance Imaging, 4*(5).

Awojoyogbe, O. B., Dada, O. M., & Faromika, O. P. (2016b). Development of magnetic resonance imaging method for computational neuro-oncology. *Journal of Neurology & Neurophysiology, 7*(Suppl), 4.

Awojoyogbe, O. B., Dada, O. M., Faromika, O. P., & Dada, O. E. (2011a). Mathematical concept of the Bloch flow equations for general magnetic resonance imaging: A review. *Concepts in Magnetic Resonance Part A, 38A*(3), 85–101.

Awojoyogbe, O. B., Faromika, O. P., Dada, M., Boubaker, K., & Ojambati, O. S. (2011b). Mathematical models of real geometrical factors in restricted blood vessels for the analysis of CAD (coronary artery diseases) using Legendre, Boubaker and Bessel polynomials. *Journal of Medical Systems, 35*(6), 1513–1520.

Awojoyogbe, O. B., Faromika, O. P., Folorunsho, O. M., Dada, M., Fuwape, I. A., & Boubaker, K. (2010). Mathematical model of the Bloch NMR flow equations for the analysis of fluid flow in restricted geometries using the Boubaker polynomials expansion scheme. *Current Applied Physics, 10*, 289–293.

Axel, L., & Dougherty, L. (1989a). Heart wall motion: Improved method of spatial modulation of magnetisation for MR imaging. *Radiology, 172*(2), 349–350.

Axel, L., & Dougherty, L. (1989b). MR imaging of motion with spatial modulation of magnetisation. *Radiology, 171*(3), 841–845.

Ayala-Peacock, D. N., Peiffer, A. M., Lucas, J. T., Isom, S., Kuremsky, J. G., Urbanic, J. J., Bourland, J. D., Laxton, A. W., Tatter, S. B., Shaw, E. G., & Chan, M. D. (2014). A nomogram for predicting distant brain failure in patients treated with gamma knife stereotactic radiosurgery without whole brain radiotherapy. *Neuro-Oncology, 16*(9), 1283–1288.

Azhari, E. E. M., Hatta, M. M., Htike, Z. Z., & Win, S. L. (2014). Glioblastoma multiforme identification from medical imaging using computer vision. *International Journal of Soft Computing, Mathematics and Control, 3*(2), 1–12.

Badve, C., Yu, A., Dastmalchian, S., Rogers, M., Ma, D., Jiang, Y., Margevicius, S., Borgelt, B., Gelber, R., Kramer, S., Brady, L. W., Chang, C. H., Davis, L. W., Perez, C. A., & Hendrickson, F. R. (1980). The palliation of brain metastases: Final results of the first two studies by the

Radiation Therapy Oncology Group. *International Journal of Radiation Oncology, Biology, Physics, 6*(1), 1–9.

Badve, C., Yu, A., Dastmalchian, S., Rogers, M., Ma, D., Jiang, Y., Margevicius, S., Pahwa, S., Lu, Z., Schluchter, M., & Sunshine, J. (2017). MR fingerprinting of adult brain tumors: Initial experience. *American Journal of Neuroradiology, 38*(3), 492–499.

Baer, T., Gore, J. C., Boyce, S., & Nye, P. W. (1987). Application of MRI to the analysis of speech production. *Magnetic Resonance Imaging, 5*(1), 1–7.

Bailey, W. J., & Ulrich, R. (2004). Molecular profiling approaches for identifying novel biomarkers. *Expert Opinion in Drug Safety, 3*, 137–151.

Balamurugan, R., & Saranya, S. (2014). Enhanced nonlocal mean filter for MRI denoising with Rician noise. *International Journal of Emerging Technology and Advanced Engineering, 4*(5), 256–261.

Balasubramanian, G., Chan, I. Y., Kolesov, R., Al-Hmoud, M., Tisler, J., Shin, C., Kim, C., Wojcik, A., Hemmer, P. R., Krueger, A., Hanke, T., Leitenstorfer, A., Bratschitsch, R., Jelezko, F., & Wrachtrup, J. (2008). Nanoscale imaging magnetometry with diamond spins under ambient conditions. *Nature, 455*(7213), 648–651.

Bammer, R., Skare, S., Newbould, R., Liu, C., Thijs, V., Ropele, S., Clayton, D. B., Krueger, G., Moseley, M. E., & Glover, G. H. (2005). Foundations of advanced magnetic resonance imaging. *NeuroRx, 2*(2), 167–196.

Bao, W. (2007). The nonlinear Schrödinger equation and applications in Bose-Einstein condensation and plasma physics. *Dynamics in Models of Coarsening, Coagulation, Condensation and Quantization, 9*, 141–239.

Bao, W., Tang, Q., & Xu, Z. (2013). Numerical methods and comparison for computing dark and bright solitons in the nonlinear Schrödinger equation. *Journal of Computational Physics, 235*, 423–445.

Barnholtz-Sloan, J. S., Sloan, A. E., Davis, F. G., Vigneau, F. D., Lai, P., & Sawaya, R. E. (2004). Incidence proportions of brain metastases in patients diagnosed (1973 to 2001) in the Metropolitan Detroit Cancer Surveillance System. *Journal of Clinical Oncology, 22*(14), 2865–2872.

Barrie, P. J. (1995). NMR applications to porous solids. *Annual Reports on NMR Spectroscopy, 30*, 37–96.

Basser, P. J. (2008). Diffusion and diffusion tensor MR imaging: Fundamentals. In *Magnetic Resonance Imaging of the Brain and Spine* (pp. 1752–1767).

Basser, P. J., & Jones, D. K. (2002). Diffusion-tensor MRI: Theory, experimental design and data analysis – A technical review. *NMR in Biomedicine, 15*, 456–467.

Basser, P. J., Mattiello, J., & LeBihan, D. (1994). Estimation of the effective self-diffusion tensor from the NMR spin echo. *Journal of Magnetic Resonance. Series B, 103*(3), 247–254.

Batchelor, G. K. (1967). *An introduction to fluid dynamics.* Cambridge University Press.

Bax, A. (1981). *Two-dimensional nuclear magnetic resonance in liquids.* (Doctoral dissertation, TU Delft, Delft University of Technology).

Becherer, A., Karanikas, G., Szabó, M., Zettinig, G., Asenbaum, S., Marosi, C., Henk, C., Wunderbaldinger, P., Czech, T., Wadsak, W., & Kletter, K. (2003). Brain tumour imaging with PET: A comparison between [18F] fluorodopa and [11C] methionine. *European Journal of Nuclear Medicine and Molecular Imaging, 30*(11), 1561–1567.

Beer, A. J., & Schwaiger, M. (2011). PET imaging of $\alpha v \beta 3$ expression in cancer patients. In K. Shah (Ed.), *Molecular imaging: Methods and protocols* (pp. 183–200). Humana Press.

Belliveau, J. W., Kennedy, D. N., McKinstry, R. C., Buchbinder, B. R., Weisskoff, R. M., Cohen, M. S., Vevea, J. M., Brady, T. J., & Rosen, B. R. (1991). Functional mapping of the human visual cortex by magnetic resonance imaging. *Science, 254*(5032), 716–719.

Benoit, M. (2010). Force spectroscopy on cells. In K. D. Sattler (Ed.), *Handbook of nanophysics: Nanomedicine and nanorobotics.* CRC Press.

Berg-Sorenson, K., Castin, Y., Bonderup, E., & Molmer, K. (1992). Momentum diffusion of atoms moving in laser fields. *Journal of Physics B: Atomic, Molecular and Optical Physics, 25*(20), 4195.

Bernasconi, A., Bernasconi, N., Caramanos, Z., Reutens, D. C., Andermann, F., Dubeau, F., Tampieri, D., Pike, B. G., & Arnold, D. L. (2000). T2 relaxometry can lateralize mesial temporal lobe epilepsy in patients with normal MRI. *Neuroimage, 12*(6), 739–746.

Berry, C. C. (2009). Progress in functionalisation of magnetic nanoparticles for applications in biomedicine. *Journal of Physics D: Applied Physics, 42*(22), 224003.

Bhosale, P., Lalge, P., Dhandekar, A., Gaykar, P., & Pate, P. (2017). Brain tumor automated detection and segmentation. *International Journal of Innovative Research in Science, Engineering and Technology, 1*(2), 1–8.

Bjornerud, A., & Johansson, L. (2004). The utility of superparamagnetic contrast agents in MRI: Theoretical consideration and applications in the cardiovascular system. *NMR in Biomedicine, 17*, 465–477.

Blanchet, K. D. (2010). Redefining personalized medicine in the postgenomic era: Developing bladder cancer therapeutics with proteomics. *BJU International, 105*(2), i–iii.

Blinc, R., & Zeks, B. (1974). *Soft modes in ferroelectrics and antiferroelectrics*. North Holland.

Block, K. T., Uecker, M., & Frahm, J. (2009). Model-based iterative reconstruction for radial fast spin-echo MRI. *IEEE Transactions on Medical Imaging, 28*(11), 1759–1769.

Boisseau, P., & Loubaton, B. (2011). Nanomedicine, nanotechnology in medicine. *Comptes Rendus Physique, 12*(7), 620–636.

Bolger, A. F., Heiberg, E., Karlsson, M., Wigstrom, L., Engvall, J., Sigfridsson, A., Ebbers, T., Kvitting, J. E., Carlhall, C. J., & Wranne, B. (2007). Transit of blood flow through the human left ventricle mapped by cardiovascular magnetic resonance. *Journal of Cardiovascular Magnetic Resonance, 9*, 741–747.

Bonnemain, B. (1998). Superparamagnetic agents in magnetic resonance imaging: Physicochemical characteristics and clinical applications. A review. *Journal of Drug Targeting, 6*, 167–174.

Borgonovi, F., Gorshkov, V. N., & Tsifrinovich, V. I. I. (2006). *Magnetic resonance force microscopy and a single-spin measurement*. World Scientific.

Borole, V. Y., Nimbhore, S. S., & Kawthekar, D. S. S. (2015). Image processing techniques for brain tumor detection: A review. *International Journal of Emerging Trends & Technology in Computer Science, 4*(5), 2.

Botnar, R., Scheidegger, M. B., & Boesiger, P. (1996). Quantification of blood flow patterns in human vessels by magnetic resonance imaging. *Technology and Health Care, 4*, 97–112.

Bottomley, P. A., Foster, T. H., Argersinger, R. E., & Pfeifer, L. M. (1984). A review of normal tissue NMR relaxation times and relaxation mechanisms from 1 to 100 MHz: Dependence on tissue type, NMR frequency, temperature, species, excision, and age. *Medical Physics, 11*, 425–448.

Bracewell, R. N. (1986). *The Fourier transform and its applications* (2nd ed., pp. 40–42). McGraw-Hill Publishing Company.

Bransden, B. H., & Joachain, C. J. (2000). *Quantum mechanics* (2nd ed.). Pearson Education.

Brix, G., Kolem, H., Nitz, W. R., Bock, M., Huppertz, A., Zech, C. J., & Dietrich, O. (2008). Basics of magnetic resonance imaging and magnetic resonance spectroscopy. In *Magnetic resonance tomography* (pp. 3–167). Springer.

Brix, G., Seebass, M., Hellwig, G., & Griebel, J. (2002). Estimation of heat transfer and temperature rise in partial-body regions during MR procedures: An analytical approach with respect to safety considerations. *Magnetic Resonance Imaging, 20*, 65–76.

Bryant, D. J., Payne, J. A., Firmin, D. N., & Longmore, D. B. (1984). Measurement of flow with NMR imaging using a gradient pulse and phase difference technique. *Journal of Computer Assisted Tomography, 8*(4), 588–593.

Buckley, D. L., & Parker, G. J. (2005). Measuring contrast agent concentration in T 1-weighted dynamic contrast-enhanced MRI. In *Dynamic contrast-enhanced magnetic resonance imaging in oncology* (pp. 69–79). Springer.

Buckner, J. C., Brown, P. D., O'Neill, B. P., Meyer, F. B., Wetmore, C. J., & Uhm, J. H. (2007, October). Central nervous system tumors. *Mayo Clinic Proceedings, 82*(10), 1271–1286.

Bulte, J. W., Brooks, R. A., Moskowitz, B. M., Bryant, L. H., & Frank, J. A. (1999). Relaxometry and magnetometry of the MR contrast agent MION-46L. *Magnetic Resonance in Medicine, 42*, 379–384.

Bulte, J. W., Vymazal, J., Brooks, R. A., Pierpaoli, C., & Frank, J. A. (1993). Frequency dependence of MR relaxation times II. Iron oxides. *Journal of Magnetic Resonance Imaging, 3*, 641–648.

Bulte, J. W. M., & Kraitchman, D. L. (2004). Iron oxide MR contrast agents for molecular and cellular imaging. *NMR in Biomedicine, 17*, 484–499.

Burkhardt, C. E., & Leventhal, J. J. (2008). *Foundations of quantum physics*. Springer-Verlag.

Cabal, C., Darias, D., González, E., & Musacchio, A. (2013). Theranostics and molecular imaging: New concepts and technologies for drug development. *Biotecnologia Aplicada, 30*(3), 172.

Cacheris, W. P., Quay, S. C., & Rocklage, S. M. (1990). The relationship between thermodynamics and the toxicity of gadolinium complexes. *Magnetic Resonance Imaging, 8*(4), 467–481.

Cairncross, J. G., Kim, J. H., & Posner, J. B. (1980). Radiation therapy for brain metastases. *Annals of Neurology: Official Journal of the American Neurological Association and the Child Neurology Society, 7*(6), 529–541.

Calamante, F., Gadian, D. G., & Connelly, A. (2002). Quantification of perfusion using bolus tracking magnetic resonance imaging in stroke: Assumptions, limitations, and potential implications for clinical use. *Stroke, 33*(4), 1146–1151.

Callaghan, P. T. (1994). *Principles of nuclear magnetic resonance microscopy*. Oxford United Kingdom: Oxford University Press.

Calvo, B. F., & Semelka, R. C. (1999). Beyond anatomy: MR imaging as a molecular diagnostic tool. *Surgical Oncology Clinics of North America, 8*(1), 171–183.

Camazine, S. (2003). *Self-organization in biological systems*. Princeton University Press.

Caravan, P. (2007). Physicochemical principles of MR contrast agents. In M. J. M. Modo & J. W. M. Bulte (Eds.), *Molecular and cellular MR imaging* (pp. 1–9). CRC Press - Taylor & Francis Group.

Caravan, P., Comuzzi, C., Crooks, W., McMurry, T. J., Choppin, G. R., & Woulfe, S. R. (2001). Thermodynamic stability and kinetic inertness of MS-325, a new blood pool agent for magnetic resonance imaging. *Inorganic Chemistry, 40*, 2170–2176.

Caravan, P., Ellison, J. J., McMurry, T. J., & Lauffer, R. B. (1999). Gadolinium (III) chelates as MRI contrast agents: Structure, dynamics, and applications. *Chemical Reviews, 99*, 2293–2352.

Castelli, D. D., Gianolio, E., & Aime, S. (2011). MRI contrast agents: State of the art and new trends. In E. Alessio (Ed.), *Bioinorganic medicinal chemistry* (pp. 223–251). Wiley-VCH Verlag GmbH & Co. KGaA.

Cha, S., Lupo, J. M., Chen, M. H., Lamborn, K. R., McDermott, M. W., Berger, M. S., Nelson, S. J., & Dillon, W. P. (2007). Differentiation of glioblastoma multiforme and single brain metastasis by peak height and percentage of signal intensity recovery derived from dynamic susceptibility-weighted contrast-enhanced perfusion MR imaging. *American Journal of Neuroradiology, 28*(6), 1078–1084.

Chandra, G. R., & Rao, K. R. H. (2016). Tumor detection in brain using genetic algorithm. *Procedia Computer Science, 79*, 449–457.

Chang, D. C., Hazlewood, C. F., Nichols, B. L., & Rorschach, H. E. (1972). Spin echo studies on cellular water. *Nature, 235*(5334), 170–171.

Chang, D. C., Rorschach, H. E., Nichols, B. L., & Hazlewood, C. F. (1973). Implications of diffusion coefficient measurements for the structure of cellular water. *Annals of the New York Academy of Sciences, 204*, 434–443.

Chang, S., & Minogin, V. (2002). Density-matrix approach to dynamics of multilevel atoms in laser fields. *Physics Reports, 365*(2), 65–143.

Charles, R. N., Norman, L. S., Robert, J. D., & David, A. R. (2005). *The human nervous system: Structure and function* (6th ed.). Humana Press. www.teamrads.com/Cases/Neuroanatomy/MRI

Chatham, J. C., & Blackband, S. J. (2001). Nuclear magnetic resonance spectroscopy and imaging in animal research. *ILAR Journal, 42*(3), 189–208.

Chen, C., & Xu, L. X. (2002). Tissue temperature oscillations in an isolated pig kidney during surface heating. *Annals of Biomedical Engineering, 30*, 1162–1171.

Chen, M., Li, Y., Yang, M., Chen, X., Chen, Y., Yang, F., Lu, S., Yao, S., Zhou, T., Liu, J., & Wu, J. Y. (2016). A new method for quantifying mitochondrial axonal transport. *Protein & Cell, 7* (11), 804–819.

Chen, S. H., Broccio, M., Liu, Y., Fratini, E., & Baglioni, P. (2007). The two-Yukawa model and its applications: The cases of charged proteins and copolymer micellar solutions. *Applied Crystallography, 40*(s1), s321–s326.

Chen, Y. (2013). Macroscopic quantum mechanics: Theory and experimental concepts of optomechanics. *Journal of Physics B: Atomic, Molecular and Optical Physics, 46*(10), 104001.

Chiagozie, O. G., & Nwaji, O. G. (2012). Radio frequency identification (RFID) based attendance system with automatic door unit. *Academic Research International, 2*(2), 168.

Chopra, R., Wachsmuth, J., Burtnyk, M., Haider, M. A., & Bronskill, M. J. (2006). Analysis of factors important for transurethral ultrasound prostate heating using MR temperature feedback. *Physics in Medicine and Biology, 51*, 827–844.

Chu, C. H., Sarangadharan, I., Regmi, A., Chen, Y. W., Hsu, C. P., Chang, W. H., Lee, G. Y., Chyi, J. I., Chen, C. C., Shiesh, S. C., & Lee, G. B. (2017). Beyond the Debye length in high ionic strength solution: Direct protein detection with field-effect transistors (FETs) in human serum. *Scientific Reports, 7*(1), 1–15.

Chu, W. C. W., Lam, W. W. M., & Metreweli, C. (2000). Incidence of adverse events after IV injection of MR contrast agents in a Chinese population: A comparison between gadopentetate and gadodiamide. *Acta Radiologica, 41*(6), 662–666.

Ciuciu, P., Poline, J. B., Marrelec, G., Idier, J., Pallier, C., & Benali, H. (2003). Unsupervised robust nonparametric estimation of the hemodynamic response function for any fMRI experiment. *IEEE Transactions on Medical Imaging, 22*(10), 1235–1251.

Clare, S. (1997). *Functional magnetic resonance imaging: Methods and applications.* Unpublished doctoral dissertation. The University of Nottingham, Nottingham, United Kingdom.

Clarke, J. L., & Chang, S. (2009). Pseudoprogression and pseudoresponse: Challenges in brain tumor imaging. *Current Neurology and Neuroscience Reports, 9*(3), 241–246.

Clement, O., Siauve, N., Lewin, M., de Kerviler, E., Cuenod, C. A., & Frija, G. (1998). Contrast agents in magnetic resonance imaging of the liver: Present and future. *Biomedicine and Pharmacotherapy, 52*, 51–58.

Cline, H. E., Dumoulin, C. L., Hart, H. R., Lorensen, W. E., & Ludke, S. (1987). 3D reconstruction of the brain from magnetic resonance images using a connectivity algorithm. *Magnetic Resonance Imaging, 5*(5), 345–352.

Cohen, B., Dafni, H., Meir, G., & Neeman, M. (2004). Ferritin as novel MR-reporter for molecular imaging of gene expression. *Proceedings of the International Society of Magnetic Resonance in Medicine, 11*, 1707.

Cohen, M. S., & Bookheimer, S. Y. (1994). Localization of brain function using magnetic resonance imaging. *Trends in Neurosciences, 17*(7), 268–277.

Cohen-Tannoudji, C., Diu, B., & Laloe, F. (1977). *Quantum mechanics* (Vol. 1). Wiley.

Cohen-Tannoudji, C., Dupont-Roc, J., & Grynberg, G. (1998). *Atom-photon interactions: Basic processes and applications* (2nd ed.). Wiley-VCH Verlag GmbH and Co..

Conti, P. S. (1995). Introduction to imaging brain tumor metabolism with positron emission tomography (PET). *Cancer Investigation, 13*, 244–259.

Cooper, R. L., Chang, D. B., Young, A. C., Martin, C. J., & Ancker-Johnson, B. (1974). Restricted diffusion in biophysical systems: Experiment. *Biophysical Journal, 14*(3), 161.

Cowan, B. P. (1997). *Nuclear magnetic resonance and relaxation* (1st ed.). Cambridge University Press.

Cristini, V., Frieboes, H. B., Li, X., Lowengrub, J. S., Macklin, P., Sanga, S., Wise, S. M., & Zheng, X. (2008). Nonlinear modelling and simulation of tumor growth. In N. Bellomo, M. Caplain, & E. De Angelis (Eds.), *Selected topics in cancer modelling: Genesis, evolution, immune competition and simulation* (pp. 1–69). Birkhäuser.

Cui, B., Zhang, S., Chen, L., Yu, J., Widhopf, G. F., Fecteau, J. F., Rassenti, L. Z., & Kipps, T. J. (2013). Targeting ROR1 inhibits epithelial–mesenchymal transition and metastasis. *Cancer Research, 73*(12), 3649–3660.

Cuyt, A. A., Petersen, V., Verdonk, B., Waadeland, H., & Jones, W. B. (2008). *Handbook of continued fractions for special functions*. Springer Science & Business Media.

Czernin, J., & Phelps, M. E. (2002). Positron emission tomography scanning: Current and future applications. *Annual Review of Medicine, 53*, 89–112.

Dada, M., Awojoyogbe, O. B., Boubaker, K., & Ojambati, O. S. (2010). BPES analyses of a new diffusion-advection equation for fluid flow in blood vessels under different bio-physico-geometrical conditions. *Journal of Biophysics and Structural Biology, 2*(3), 28–34.

Dada, M., Awojoyogbe, O. B., Moses, O. F., Ojambati, O. S., & De, D. K. (2009). A mathematical analysis of stenosis geometry, NMR magnetizations and signals based on the Bloch NMR flow equations, Bessel and Boubaker polynomial expansions. *Journal of Biological Physics and Chemistry, 9*(3), 101–106.

Dada, M. O., Jayeoba, B., Awojoyogbe, B. O., Uno, U. E., & Awe, O. E. (2017). Mathematical development and computational analysis of harmonic phase-magnetic resonance imaging (HARP-MRI) based on Bloch nuclear magnetic resonance (NMR) diffusion model for myocardial motion. *Journal of Medical Systems, 41*(10), 168.

Dada, O. M., & Awojoyogbe, O. B. (2015). Analytical solutions to Bloch NMR flow equation in porous system: Future and emerging magnetic resonance computational imaging for medical and biomedical engineering. *Recent Patents and Topics on Imaging, 5*(1), 31–43.

Dada, O. M., Awojoyogbe, O. B., Adesola, O. A., & Boubaker, K. (2013a). Magnetic resonance imaging derived flow parameters for the analysis of cardiovascular diseases and drug development. *Magnetic Resonance Insights, 6*, 83–93.

Dada, O. M., Awojoyogbe, O. B., Baroni, S., & Aweda, M. A. (2013b). *Application of Bloch NMR equation and Pennes bioheat equation to theranostics*. Paper presented at the World Molecular Imaging Congress, Savannah, Georgia, USA, 18–21 September 2013.

Dada, O. M., Awojoyogbe, O. B., & Ukoha, A. C. (2015). A computational analysis for quantitative evaluation of petrol-physical properties of rock fluids based on Bloch NMR diffusion model for porous media. *Journal of Petroleum Science and Engineering, 127*, 137–147.

Dada, O. M., Baroni, S., & Awojoyogbe, O. B. (2014). Computational model of Bloch flow equation for in-vivo assessment of contrast agents. *Journal of Cardiovascular Magnetic Resonance, 16*(Suppl 1), O93.

Dahab, D. A., Ghoniemy, S. S., & Selim, G. M. (2012). Automated brain tumor detection and identification using image processing and probabilistic neural network techniques. *International Journal of Image Processing and Visual Communication, 1*(2), 1–8.

Dahlem, M. A., & Müller, S. C. (2004). Reaction-diffusion waves in neuronal tissue and the window of cortical excitability. *Annalen der Physik, 13*(7–8), 442–449.

Damadian, R. (1971). Tumor detection by nuclear magnetic resonance. *Science, 171*(3976), 1151.

Damadian, R., Goldsmith, M., & Minkoff, L. (1977). NMR in cancer: XVI. FONAR image of the live human body. *Physiological Chemistry and Physics, 9*(1), 97–100.

Damodharan, S., & Raghavan, D. (2015). Combining tissue segmentation and neural network for brain tumor detection. *International Arab Journal of Information Technology, 12*(1), 42–52.

Dastjerdi, A. V., & Buyya, R. (2016). Fog computing: Helping the internet of things realize its potential. *Computer, 49*(8), 112–116.

Dawson, J., & Lauterbur, P. C. (2008). Magnetic resonance imaging. *Scholarpedia, 3*(7), 3381. revision #123947.

De Marsily, G. (1986). *Quantitative hydrogeology: Groundwater hydrology for engineers*. Academic Press.

De Vries, A., Moonen, R., Yildirim, M., Langereis, S., Lamerichs, R., Pikkemaat, J. A., Baroni, S., Terreno, E., Nicolay, K., Strijkers, G. J., & Grüll, H. (2014). Relaxometric studies of gadolinium-functionalized perfluorocarbon nanoparticles for MR imaging. *Contrast Media & Molecular Imaging, 9*(1), 83–91.

Degen, C. L., Poggio, M., Mamin, H. J., Rettner, C. T., & Rugar, D. (2009). Nanoscale magnetic resonance imaging. *Proceedings of the National Academy of Sciences of the United States of America, 106*(5), 1313–1317.

Delikatny, E. J., & Poptani, H. (2005). MR techniques for in vivo molecular and cellular imaging. *Radiologic Clinics, 43*, 205–220.

Demir, H., & Süngü, İ. Ç. (2009). Numerical solution of a class of nonlinear Emden-Fowler equations by using differential transform method. *Cankaya University Journal of Science and Engineering, 12*(2), 75.

Denney, T. S. (1997, June). Identification of myocardial tags in tagged MR images without prior knowledge of myocardial contours. In *Biennial International Conference on Information Processing in Medical Imaging* (pp. 327–340). Springer.

Detre, J. A., & Wang, J. (2002). Technical aspects and utility of fMRI using BOLD and ASL. *Clinical Neurophysiology, 113*(5), 621–634.

Devkota, B. H., & Imberger, J. (2009). Lagrangian modelling of advection-diffusion transport in open channel flow. *Water Resources Research, 45*(12), W12406.

Dickerson, B. C., & Sperling, R. A. (2005). Neuroimaging biomarkers for clinical trials of disease-modifying therapies in Alzheimer's disease. *NeuroRx, 2*(2), 348–360.

Dietel, M., & Sers, C. (2006). Personalized medicine and development of targeted therapies: The upcoming challenge for diagnostic molecular pathology. A review. *Virchows Archiv, 448*(6), 744–755.

Divya, A., Raj, S. N., & Ramesh, S. (2020). A review of MRI Brain tumor noise removal, segmentation, and classification. *Journal of Critical Reviews, 7*(12), 2072–2081.

Dixit, S., Kini, M. R., Patil, M. D., & Chaithra, S. (2016). Brain tumor detection from clinical CT and MRI using wavelet based image fusion technique. *PARIPEX-Indian Journal of Research, 5*(6), 284–286.

Dobre, M. C., Ugurbil, K., & Marjanska, M. (2007). Determination of blood longitudinal relaxation time (T_1) at high magnetic field strengths. *Magnetic Resonance Imaging, 25*, 733–735.

Donahue, K. M., Weisskoff, R. M., & Burstein, D. (1997). Water diffusion and exchange as they influence contrast enhancement. *Journal of Magnetic Resonance Imaging, 7*, 102–110.

Dumas, S., Jacques, V., Sun, W. C., Troughton, J. S., Welch, J. T., Chasse, J. M., Schmitt-Willich, H., & Caravan, P. (2010). High relaxivity MRI contrast agents Part 1: Impact of single donor atom substitution on relaxivity of serum albumin-bound gadolinium complexes. *Investigative Radiology, 45*(10), 600.

Eberly, J. H., & Stroud, C. R. (2006). Coherent transients. In G. W. F. Drake (Ed.), *Springer handbooks of atomic, molecular, and optical physics* (pp. 1065–1076). Springer Science.

Edvardsen, T., Rosen, B. D., Pan, L., Jerosch-Herold, M., Lai, S., Hundley, W. G., Sinha, S., Kronmal, R. A., Bluemke, D. A., & Lima, J. A. (2006). Regional diastolic dysfunction in individuals with left ventricular hypertrophy measured by tagged magnetic resonance imaging—The Multi-Ethnic Study of Atherosclerosis (MESA). *American Heart Journal, 151*(1), 109–114.

Edwards, J. P., Gerber, U., Schubert, C., Trejo, M. A., & Weber, A. (2017). The Yukawa potential: Ground state energy and critical screening. *Progress of Theoretical and Experimental Physics, 2017*(8), 083A01.

Eisberg, R., & Resnick, R. (1985). *Quantum physics of atoms, molecules, solids, nuclei, and particles*. John Wiley and Sons.

El-Dahshan, E. S. A., Mohsen, H. M., Revett, K., & Salem, A. B. M. (2014). Computer-aided diagnosis of human brain tumor through MRI: A survey and a new algorithm. *Expert Systems with Applications, 41*(11), 5526–5545.

Engler, N., Ostermann, A., Niimura, N., & Parak, F. G. (2003). Hydrogen atoms in proteins: Positions and dynamics. *Proceedings of the National Academy of Sciences, 100*(18), 10243–10248.

Ernst, R. R., Bodenhausen, G., & Wokaun, A. (1987). *Principles of nuclear magnetic resonance in one and two dimensions* (Vol. 14). Clarendon Press.

Falk, J. (2008). Epigenetics & Sequencing-Gene Expression Systems' Second International Meeting. Chromatin methylation to disease biology & theranostics. *IDrugs: The Investigational Drugs Journal, 11*(9), 650.

Falk, M. H., & Issels, R. D. (2001). Hyperthermia in oncology. *International Journal of Hyperthermia, 17*(1), 1–18.

Farouki, R. T., & Hamaguchi, S. (1994). Thermodynamics of strongly-coupled Yukawa systems near the one-component-plasma limit. II. Molecular dynamics simulations. *The Journal of Chemical Physics, 101*(11), 9885–9893.

Farraher, S. W., Jara, H., Chang, K. J., Ozonoff, A., & Soto, J. A. (2006). Differentiation of hepatocellular carcinoma and hepatic metastasis from cysts and hemangiomas with calculated T2 relaxation times and the T1/T2 relaxation times ratio. *Journal of Magnetic Resonance Imaging: An Official Journal of the International Society for Magnetic Resonance in Medicine, 24*(6), 1333–1341.

Ferlay, J. S. H. R., Shin, H. R., Bray, F., Forman, D., Mathers, C., & Parkin, D. M. (2010). *Cancer incidence and mortality worldwide*. International Agency for Research on Cancer.

Field, J. H. (2004). Relationship of quantum mechanics to classical electromagnetism and classical relativistic mechanics. *European Journal of Physics, 25*(3), 385.

Firouzi, F., Rahmani, A. M., Mankodiya, K., Badaroglu, M., Merrett, G. V., Wong, P., & Farahani, B. (2018). Internet-of-Things and big data for smarter healthcare: From device to architecture, applications and analytics. *Future Generation Computer Systems, 78*, 583–586.

Flickinger, J. C., Kondziolka, D., Lunsford, L. D., Coffey, R. J., Goodman, M. L., Shaw, E. G., Hudgins, W. R., Weiner, R., Harsh, G. R., IV, Sneed, P. K., & Larson, D. A. (1994). A multi-institutional experience with stereotactic radiosurgery for solitary brain metastasis. *International Journal of Radiation Oncology, Biology, Physics, 28*(4), 797–802.

Fox, B. D., Cheung, V. J., Patel, A. J., Suki, D., & Rao, G. (2011). Epidemiology of metastatic brain tumors. *Neurosurgery Clinics, 22*(1), 1–6.

Frank, J. A., Anderson, S. A., & Arbab, A. S. (2007). Methods for labelling nonphagocytic cells with MR contrast agents. In M. J. M. Modo & J. W. M. Bulte (Eds.), *Molecular and cellular MR imaging* (pp. 295–324). CRC Press - Taylor & Francis Group.

Frauenfelder, H., Sligar, S. G., & Wolynes, P. G. (1991). The energy landscapes and motions of proteins. *Science, 254*(5038), 1598–1603.

Freitas, R. A., Jr. (2005). What is nanomedicine? *Nanomedicine: Nanotechnology, Biology and Medicine, 1*(1), 2–9.

Fritsche, L., & Haugk, M. (2003). A new look at the derivation of the Schrödinger equation from Newtonian mechanics. *Annalen der Physik, 12*(6), 371–402.

Frost, J. J. (2003). Molecular imaging of the brain: A historical perspective. *Neuroimaging Clinics, 13*, 653–658.

Furlani, E. P. (2010). Nanoscale magnetic biotransport. In K. D. Sattler (Ed.), *Handbook of nanophysics: Nanomedicine and nanorobotics* (pp. 10-1–10-19). CRC Press - Taylor & Francis Group.

Fusco, A., & Fedele, M. (2007). Roles of HMGA proteins in cancer. *Nature Reviews Cancer, 7*(12), 899–910.

Gamper, U., Boesiger, P., & Kozerke, S. (2008). Compressed sensing in dynamic MRI. *Magnetic Resonance in Medicine, 59*(2), 365–373.

Garavelli, S. L., & Oliveira, F. A. (1991). Analytical solution for a Yukawa-type potential. *Physical Review Letters, 66*(10), 1310.

Gawad, C. (2005). Towards molecular medicine: A case for a biological periodic table. *American Journal of Pharmacogenomics, 5*, 207–211.

Geraldes, C. F., & Laurent, S. (2009). Classification and basic properties of contrast agents for magnetic resonance imaging. *Contrast Media & Molecular Imaging, 4*, 1–23.

Geurts, P., Ernst, D., & Wehenkel, L. (2006). Extremely randomized trees. *Machine Learning, 63*(1), 3–42.

Ghare, S., Gaikwad, N., Kulkarni, N., & Nerkar, M. (2015). Detection of possibility of brain tumor using image segmentation. *International Journal of Innovative Research in Computer and Communication Engineering, 3*(4), 2701–2704.

Gheysens, O., & Gambhir, S. S. (2005). Studying molecular and cellular processes in the intact organism. *Progress in Drug Research, 62*, 117–150.

Gibby, W. A. (2005). Basic principles of magnetic resonance imaging. *Neurosurgery Clinics of North America, 16*(1), 1–64.

Gibson, R. E., Burns, H. D., Hamill, T. G., Eng, W. S., Francis, B. E., & Ryan, C. (2000). Non-invasive radiotracer imaging as a tool for drug development. *Current Pharmaceutical Design, 6*, 973–989.

Giese, A., Bjerkvig, R., Berens, M. E., & Westphal, M. (2003). Cost of migration: Invasion of malignant gliomas and implications for treatment. *Journal of Clinical Oncology, 21*(8), 1624–1636.

Glaser, R. (2012). *Biophysics: An introduction* (p. 233). Springer Science & Business Media.

Goose, J. E., & Sansom, M. S. (2013). Reduced lateral mobility of lipids and proteins in crowded membranes. *PLoS Computational Biology, 9*(4), e1003033.

Gordillo, N., Montseny, E., & Sobrevilla, P. (2013). State of the art survey on MRI brain tumor segmentation. *Magnetic Resonance Imaging, 31*(8), 1426–1438.

Gries, H. (2002). Extracellular MRI contrast agents based on gadolinium. *Topics in Current Chemistry, 221*, 1–24.

Griffiths, D. (2019). *Quantum physics* (3rd ed.). Cambridge University Press.

Griffiths, H. J. (1986). Manual of clinical magnetic resonance imaging. *Radiology, 160*(1), 226–226.

Grolemund, G. (2014). *Hands-on programming with R: Write your own functions and simulations.* O'Reilly Media, Inc..

Grotz, B., Beck, J., Neumann, P., Naydenov, B., Reuter, R., Reinhard, F., Jelezko, F., Wrachtrup, J., Schweinfurth, D., Sarkar, B., & Hemmer, P. (2011). Sensing external spins with nitrogen-vacancy diamond. *New Journal of Physics, 13*, 055004.

Gruber, M. L., & Buster, W. P. (2004). Temozolomide in combination with irinotecan for treatment of recurrent malignant glioma. *American Journal of Clinical Oncology, 27*(1), 33–38.

Gubbi, J., Buyya, R., Marusic, S., & Palaniswami, M. (2013). Internet of Things (IoT): A vision, architectural elements, and future directions. *Future Generation Computer Systems, 29*(7), 1645–1660.

Guccione, S., Li, K. C., & Bednarski, M. D. (2003). Molecular imaging and therapy directed at the neovasculature in pathologies. How imaging can be incorporated into vascular-targeted delivery systems to generate active therapeutic agents. *IEEE Engineering in Medicine and Biology Magazine: The Quarterly Magazine of the Engineering in Medicine & Biology Society, 23*(5), 50–56.

Guimaraes, M. D., Schuch, A., Hochhegger, B., Gross, J. L., Chojniak, R., & Marchiori, E. (2014). Functional magnetic resonance imaging in oncology: State of the art. *Radiologia Brasileira, 47*(2), 101–111.

Gupta, A., Stait-Gardner, T., Ghadirian, B., Price, W. S., Dada, O. M., & Awojoyogbe, O. B. (2014). *Theory, dynamics and applications of magnetic resonance imaging-I.* Science PG.

Guttman, M. A., Prince, J. L., & McVeigh, E. R. (1994). Tag and contour detection in tagged MR images of the left ventricle. *IEEE Transactions on Medical Imaging, 13*(1), 74–88.

Haacke, E. M., Brown, R. W., Thompson, M. R., & Venkatesan, R. (1999). *Magnetic resonance imaging: Physical principles and sequence design* (pp. 39–40). A John Wiley and Sons.

Haacke, E. M., Lin, W., & Li, D. (2000). Whole body magnetic resonance angiography. In I. R. Young (Ed.), *Methods in biomedical magnetic resonance imaging and spectroscopy* (pp. 488–500). John Wiley & Sons.

Haacke, E. M., Xu, Y., Cheng, Y. C. N., & Reichenbach, J. R. (2004). Susceptibility weighted imaging (SWI). *Magnetic Resonance in Medicine, 52*(3), 612–618.

Hahn, E. L. (1980). Heritage of the Bloch equations in quantum optics. In *Felix Bloch Festschrift* (p. 75).

Halldin, C., Gulyas, B., Langer, O., & Farde, L. (2001). Brain radioligands--State of the art and new trends. *The Quarterly Journal of Nuclear Medicine and Molecular Imaging, 45*(2), 139.

Handore, S., & Kokare, D. (2015, January). Performance analysis of various methods of tumour detection. In *2015 International Conference on Pervasive Computing (ICPC)* (pp. 1–4). IEEE.

Handore, S. S. V., & Patil, P. M. (2010). Brain tumor segmentation from MRI using GLCM. *International Journal of Emerging Technologies in Engineering, 2*(4), 591–595.

Hanson, L. G. (2008). Is quantum mechanics necessary for understanding magnetic resonance? *Concepts in Magnetic Resonance Part A, 32*(5), 329–340.

Hanusch, K. U., Janssen, C. H., Billheimer, D., Jenkins, I., Spurgeon, E., Lowry, C. A., & Raison, C. L. (2013). Whole-body hyperthermia for the treatment of major depression: Associations with thermoregulatory cooling. *The American Journal of Psychiatry, 170*(7), 802.

Haralick, R. M., Shanmugam, K., & Dinstein, I. H. (1973). Textural features for image classification. *IEEE Transactions on Systems, Man, and Cybernetics, SMC-3*(6), 610–621.

Hardy, P., & Henkelman, R. M. (1991). On the transverse relaxation rate enhancement induced by diffusion of spins through inhomogeneous fields. *Magnetic Resonance in Medicine, 17*(2), 348–356.

Harris, R. K. (1986). *Nuclear magnetic resonance spectroscopy*. John Wiley and Sons.

Hartung, M. P., Grist, T. M., & François, C. J. (2011). Magnetic resonance angiography: Current status and future directions. *Journal of Cardiovascular Magnetic Resonance, 13*(1), 19.

Havaei, M., Davy, A., Warde-Farley, D., Biard, A., Courville, A., Bengio, Y., Pal, C., Jodoin, P. M., & Larochelle, H. (2017). Brain tumor segmentation with deep neural networks. *Medical Image Analysis, 35*, 18–31.

Hazlewood, C. F., Chang, D. C., Nichols, B. L., & Woessner, D. E. (1974). Nuclear magnetic resonance transverse relaxation times of water protons in skeletal muscle. *Biophysical Journal, 14*(8), 583.

Hazlewood, U. F., Rorschach, H. E., & Lin, C. (1991). Diffusion of water in tissues and MRI. *Magnetic Resonance in Medicine, 19*(2), 214–216.

Heckl, S., Pipkorn, R., Nägele, T., Vogel, U., Küker, W., & Voight, K. (2004). Molecular imaging: Bridging the gap between neuroradiology and neurohistology. *Histology and Histopathology, 19*(2), 651–668.

Heiken, J. P., Weyman, P. J., Lee, J. K., Balfe, D. M., Picus, D., Brunt, E. M., & Flye, M. W. (1989). Detection of focal hepatic masses: Prospective evaluation with CT, delayed CT, CT during arterial portography, and MR imaging. *Radiology, 171*(1), 47–51.

Hemmer, P. (2013). Toward molecular-scale MRI. *Science, 339*(6119), 529–530.

Hendrick, R. E. (2007). *Breast MRI: Fundamentals and technical aspects*. Springer Science & Business Media.

Henson, J. W., Ulmer, S., & Harris, G. J. (2008). Brain tumor imaging in clinical trials. *American Journal of Neuroradiology, 29*(3), 419–424.

Hernando, C. G., Esteban, L., Cañas, T., Van den Brule, E., & Pastrana, M. (2010). The role of magnetic resonance imaging in oncology. *Clinical and Translational Oncology, 12*(9), 606–613.

Herschman, H. R. (2003). Molecular imaging: Looking at problems, seeing solutions. *Science, 302* (5645), 605–608.

Heyn, C., Ronald, J. A., Mackenzie, L. T., MacDonald, I. C., Chambers, A. F., Rutt, B. K., & Foster, P. J. (2006). In vivo magnetic resonance imaging of single cells in mouse brain with optical validation. *Magnetic Resonance in Medicine, 55*(1), 23–29.

Heywang-Köbrunner, S. H., & Oellinger, H. (1996). Breast MRI. In D. M. Grant & R. K. Harris (Eds.), *The encyclopaedia of nuclear magnetic resonance*. Wiley.

Higgins, L. J., & Pomper, M. G. (2011, February). The evolution of imaging in cancer: Current state and future challenges. *Seminars in Oncology, 38*(1), 3–15.

Hinshaw, W. S., & Lent, A. H. (1983). An introduction to NMR imaging: From the Bloch equation to the imaging equation. *Proceedings of the IEEE, 71*(3), 338–350.

Histed, S. N., Lindenberg, M. L., Mena, E., Turkbey, B., Choyke, P. L., & Kurdziel, K. A. (2012). Review of functional/anatomic imaging in oncology. *Nuclear Medicine Communications, 33* (4), 349.

Hoefnagels, F. W., Lagerwaard, F. J., Sanchez, E., Haasbeek, C. J., Knol, D. L., Slotman, B. J., & Vandertop, W. P. (2009). Radiological progression of cerebral metastases after radiosurgery: Assessment of perfusion MRI for differentiating between necrosis and recurrence. *Journal of Neurology, 256*(6), 878.

Hoh, C. K., Schiepers, C., Seltzer, M. A., Gambhir, S. S., Silverman, D. H., Czernin, J., Maddahi, J., & Phelps, M. E. (1997, April). PET in oncology: Will it replace the other modalities? *Seminars in Nuclear Medicine, 27*(2), 94–106.

Hoit, B. D. (2011). Strain and strain rate echocardiography and coronary artery disease. *Circulation. Cardiovascular Imaging, 4*(2), 179–190.

Hopf, F. A., Meystre, P., Scully, M. O., & Louisell, W. H. (1976). Strong-signal theory of a free-electron laser. *Physical Review Letters, 37*(20), 1342.

Hosmer, D. W., Jr., Lemeshow, S., & Sturdivant, R. X. (2013). *Applied logistic regression.* John Wiley & Sons.

Hu, H., Wen, Y., Chua, T. S., & Li, X. (2014). Toward scalable systems for big data analytics: A technology tutorial. *IEEE Access, 2*, 652–687.

Huang, C., Graff, C. G., Clarkson, E. W., Bilgin, A., & Altbach, M. I. (2012). T2 mapping from highly undersampled data by reconstruction of principal component coefficient maps using compressed sensing. *Magnetic Resonance in Medicine, 67*(5), 1355–1366.

Işın, A., Direkoğlu, C., & Şah, M. (2016). Review of MRI-based brain tumor image segmentation using deep learning methods. *Procedia Computer Science, 102*, 317–324.

Isin, E. M., Elmore, C. S., Nilsson, G. N., Thompson, R. A., & Weidolf, L. (2012). Use of radiolabeled compounds in drug metabolism and pharmacokinetic studies. *Chemical Research in Toxicology, 25*(3), 532–542.

Jackson, G. D., Connelly, A., Duncan, J. S., Grünewald, R. A., & Gadian, D. G. (1993). Detection of hippocampal pathology in intractable partial epilepsy: Increased sensitivity with quantitative magnetic resonance T2 relaxometry. *Neurology, 43*(9), 1793–1793.

Jacobs, A. H., Kracht, L. W., Gossmann, A., Rüger, M. A., Thomas, A. V., Thiel, A., & Herholz, K. (2005). Imaging in neurooncology. *NeuroRx, 2*(2), 333–347.

Jacobs, R. E., & Cherry, S. R. (2001). Complementary emerging techniques: High-resolution PET and MRI. *Current Opinion in Neurobiology, 11*(5), 621–629.

Jacques, V., Dumas, S., Sun, W. C., Troughton, J. S., Greenfield, M. T., & Caravan, P. (2010). High relaxivity MRI contrast agents part 2: Optimization of inner-and second-sphere relaxivity. *Investigative Radiology, 45*(10), 613.

Jager, P. L., de Korte, M. A., Lub-de Hooge, M. N., Van Waarde, A., Koopmans, K. P., Perik, P. J., & De Vries, E. G. (2005). Molecular imaging: What can be used today. *Cancer Imaging, 5 (Spec No A)*, S27–S32.

Jain, K. K. (2008). Nanomedicine: Application of nanobiotechnology in medical practice. *Medical Principles and Practice, 17*(2), 89–101.

Jasanoff, A. (2005). Functional MRI using molecular imaging agents. *Trends in Neurosciences, 28* (3), 120–126.

Jayaraman, M., Vellingiri, K., & Guo, Y. (2019). Neutrosophic set in medical image denoising. In *Neutrosophic set in medical image analysis* (pp. 77–100). Academic Press.

Jellison, B. J., Field, A. S., Medow, J., Lazar, M., Salamat, M. S., & Alexander, A. L. (2004). Diffusion tensor imaging of cerebral white matter: A pictorial review of physics, fiber tract anatomy, and tumor imaging patterns. *American Journal of Neuroradiology, 25*(3), 356–369.

Jiang, S., Gnanasammandhan, M. K., & Zhang, Y. (2010). Optical imaging-guided cancer therapy with fluorescent nanoparticles. *Journal of the Royal Society Interface, 7*(42), 3–18.

Jiang, S. C., Ma, N., Li, H. J., & Zhang, X. X. (2002). Effects of thermal properties and geometrical dimensions on skin burn injuries. *Burns, 28*(8), 713–717.

Jorgensen, W. L. (2004). The many roles of computation in drug discovery. *Science, 303*(5665), 1813–1818.

Joseph, R. P., Singh, C. S., & Manikandan, M. (2014). Brain tumor MRI image segmentation and detection in image processing. *International Journal of Research in Engineering and Technology, 3*(1), 1–5.

Jung, C. W., & Jacobs, P. (1995). Physical and chemical properties of superparamagnetic iron oxide MR contrast agents: Ferumoxides, ferumoxtran, ferumoxsil. *Magnetic Resonance Imaging, 13* (5), 661–674.

Kalaiselvi, T., Nagaraja, P., & Sriramakrishnan, P. (2016). A simple image processing approach to abnormal slices detection from MRI tumor volumes. *International Journal of Multimedia & Its Applications, 8*(1), 55–64.

Kaplan, W. (1991). *Advanced calculus* (pp. 189–191). Addison-Wesley.

Kapoor, L., & Thakur, S. (2017, January). A survey on brain tumor detection using image processing techniques. In *2017 7th International Conference on Cloud Computing, Data Science & Engineering-Confluence* (pp. 582–585). IEEE.

Kegeles, L. S., & Mann, J. J. (1997). In vivo imaging of neurotransmitter systems using radiolabeled receptor ligands. *Neuropsychopharmacology, 17*(5), 293–307.

Khalifa, N. E. M., Taha, M. H. N., Hassanien, A. E., & Mohamed, H. N. E. T. (2019). Deep iris: Deep learning for gender classification through iris patterns. *Acta Informatica Medica, 27*(2), 96.

Khalil, H. K., & Grizzle, J. W. (1996). *Nonlinear systems* (Vol. 3). Prentice hall.

Khanafer, K., & Vafai, K. (2006). The role of porous media in biomedical engineering as related to magnetic resonance imaging and drug delivery. *Heat and Mass Transfer, 42*(10), 939–953.

Kiessling, F., Boese, J., Corvinus, C., Ederle, J. R., Zuna, I., Schoenberg, S. O., Brix, G., Schmähl, A., Tuengerthal, S., Herth, F., Kauczor, H. U., & Essig, M. (2004). Perfusion CT in patients with advanced bronchial carcinomas: A novel chance for characterization and treatment monitoring. *European Radiology, 14*(7), 1226–1233.

Kim, T. K., Choi, B. I., Han, J. K., Jang, H. J., & Han, M. C. (1997). Optimal MR protocol for hepatic hemangiomas: Comparison of conventional spin-echo sequences with T2-weighted turbo spin-echo and serial gradient-echo (FLASH) sequences with gadolinium enhancement. *Acta Radiologica, 38*(4), 565–571.

Kimmich, R., & Anoardo, E. (2004). Field-cycling NMR relaxometry. *Progress in Nuclear Magnetic Resonance Spectroscopy, 44*(3), 257–320.

Kinoshita, M., Goto, T., Okita, Y., Kagawa, N., Kishima, H., Hashimoto, N., & Yoshimine, T. (2010). Diffusion tensor-based tumor infiltration index cannot discriminate vasogenic edema from tumor-infiltrated edema. *Journal of Neuro-Oncology, 96*(3), 409–415.

Kirik, D., Breysse, N., Björklund, T., Besret, L., & Hantraye, P. (2005). Imaging in cell-based therapy for neurodegenerative diseases. *European Journal of Nuclear Medicine and Molecular Imaging, 32*(2), S417–S434.

Krabbe, K., Gideon, P., Wagn, P., Hansen, U., Thomsen, C., & Madsen, F. (1997). MR diffusion imaging of human intracranial tumours. *Neuroradiology, 39*(7), 483–489.

Kreyszig, E. (1996). *Advanced engineering mathematics*. John Wiley and Sons.

Kuffler, S. W., & Potter, D. D. (1964). Glia in the leech central nervous system: Physiological properties and neuron-glia relationship. *Journal of Neurophysiology, 27*(2), 290–320.

Kundu, P. K., & Cohen, I. M. (2004). *Fluid mechanics*. Elsevier Academic Press.

Kwock, L. (1998). Localized MR spectroscopy: Basic principles. *Neuroimaging Clinics of North America, 8*(4), 713–731.

Kwong, K. K., Belliveau, J. W., Chesler, D. A., Goldberg, I. E., Weisskoff, R. M., Poncelet, B. P., Kennedy, D. N., Hoppel, B. E., Cohen, M. S., & Turner, R. (1992). Dynamic magnetic resonance imaging of human brain activity during primary sensory stimulation. *Proceedings of the National Academy of Sciences, 89*(12), 5675–5679.

Laddha, M., & Ladhake, S. A. (2014). Brain tumor detection using morphological and watershed operators. *International Journal of Application or Innovation in Engineering and Management, 3*(3), 383–387.

Lakshmi, S. V. S. S., & Kavila, S. D. (2018). Machine learning for credit card fraud detection system. *International Journal of Applied Engineering Research, 13*(24), 16819–16824.

Laub, G. (2000). Time-of-flight method of MRA. In I. R. Young (Ed.), *Methods in biomedical magnetic resonance imaging and spectroscopy* (pp. 500–504). John Wiley & Sons.

Laurent, S., Forge, D., Port, M., Roch, A., Robic, C., Vander Elst, L., & Muller, R. N. (2008). Magnetic iron oxide nanoparticles: Synthesis, stabilization, vectorization, physicochemical characterizations, and biological applications. *Chemical Reviews, 108*(6), 2064–2110.

Lauterbur, P. C. (1979). Medical imaging by nuclear magnetic resonance zeugmatography. *IEEE Transactions on Nuclear Science, 26*(2), 2807–2811.

Le Bihan, D., & Iima, M. (2015). Diffusion magnetic resonance imaging: What water tells us about biological tissues. *PLoS Biology, 13*(7), e1002203.

Le Bihan, D., Mangin, J. F., Poupon, C., Clark, C. A., Pappata, S., Molko, N., & Chabriat, H. (2001). Diffusion tensor imaging: Concepts and applications. *Journal of Magnetic Resonance Imaging, 13*(4), 534–546.

Le Bihan, D., Urayama, S. I., Aso, T., Hanakawa, T., & Fukuyama, H. (2006). Direct and fast detection of neuronal activation in the human brain with diffusion MRI. *Proceedings of the National Academy of Sciences, 103*(21), 8263–8268.

Lee, M. H., Smyser, C. D., & Shimony, J. S. (2013). Resting-state fMRI: A review of methods and clinical applications. *American Journal of Neuroradiology, 34*(10), 1866–1872.

Lee, S. P., Silva, A. C., Ugurbil, K., & Kim, S. G. (1999). Diffusion-weighted spin-echo fMRI at 9.4T: Microvascular/tissue contribution to BOLD signal changes. *Magnetic Resonance in Medicine, 42*(5), 919–928.

Leggett, A. J. (1980). Macroscopic quantum systems and the quantum theory of measurement. *Progress of Theoretical Physics Supplement, 69*, 80–100.

Lévesque, M. F., Nakasato, N., Vinters, H. V., & Babb, T. L. (1991). Surgical treatment of limbic epilepsy associated with extrahippocampal lesions: The problem of dual pathology. *Journal of Neurosurgery, 75*(3), 364–370.

Levin, C. S. (2005). Primer on molecular imaging technology. *European Journal of Nuclear Medicine and Molecular Imaging, 32*(2), S325–S345.

Li, A., Wong, C. S., Wong, M. K., Lee, C. M., & Au Yeung, M. C. (2006). Acute adverse reactions to magnetic resonance contrast media–gadolinium chelates. *The British Journal of Radiology, 79*(941), 368–371.

Li, W. H., Parigi, G., Fragai, M., Luchinat, C., & Meade, T. J. (2002). Mechanistic studies of a calcium-dependent MRI contrast agent. *Inorganic Chemistry, 41*(15), 4018–4024.

Li, X., Li, L., Lu, H., Chen, D., & Liang, Z. (2003, May). Inhomogeneity correction for magnetic resonance images with fuzzy C-mean algorithm. In *Medical Imaging 2003: Image Processing* (Vol. 5032, pp. 995–1005). International Society for Optics and Photonics.

Liang, Z. P., & Lauterbur, P. C. (2000). *Principles of magnetic resonance imaging*. SPIE Optical Engineering Press.

Libby, P. (1995). Molecular bases of the acute coronary syndromes. *Circulation, 91*(11), 2844–2850.

Lieb, J. P., & Babb, T. L. (1986). Interhemispheric propagation time of human hippocampal seizures: II. Relationship to pathology and cell density. *Epilepsia, 27*(3), 294–300.

Liotta, L. A., Luchini, A., Petricoin, E. F., & Espina, V. (2015). *Hydrogel nanoparticle based immunoassay (No. 9,012,240)*. George Mason Research Foundation.

Liu, E. H., Saidel, G. M., & Harasaki, H. (2003). Model analysis of tissue responses to transient and chronic heating. *Annals of Biomedical Engineering, 31*(8), 1007–1014.

Liu, X., Kim, C. N., Yang, J., Jemmerson, R., & Wang, X. (1996). Induction of apoptotic program in cell-free extracts: Requirement for dATP and cytochrome c. *Cell, 86*(1), 147–157.

Liu, Y., Chen, W. R., & Chen, S. H. (2005). Cluster formation in two-Yukawa fluids. *The Journal of Chemical Physics, 122*(4), 044507.

Lodish, H. F. (2004). *Molecular cell biology*. W. H. Freeman & Company.

Logothetis, N. K., Pauls, J., Augath, M., Trinath, T., & Oeltermann, A. (2001). Neurophysiological investigation of the basis of the fMRI signal. *Nature, 412*(6843), 150.

Loman, A., Gregor, I., Stutz, C., Mund, M., & Enderlein, J. (2010). Measuring rotational diffusion of macromolecules by fluorescence correlation spectroscopy. *Photochemical & Photobiological Sciences, 9*(5), 627–636.

Lon, W. J., Ananta, N. J. S., Balaji, S., & Bennett, H. K. (2010). *Carbon nanotube based magnetic resonance imaging contrast agents*. WO2010017546.

Lorusso, V., Pascolo, L., Fernetti, C., Anelli, P. L., Uggeri, F., & Tiribelli, C. (2005). Magnetic resonance contrast agents: From the bench to the patient. *Current Pharmaceutical Design, 11*(31), 4079–4098.

Louie, A. Y., Hüber, M. M., Ahrens, E. T., Rothbächer, U., Moats, R., Jacobs, R. E., Fraser, S. E., & Meade, T. J. (2000). In vivo visualization of gene expression using magnetic resonance imaging. *Nature Biotechnology, 18*(3), 321–325.

Louis, D. N., Ohgaki, H., Wiestler, O. D., Cavenee, W. K., Burger, P. C., Jouvet, A., Scheithauer, B. W., & Kleihues, P. (2007). The 2007 WHO classification of tumours of the central nervous system. *Acta Neuropathologica, 114*(2), 97–109.

Lowe, M. P. (2004). Activated MR contrast agents. *Current Pharmaceutical Biotechnology, 5*(6), 519–528.

Lu, H., Nagae-Poetscher, L. M., Golay, X., Lin, D., Pomper, M., & van Zijl, P. (2005). Routine clinical brain MRI sequences for use at 3.0 Tesla. *Journal of Magnetic Resonance Imaging, 22*(1), 13–22.

Lustig, M., Donoho, D., & Pauly, J. M. (2007). Sparse MRI: The application of compressed sensing for rapid MR imaging. *Magnetic Resonance in Medicine, 58*(6), 1182–1195.

Lustig, M., Donoho, D. L., Santos, J. M., & Pauly, J. M. (2008). Compressed sensing MRI. *IEEE Signal Processing Magazine, 25*(2), 72–82.

Ma, C., Wang, X., & Varghese, T. (2016). Segmental analysis of cardiac short-Axis views using Lagrangian radial and circumferential strain. *Ultrasonic Imaging, 38*(6), 363–383.

Ma, D., Gulani, V., Seiberlich, N., Liu, K., Sunshine, J. L., Duerk, J. L., & Griswold, M. A. (2013). Magnetic resonance fingerprinting. *Nature, 495*(7440), 187–192.

Macdonald, D. R., Cascino, T. L., Schold, S. C., Jr., & Cairncross, J. G. (1990). Response criteria for phase II studies of supratentorial malignant glioma. *Journal of Clinical Oncology, 8*(7), 1277–1280.

Mackay, C. E., Webb, J. A., Eldridge, P. R., Chadwick, D. W., Whitehouse, G. H., & Roberts, N. (2000). Quantitative magnetic resonance imaging in consecutive patients evaluated for surgical treatment of temporal lobe epilepsy. *Magnetic Resonance Imaging, 18*(10), 1187–1199.

Mahoney, B. P., Raghunand, N., Baggett, B., & Gillies, R. J. (2003). Tumor acidity, ion trapping and chemotherapeutics: I. Acid pH affects the distribution of chemotherapeutic agents in vitro. *Biochemical Pharmacology, 66*(7), 1207–1218.

Mallikarjuna, E. (2015). Brain tumor detection using segmentation based object labeling algorithm. *Journal of Computation in Biosciences and Engineering, 3*(2), 1–3.

Mamin, H. J., Kim, M., Sherwood, M. H., Rettner, C. T., Ohno, K., Awschalom, D. D., & Rugar, D. (2013). Nanoscale nuclear magnetic resonance with a nitrogen-vacancy spin sensor. *Science, 339*(6119), 557–560.

Marieb, E. N., & Hoehn, K. (2013). *The cardiovascular system: Blood vessels. Human anatomy & physiology* (p. 712). Pearson.

Marsden, P. K., Strul, D., Keevil, S. F., Williams, S. C., & Cash, D. (2002). Simultaneous PET and NMR. *The British Journal of Radiology, 75*(Suppl_9), S53–S59.

Martell, A. E., & Motekaitis, R. J. (1992). *Determination and use of stability constants*. VCH Publishers.

Martínez, M. C., & de la Torre, J. G. (1987). Brownian dynamics simulation of restricted rotational diffusion. *Biophysical Journal, 52*(2), 303–310.

Martins, A. N., Johnston, J. S., Henry, J. M., Stoffel, T. J., & Di Chiro, G. (1977). Delayed radiation necrosis of the brain. *Journal of Neurosurgery, 47*(3), 336–345.

Massoud, T. F., & Gambhir, S. S. (2003). Molecular imaging in living subjects: Seeing fundamental biological processes in a new light. *Genes & Development, 17*(5), 545–580.

Maurer, P. C., Kucsko, G., Latta, C., Jiang, L., Yao, N. Y., Bennett, S. D., Pastawski, F., Hunger, D., Chisholm, N., Markham, M., Twitchen, D. J., Cirac, J. I., & Lukin, M. D. (2012). Room-temperature quantum bit memory exceeding one second. *Science, 336*(6086), 1283–1286.

Maze, J. R., Stanwix, P. L., Hodges, J. S., Hong, S., Taylor, J. M., Cappellaro, P., Jiang, L., Gurudev Dutt, M. V., Togan, E., Zibrov, A. S., Yacoby, A., Walsworth, R. L., & Lukin, M. D. (2008). Nanoscale magnetic sensing with an individual electronic spin in diamond. *Nature, 455* (7213), 644–647.

Mehrabian, H., Desmond, K. L., Soliman, H., Sahgal, A., & Stanisz, G. J. (2017). Differentiation between radiation necrosis and tumor progression using chemical exchange saturation transfer. *Clinical Cancer Research, 23*(14), 3667–3675.

Mehrotra, R., & Ansari, M. A. (2020). Feature extraction for brain tumour analysis and classification: A review. *International Journal of Digital Signals and Smart Systems, 4*(1–3), 199–218.

Menze, B. H., Jakab, A., Bauer, S., Kalpathy-Cramer, J., Farahani, K., Kirby, J., et al. (2014). The multimodal brain tumor image segmentation benchmark (BRATS). *IEEE Transactions on Medical Imaging, 34*(10), 1993–2024.

Merzbacher, E. (1998). *Quantum mechanics*. John Wiley & Sons.

Messiah, A. (1999). *Quantum mechanics: Two volumes bound as one*. Dover. (Two vol. bound as one, unabridged reprint ed.).

Metcalf, H. J., & Straten, P. (2007). *Laser cooling and trapping of neutral atoms*. Wiley-VCH Verlag GmbH & Co. KGaA.

Meyer, M. A. (2008). Malignant gliomas in adults. *The New England Journal of Medicine, 359*(17), 1850.

Meystre, P. (2006). Light-matter interaction. In G. W. F. Drake (Ed.), *Springer handbooks of atomic, molecular, and optical physics* (pp. 997–1007). Springer Science.

Mitchell, D. G., & Stark, D. D. (1992). *Hepatobiliary MRI: A text-atlas at mid and high field*. Mosby Inc..

Modo, M., Hoehn, M., & Bulte, J. W. (2005). Cellular MR imaging. *Molecular Imaging, 4*(3), 143–164.

Modo, M. J. M., & Bulte, J. W. M. (2007). What is molecular and cellular imaging. In M. J. M. Modo & J. W. M. Bulte (Eds.), *Molecular and cellular MR imaging* (pp. 1–9). CRC Press-Taylor & Francis Group.

Moghimi, S. M., Hunter, A. C., & Murray, J. C. (2005). Nanomedicine: Current status and future prospects. *The FASEB Journal, 19*(3), 311–330.

Mohanty, S. (2012). Optically-actuated translational and rotational motion at the microscale for microfluidic manipulation and characterization. *Lab on a Chip, 12*(19), 3624–3636.

Mohanty, S. K., Sharma, M., Panicker, M. M., & Gupta, P. K. (2005). Controlled induction, enhancement, and guidance of neuronal growth cones by use of line optical tweezers. *Optics Letters, 30*(19), 2596–2598.

Mohanty, S. K., Uppal, A., & Gupta, P. K. (2004). Self-rotation of red blood cells in optical tweezers: Prospects for high throughput malaria diagnosis. *Biotechnology Letters, 26*(12), 971–974.

Mohanty, S. K., Uppal, A., & Gupta, P. K. (2008). Optofluidic stretching of RBCs using single optical tweezers. *Journal of Biophotonics, 1*(6), 522–525.

Moore, A., Weissleder, R., & Bogdanov, A. (1997). Uptake of dextran-coated monocrystalline iron oxides in tumor cells and macrophages. *Journal of Magnetic Resonance Imaging, 7*(6), 1140–1145.

368 Bibliography

Moore, C. C., McVeigh, E. R., & Zerhouni, E. A. (1999). Noninvasive measurement of three-dimensional myocardial deformation with tagged magnetic resonance imaging during graded local ischemia. *Journal of Cardiovascular Magnetic Resonance, 1*(3), 207–222.

Moratal, D., Brummer, M. E., Martí-Bonmatí, L., & Vallés-Lluch, A. (2006). NMR imaging. In *Wiley encyclopedia of biomedical engineering.*

Mori, S., & Barker, P. B. (1999). Diffusion magnetic resonance imaging: Its principle and applications. *The Anatomical Record, 257*(3), 102–109.

Mori, S., & Tournier, J. D. (2013). *Introduction to diffusion tensor imaging: And higher order models.* Academic Press.

Moroz, P., Jones, S. K., & Gray, B. N. (2001). Status of hyperthermia in the treatment of advanced liver cancer. *Journal of Surgical Oncology, 77*(4), 259–269.

Morrisett, J., Vick, W., Sharma, R., Lawrie, G., Reardon, M., Ezell, E., Schwartz, J., & Gorenstein, D. (2003). Discrimination of components in atherosclerotic plaques from human carotid endarterectomy specimens by magnetic resonance imaging ex vivo. *Magnetic Resonance Imaging, 21* (5), 465–474.

Morriss, C. E., Freedman, R., Straley, C., Johnston, M., Vinegar, H. J., & Tutunjian, P. N. (1994). Hydrocarbon saturation and viscosity estimation from NMR logging in the Belridge Diatomite. In *SPWLA 35th Annual Logging Symposium. Society of Petrophysicists and Well-Log Analysts.*

Mourier-Robert, V., Bibette, J., Goutayer, M., Garcia, F. N. Y., & Texier-Nogues, I. (2018). *U.S. Patent No. 10,092,506.* U.S. Patent and Trademark Office.

Muller, R. N., Roch, A., Colet, J. M., Ouakssim, A., & Gillis, P. (2001). Particulate magnetic contrast agents. In E. Toth & A. E. Merbach (Eds.), *Chemistry of contrast agents in medical magnetic resonance imaging* (pp. 417–435). Wiley.

Mustaqeem, A., Javed, A., & Fatima, T. (2012). An efficient brain tumor detection algorithm using watershed & thresholding-based segmentation. *International Journal of Image, Graphics and Signal Processing, 4*(10), 34.

Nabizadeh, N., & Kubat, M. (2015). Brain tumors detection and segmentation in MR images: Gabor wavelet vs. statistical features. *Computers and Electrical Engineering, 45,* 286–301.

Naik, A., & Samant, L. (2016). Correlation review of classification algorithm using data mining tool: WEKA, Rapidminer, Tanagra, Orange and Knime. *Procedia Computer Science, 85,* 662–668.

Naik, D., Viswamitra, S., Kumar, A. A., & Srinath, M. G. (2014). Susceptibility weighted magnetic resonance imaging of brain: A multifaceted powerful sequence that adds to understanding of acute stroke. *Annals of Indian Academy of Neurology, 17*(1), 58–61.

Naik, J., & Patel, S. (2014). Tumor detection and classification using decision tree in brain MRI. *International Journal of Computer Science and Network Security, 14*(6), 87.

Nakhmani, A., Kikinis, R., & Tannenbaum, A. (2014). MRI brain tumor segmentation and necrosis detection using adaptive Sobolev snakes. In *Medical Imaging 2014: Image Processing* (Vol. 9034, p. 903442). International Society for Optics and Photonics.

Nandi, A. (2015, November). Detection of human brain tumour using MRI image segmentation and morphological operators. In *2015 IEEE International Conference on Computer Graphics, Vision and Information Security (CGVIS)* (pp. 55–60). IEEE.

Naz, S., & Hameed, I. A. (2017, October). Automated techniques for brain tumor segmentation and detection: A review study. In *2017 International Conference on Behavioral, Economic, Socio-cultural Computing (BESC)* (pp. 1–6). IEEE.

Nelson, D. L., & Cox, M. M. (2008). *Lehninger principles of biochemistry.* W. H. Freeman.

Nelson, E. (1966). Derivation of the Schrödinger equation from Newtonian mechanics. *Physical Review, 150*(4), 1079.

Neuman, K. C., & Block, S. M. (2004). Optical trapping. *Review of Scientific Instruments, 75*(9), 2787–2809.

Ng, E. Y. K., & Chua, L. T. (2002). Comparison of one-and two-dimensional programmes for predicting the state of skin burns. *Burns, 28*(1), 27–34.

Nicholson, C., & Rice, M. E. (1991). Diffusion of ions and transmitters in the brain cell microenvironment. In K. Flux & L. F. Agnati (Eds.), *Volume transmission in the brain: Novel mechanisms for neural transmission* (pp. 279–294). Raven Press.

Nieto, M. M. (1979). Hydrogen atom and relativistic pi-mesic atom in N-space dimensions. *American Journal of Physics, 47*(12), 1067–1072.

Nisar, S., Khan, M. M., Ibrahim, M., & Tariq, M. (2015). Efficient detection of brain tumor using normalized histogram. *Journal of Basic and Applied Scientific Research, 5*(4), 34–43.

Nishimura, K., Okayama, H., Inoue, K., Saito, M., Yoshii, T., Hiasa, G., Sumimoto, T., Inaba, S., Ogimoto, A., Funada, J. I., & Higaki, J. (2012). Direct measurement of radial strain in the inner-half layer of the left ventricular wall in hypertensive patients. *Journal of Cardiology, 59*(1), 64–71.

Novelline, R. A., & Squire, L. F. (2004). *Squire's fundamentals of radiology*. La Editorial, UPR.

O'Reilly, R. C. (2000). *General issues in computational modeling*. http://psych.colorado.edu/~oreilly/cecn/node16.html

O'Reilly, R. C., Herd, S. A., & Pauli, W. M. (2010). Computational models of cognitive control. *Current Opinion in Neurobiology, 20*(2), 257–261.

O'Reilly, R. C., & Munakata, Y. (2000). *Computational explorations in cognitive neuroscience: Understanding the mind by simulating the brain*. MIT press.

Occhiuzzi, C., Vallese, C., Amendola, S., Manzari, S., & Marrocco, G. (2014). NIGHT-Care: A passive RFID system for remote monitoring and control of overnight living environment. *Procedia Computer Science, 32*, 190–197.

Ogawa, S., Lee, T. M., Kay, A. R., & Tank, D. W. (1990). Brain magnetic resonance imaging with contrast dependent on blood oxygenation. *Proceedings of the National Academy of Sciences, 87*(24), 9868–9872.

Ogawa, S., Menon, R. S., Kim, S. G., & Ugurbil, K. (1998). On the characteristics of functional magnetic resonance imaging of the brain. *Annual Review of Biophysics and Biomolecular Structure, 27*(1), 447–474.

Okada, H., Kohanbash, G., Zhu, X., Kastenhuber, E. R., Hoji, A., Ueda, R., & Fujita, M. (2009). Immunotherapeutic approaches for glioma. *Critical Reviews in Immunology, 29*(1), 42.

Onate, C. A., & Ojonubah, J. O. (2016). Eigensolutions of the Schrödinger equation with a class of Yukawa potentials via supersymmetric approach. *Journal of Theoretical and Applied Physics, 10*(1), 21–26.

Osman, N. F., Kerwin, W. S., McVeigh, E. R., & Prince, J. L. (1999). Cardiac motion tracking using CINE harmonic phase (HARP) magnetic resonance imaging. *Magnetic Resonance in Medicine: An Official Journal of the International Society for Magnetic Resonance in Medicine, 42*(6), 1048–1060.

Osman, N. F., McVeigh, E. R., & Prince, J. L. (2000). Imaging heart motion using harmonic phase MRI. *IEEE Transactions on Medical Imaging, 19*(3), 186–202.

Ovsiannikov, L. V. (1982). *Group analysis of differential equations*. (W. F. Ames, Trans.). Academic.

Ozbas, E. E., Aksu, D., Ongen, A., Aydin, M. A., & Ozcan, H. K. (2019). Hydrogen production via biomass gasification, and modeling by supervised machine learning algorithms. *International Journal of Hydrogen Energy, 44*(32), 17260–17268.

Padhani, A. R. (2011, April). Diffusion magnetic resonance imaging in cancer patient management. *Seminars in Radiation Oncology, 21*(2), 119–140.

Palanisamy, T. S. C. A., Jayaraman, M., Vellingiri, K., & Guo, Y. (2019). Optimization-based neutrosophic set for medical image processing. In *Neutrosophic set in medical image analysis* (pp. 189–206). Academic Press.

Panayotounakos, D. E., & Kravvaritis, D. C. (2006). Exact analytic solutions of the Abel, Emden–Fowler and generalized Emden–Fowler nonlinear ODEs. *Nonlinear Analysis: Real World Applications, 7*(4), 634–650.

Pape, H., Clauser, C., & Iffland, J. (1998). Permeability prediction for reservoir sandstones and basement rocks based on fractal pore space geometry. In *Proceedings of SEG Annual Meeting: New Orleans.*

Parker, S. P. (1997). *Encyclopedia of science and technology* (Vol. 21). McGraw-Hill.

Parthasarathy, V. (2006). *Characterization of harmonic phase MRI: Theory, simulations, and applications.* (Doctoral dissertation, Johns Hopkins University).

Parvez, K., Parvez, A., & Zadeh, G. (2014). The diagnosis and treatment of pseudoprogression, radiation necrosis and brain tumor recurrence. *International Journal of Molecular Sciences, 15* (7), 11832–11846.

Paschotta, R. (2008). *Encyclopedia of laser physics and technology* (Vol. 1). Wiley-VCH.

Patchell, R. A. (2003). The management of brain metastases. *Cancer Treatment Reviews, 29*(6), 533–540.

Patel, L., Shukla, T., Huang, X., Ussery, D. W., & Wang, S. (2020). Machine learning methods in drug discovery. *Molecules, 25*(22), 5277.

Patil, R. C., & Bhalchandra, A. S. (2012). Brain tumour extraction from MRI images using MATLAB. *International Journal of Electronics, Communication and Soft Computing Science & Engineering, 2*(1), 1.

Pauleit, D., Stoffels, G., Bachofner, A., Floeth, F. W., Sabel, M., Herzog, H., Tellmann, L., Jansen, P., Reifenberger, G., Hamacher, K., & Coenen, H. H. (2009). Comparison of 18F-FET and 18F-FDG PET in brain tumors. *Nuclear Medicine and Biology, 36*(7), 779–787.

Peleg, Y., Pnini, R., & Zaarur, E. (1998). *Schaum's outline of theory and problems of quantum mechanics* (pp. 123–124). McGraw-Hill.

Perkuhn, M., Stavrinou, P., Thiele, F., Shakirin, G., Mohan, M., Garmpis, D., Kabbasch, C., & Borggrefe, J. (2018). Clinical evaluation of a multiparametric deep learning model for glioblastoma segmentation using heterogeneous magnetic resonance imaging data from clinical routine. *Investigative Radiology, 53*(11), 647.

Perez, J. M., Josephson, L., O'Loughlin, T., Högemann, D., & Weissleder, R. (2002a). Magnetic relaxation switches capable of sensing molecular interactions. *Nature Biotechnology, 20*(8), 816–820.

Perez, J. M., O'Loughin, T., Simeone, F. J., Weissleder, R., & Josephson, L. (2002b). DNA-based magnetic nanoparticle assembly acts as a magnetic relaxation nanoswitch allowing screening of DNA-cleaving agents. *Journal of the American Chemical Society, 124*(12), 2856–2857.

Perona, P., & Malik, J. (1990). Scale-space and edge detection using anisotropic diffusion. *IEEE Transactions on Pattern Analysis and Machine Intelligence, 12*(7), 629–639.

Phelps, M. E. (2000). PET: The merging of biology and imaging into molecular imaging. *Journal of Nuclear Medicine, 41*(4), 661–681.

Pintar, M. M. (1976). NMR basic principles and progress. In P. Diehl, E. Fluck, H. Günther, R. Kosfeld, & J. Seelig (Eds.), *Introductory essays* (Vol. 13). Springer-Verlag.

Piwnica-Worms, D., Schuster, D. P., & Garbow, J. R. (2004). Technoreview: Molecular imaging of host–pathogen interactions in intact small animals. *Cellular Microbiology, 6*(4), 319–331.

Podulso, J. F., Wengenack, T. M., Curran, G. L., Wisniewski, T., Sigurdsson, E. M., Macura, S. I., Borowski, B. J., & Jack, C. R. (2002). Molecular targeting of Alzheimer's amyloid plaques for contrast-enhanced magnetic resonance imaging. *Neurobiology of Disease, 2*(11), 315–329.

Polyanin, A. D., Schiesser, W. E., & Zhurov, A. I. (2008). Partial differential equation. *Scholarpedia, 3*(10), 4605.

Pomper, M. G. (2004). Translational molecular imaging for cancer. *Cancer Imaging: The Official Publication of the International Cancer Imaging Society, 5*, S16–S26.

Poonam, J. P. (2013). Review of image processing techniques for automatic detection of tumor in human brain. *International Journal of Computer Science and Mobile Computing, 2*(11), 117–122.

Price, S. J., Jena, R., Burnet, N. G., Hutchinson, P. J., Dean, A. F., Pena, A., Pickard, J. D., Carpenter, T. A., & Gillard, J. H. (2006). Improved delineation of glioma margins and regions of infiltration with the use of diffusion tensor imaging: An image-guided biopsy study. *American Journal of Neuroradiology, 27*(9), 1969–1974.

Price, W. S. (1997). Pulsed-field gradient nuclear magnetic resonance as a tool for studying translational diffusion: Part 1. Basic theory. *Concepts in Magnetic Resonance: An Educational Journal, 9*(5), 299–336.

Price, W. S. (1998). Pulsed-field gradient nuclear magnetic resonance as a tool for studying translational diffusion: Part II. Experimental aspects. *Concepts in Magnetic Resonance: An Educational Journal, 10*(4), 197–237.

Price, W. S. (2009). *NMR studies of translational motion: Principles and applications*. Cambridge University Press.

Price, W. S., & Kuchel, P. W. (1991). Effect of nonrectangular field gradient pulses in the Stejskal and Tanner (diffusion) pulse sequence. *Journal of Magnetic Resonance, 94*(1), 133–139.

Prima, S., Ayache, N., Barrick, T., & Roberts, N. (2001, October). Maximum likelihood estimation of the bias field in MR brain images: Investigating different modelings of the imaging process. In *International Conference on Medical Image Computing and Computer-Assisted Intervention* (pp. 811–819). Springer.

Prokopová-Kubinová, Š., Vargova, L., Tao, L., Ulbrich, K., Šubr, V., Syková, E., & Nicholson, C. (2001). Poly [N-(2-hydroxypropyl) methacrylamide] polymers diffuse in brain extracellular space with same tortuosity as small molecules. *Biophysical Journal, 80*(1), 542–548.

Prudnikov, A. P., Brychkov, I. A., & Marichev, O. I. (1986). *Integrals and series: Special functions* (Vol. 2). CRC Press.

Rabi, I. I. (1937). Space quantization in a gyrating magnetic field. *Physical Review, 51*(8), 652.

Rai, A. J., Yee, J., & Fleisher, M. (2010a). Biomarkers in the era of personalized medicine–A multiplexed SNP assay using capillary electrophoresis for assessing drug metabolism capacity. *Scandinavian Journal of Clinical and Laboratory Investigation, 70*(suppl 242), 15–18.

Rai, P., Mallidi, S., Zheng, X., Rahmanzadeh, R., Mir, Y., Elrington, S., Khurshid, A., & Hasan, T. (2010b). Development and applications of photo-triggered theranostic agents. *Advanced Drug Delivery Reviews, 62*(11), 1094–1124.

Rausch, M., Rudin, M., Allegrini, P. R., & Beckmann, N. (2007). Cellular imaging of macrophage activity in infection and inflammation. In M. J. M. Modo & J. W. M. Bulte (Eds.), *Molecular and cellular MR imaging* (pp. 1–9). CRC Press - Taylor & Francis Group.

Raymond, C., Dada, O. M., Musa, E., Oni, O. O., & Awojoyogbe, O. B. (2021). Development of artificial intelligence algorithm for computer aided diagnosis of brain tumour (CADbrat) using tensor flow, TFlearn library and magnetic resonance images. *African Journal of Medical Physics, 3*(2), 24–34.

Raymond, C., Dada, O. M., Adesola, O. A., & Awojoyogbe, O. B. (2019). Computational magnetic resonance diffusion tensor imaging for probing human brain microstructure. *African Journal of Medical Physics, 1*(1), 14–24.

Redfield, A. G. (1996). Relaxation theory: Density matrix formulation. In D. M. Grant & R. K. Harris (Eds.), *The encyclopaedia of nuclear magnetic resonance*. Wiley.

Reichert, D. E., Lewis, J. S., & Anderson, C. J. (1999). Metal complexes as diagnostic tools. *Coordination Chemistry Reviews, 184*(1), 3–66.

Rider, W. (1963). Radiation damage to the brain--A new syndrome. *Journal of the Canadian Association of Radiologists, 14*, 67–69.

Roberts, T. P. L., Chuang, N., & Roberts, H. C. (2000). Neuroimaging: Do we really need new contrast agents for MRI? *European Journal of Radiology, 34*(3), 166–178.

Roberts, S., Osborne, M., Ebden, M., Reece, S., Gibson, N., & Aigrain, S. (2013). Gaussian processes for time-series modelling. *Philosophical Transactions of the Royal Society A: Mathematical, Physical and Engineering Sciences, 371*(1984), 20110550.

Rohrer, M., Bauer, H., Mintorovitch, J., Requardt, M., & Weinmann, H. J. (2005). Comparison of magnetic properties of MRI contrast media solutions at different magnetic field strengths. *Investigative Radiology, 40*(11), 715–724.

Rorschach, H. E., Chang, D. C., Hazlewood, C. F., & Nichols, B. L. (1973). The diffusion of water in striated muscle. *Annals of the New York Academy of Sciences, 204*(1), 444–452.

Ruan, S., Jaggi, C., Xue, J., Fadili, J., & Bloyet, D. (2000). Brain tissue classification of magnetic resonance images using partial volume modeling. *IEEE Transactions on Medical Imaging, 19* (12), 1179–1187.

Rudie, J. D., Rauschecker, A. M., Bryan, R. N., Davatzikos, C., & Mohan, S. (2019). Emerging applications of artificial intelligence in neuro-oncology. *Radiology, 290*(3), 607–618.

Rudin, M., & Weissleder, R. (2003). Molecular imaging in drug discovery and development. *Nature Reviews Drug Discovery, 2*(2), 123–131.

Rugar, D., Budakian, R., Mamin, H. J., & Chui, B. W. (2004). Single spin detection by magnetic resonance force microscopy. *Nature, 430*(6997), 329–332.

Rugar, D., Zuger, O., Hoen, S., Yannoni, C. S., Vieth, H. M., & Kendrick, R. D. (1994). Force detection of nuclear magnetic resonance. *Science, 264*, 1560–1563.

Runge, V. M. (2000). Safety of approved MR contrast media for intravenous injection. *Journal of Magnetic Resonance Imaging, 12*(2), 205–213.

Ryan, J. M., Loy, R., & Tariot, P. N. (2001). Impact of molecular medicine on neuropsychiatry: The clinician's perspective. *Current Psychiatry Reports, 3*(5), 355–360.

Sakamoto, K., Kamiya, M., Uchida, T., Kawano, K., & Ishimori, K. (2010). Redox-controlled backbone dynamics of human cytochrome c revealed by 15N NMR relaxation measurements. *Biochemical and Biophysical Research Communications, 398*(2), 231–236.

Sakurai, J. J. (1994). Modern quantum mechanics. In S. F. Tuan (Ed.), *Modern quantum mechanics* (Rev ed.). Addison-Wesley Publishing Company, Inc..

Salam, A. K. A., & Deorankar, A. V. (2015). Systematic approach for brain tumor detection using rough sets on DICOM images. *International Journal for Research in Applied Science & Engineering Technology, 3*(3), 645–650.

Sarka, L., Burai, L., Király, R., Zékány, L., & Brücher, E. (2002). Studies on the kinetic stabilities of the Gd^{3+} complexes formed with the N-mono (methylamide), N′-mono (methylamide) and N, N″-bis (methylamide) derivatives of diethylenetriamine-N,N,N′,N″,N″-pentaacetic acid. *Journal of Inorganic Biochemistry, 91*(1), 320–326.

Sasaki, M., Inoue, T., Tohyama, K., Oikawa, H., Ehara, S., & Ogawa, A. (2003). High-field MRI of the central nervous system: Current approaches to clinical and microscopic imaging. *Magnetic Resonance in Medical Sciences: MRMS: An Official Journal of Japan Society of Magnetic Resonance in Medicine, 2*(3), 133–139.

Schmidt, M., Meier, B., & Parak, F. (1996). X-ray structure of the cambialistic superoxide dismutase from Propionibacterium shermanii active with Fe or Mn. *JBIC, Journal of Biological Inorganic Chemistry, 1*(6), 532–541.

Schwaiger, M., & Weber, W. (2004). Molecular imaging: Dream or reality? In *From morphological imaging to molecular targeting* (pp. 1–18). Berlin Heidelberg.

Schwarz, R., Schuurmans, M., Seelig, J., & Kuennecke, B. (1999). 19F-MRI of perfluorononane as a novel contrast modality for gastrointestinal imaging. *Magnetic Resonance in Medicine, 41*(1), 80–86.

Selkar, R. G., & Thakare, M. N. (2014). Brain tumor detection and segmentation by using thresholding and watershed algorithm. *International Journal of Advanced Information and Communication Technology, 1*(3), 321–324.

Shankar, R. (2012). *Principles of quantum mechanics*. Springer Science & Business Media.

Shapiro, E. M., Sharer, K., Skrtic, S., & Koretsky, A. P. (2006). In vivo detection of single cells by MRI. *Magnetic Resonance in Medicine, 55*(2), 242–249.

Shapiro, E. M., Skrtic, S., Sharer, K., Hill, J. M., Dunbar, C. E., & Koretsky, A. P. (2004). MRI detection of single particles for cellular imaging. *Proceedings of the National Academy of Sciences of the United States of America, 101*(30), 10901–10906.

Shih, T. C., Yuan, P., Lin, W. L., & Kou, H. S. (2007). Analytical analysis of the Pennes bioheat transfer equation with sinusoidal heat flux condition on skin surface. *Medical Engineering & Physics, 29*(9), 946–953.

Shirehjini, A. A. N., Yassine, A., & Shirmohammadi, S. (2012). Equipment location in hospitals using RFID-based positioning system. *IEEE Transactions on Information Technology in Biomedicine, 16*(6), 1058–1069.

Siepmann, J., & Siepmann, F. (2012). Modeling of diffusion controlled drug delivery. *Journal of Controlled Release, 161*(2), 351–362.

Silva, A. C., Lee, J. H., Aoki, I., & Koretsky, A. P. (2004). Manganese-enhanced magnetic resonance imaging (MEMRI): Methodological and practical considerations. *NMR in Biomedicine, 17*(8), 532–543.

Sindhu, A., & Meera, S. (2015). A survey on detecting brain tumor in MRI images using image processing techniques. *International Journal of Innovative Research in Computer and Communication Engineering, 3*(1), 16.

Slichter, C. P. (2013). *Principles of magnetic resonance* (Vol. 1). Springer Science & Business Media.

Smith, C., & Ironside, J. W. (2007). Diagnosis and pathogenesis of gliomas. *Current Diagnostic Pathology, 13*(3), 180–192.

Spiegel, M. R. (1974). *Fourier analysis with applications to boundary value problems*. McGraw-Hill.

Spiegel, M. R. (1983). *Schaum's outline series of theory and problems of advanced mathematics for engineers and scientists*. McGraw-Hill.

Sprawls, P. (1987). *Physical principles of medical imaging*. Aspen Publishers.

Staddon, S., Arranz, M. J., Mancama, D., Mata, I., & Kerwin, R. W. (2002). Clinical applications of pharmacogenetics in psychiatry. *Psychopharmacology, 162*(1), 18–23.

Stark, D., & Davidoff, A. (1996). Liver, pancreas, spleen, and kidney. In D. M. Grant & R. K. Harris (Eds.), *The encyclopaedia of nuclear magnetic resonance*. Wiley.

Staudacher, T., Shi, F., Pezzagna, S., Meijer, J., Du, J., Meriles, C. A., Reinhard, F., & Wrachtrup, J. (2013). Nuclear magnetic resonance spectroscopy on a (5-nanometer) 3 Sample Volume. *Science, 339*(6119), 561–563.

Steel, M. (2005). Molecular medicine: Promises, promises? *Journal of the Royal Society of Medicine, 98*(5), 197–199.

Straten, P. V. D., & Metcalf, H. (2016). *Atoms and molecules interacting with light*. Cambridge University Press.

Stupp, R. (2005). European Organisation for Research and Treatment of Cancer brain tumor and radiotherapy groups; National Cancer Institute of Canada clinical trials group. Radiotherapy plus concomitant and adjuvant temozolomide for glioblastoma. *New England Journal of Medicine, 352*, 987–996.

Stupp, R., Dietrich, P. Y., Kraljevic, S. O., Pica, A., Maillard, I., Maeder, P., Meuli, R., Janzer, R., Pizzolato, G., Miralbell, R., & Porchet, F. (2002). Promising survival for patients with newly diagnosed glioblastoma multiforme treated with concomitant radiation plus temozolomide followed by adjuvant temozolomide. *Journal of Clinical Oncology, 20*(5), 1375–1382.

Subashini, M. M., & Sahoo, S. K. (2013). Brain MR image segmentation for tumor detection using artificial neural networks. *International Journal of Engineering Technology, 5*(2), 925–933.

Subramanyam, M., & Raviraj, P. (2015). An automatic brain tumor detection and segmentation scheme for clinical brain images. *International Journal of Emerging Technologies in Computational and Applied Sciences, 8*(1), 37–42.

Sugapriya, C. (2017). Semi supervised approach based brain tumor detection with noise removal. *International Journal on Recent and Innovation Trends in Computing and Communication, 5*(5), 446–452.

Sundar, A. (2018). *An end-to-end guide to understand the math behind XGBoost*. https://medium.com/analytics-vidhya/an-end-to-end-guide-to-understand-the-math-behindxgboost-72c07acb4afb

Svetnik, V., Liaw, A., Tong, C., Culberson, J. C., Sheridan, R. P., & Feuston, B. P. (2003). Random forest: A classification and regression tool for compound classification and QSAR modeling. *Journal of Chemical Information and Computer Sciences, 43*(6), 1947–1958.

Sykova, E. (1991). Activity-related ionic and volume changes in neuronal microenvironment. In K. Flux & L. F. Agnati (Eds.), *Volume transmission in the brain: Novel mechanisms for neural transmission* (pp. 217–336). Raven Press.

Symon, K. R. (1971). *Mechanics*. Addison-Wesley Publishing Company.

Tahayori, B., Johnston, L. A., Mareels, I. M., & Farrell, P. M. (2009). Novel insight into magnetic resonance through a spherical coordinate framework for the Bloch equation. In *SPIE Medical Imaging* (p. 72580Y). International Society for Optics and Photonics.

Tamaki, N., & Kuge, Y. (2010). Advances in molecular imaging. In N. Tamaki & Y. Kuge (Eds.), *Molecular imaging for integrated medical therapy and drug development* (pp. 1–6). Springer.

Tang, C. L. (2005). *Fundamentals of quantum mechanics for solid state electronics and optics*. Cambridge University Press.

Tanner, J. E. (1978). Transient diffusion in a system partitioned by permeable barriers. Application to NMR measurements with a pulsed field gradient. *The Journal of Chemical Physics, 69*(4), 1748–1754.

Tarn, M. D., Lopez-Martinez, M. J., & Pamme, N. (2014). On-chip processing of particles and cells via multilaminar flow streams. *Analytical and Bioanalytical Chemistry, 406*(1), 139–161.

Taylor, D. G., & Bushell, M. C. (1985). The spatial mapping of translational diffusion coefficients by the NMR imaging technique. *Physics in Medicine and Biology, 30*(4), 345.

Terreno, E., Castelli, D. D., & Aime, S. (2013). Paramagnetic CEST MRI contrast agents. In *The chemistry of contrast agents in medical magnetic resonance imaging* (2nd ed., pp. 387–425).

Thirring, W. (2013). *Quantum mathematical physics: Atoms, molecules and large systems*. Springer Science & Business Media.

Thulborn, K. R., Waterton, J. C., Matthews, P. M., & Radda, G. K. (1982). Oxygenation dependence of the transverse relaxation time of water protons in whole blood at high field. *Biochimica et Biophysica Acta (BBA) - General Subjects, 714*(2), 265–270.

Tien, R. D., Felsberg, G. J., Friedman, H., Brown, M., & MacFall, J. (1994). MR imaging of high-grade cerebral gliomas: Value of diffusion-weighted echoplanar pulse sequences. *AJR. American Journal of Roentgenology, 162*(3), 671–677.

Tilton, N., & Cortelezzi, L. (2015). Stability of boundary layers over porous walls with suction. *AIAA Journal, 53*(10), 2856–2868.

Torrey, H. C. (1956). Bloch equations with diffusion terms. *Physical Review, 104*(3), 563.

Tortora, G. J., & Derrickson, B. (2012). *The cardiovascular system: Blood vessels and hemodynamics. Principles of anatomy and physiology* (p. 817). John Wiley & Sons.

Tritton, D. J. (1988). *Physical fluid dynamics* (pp. 53–55). Clarendon Press.

Tung, C. K., Krupa, O., Apaydin, E., Liou, J. J., Diaz-Santana, A., Kim, B. J., & Wu, M. (2013). A contact line pinning based microfluidic platform for modelling physiological flows. *Lab on a Chip, 13*(19), 3876–3885.

Turkyilmazoglu, M. (2012). Solution of the Thomas–Fermi equation with a convergent approach. *Communications in Nonlinear Science and Numerical Simulation, 17*(11), 4097–4103.

Turner, R., Bihan, D. L., Moonen, C. T., Despres, D., & Frank, J. (1991). Echo-planar time course MRI of cat brain oxygenation changes. *Magnetic Resonance in Medicine, 22*(1), 159–166.

Underhill, H. R., Yuan, C., & Yarnykh, V. L. (2009). Direct quantitative comparison between cross-relaxation imaging and diffusion tensor imaging of the human brain at 3.0 T. *NeuroImage, 47*(4), 1568–1578.

Ungar, P. J., Weiss, D. S., Riis, E., & Chu, S. (1989). Optical molasses and multilevel atoms: Theory. *Journal of the Optical Society of America B: Optical Physics, 6*, 2058–2071.

Valverde, S., Oliver, A., Cabezas, M., Roura, E., & Lladó, X. (2015). Comparison of 10 brain tissue segmentation methods using revisited IBSR annotations. *Journal of Magnetic Resonance Imaging, 41*(1), 93–101.

Vander Elst, L., Maton, F., Laurent, S., Seghi, F., Chapelle, F., & Muller, R. N. (1997). A multinuclear MR study of Gd-EOB-DTPA: Comprehensive preclinical characterization of an organ specific MRI contrast agent. *Magnetic Resonance in Medicine, 38*(4), 604.

Vander Elst, L., Van Haverbeke, Y., Goudemant, J. F., & Muller, R. N. (1994). Stability assessment of gadolinium complexes by P-31 and H-1 relaxometry. *Magnetic Resonance in Medicine, 31* (4), 437.

Velikina, J., Alexander, A. L., & Samsonov, A. A. (2010). A novel approach for T1 relaxometry using constrained reconstruction in parametric dimension. In *Proceedings of the International Society for Magnetic Resonance in Medicine* (Vol. 18, p. 350).

Vincent, T., Risser, L., & Ciuciu, P. (2010). Spatially adaptive mixture modeling for analysis of fMRI time series. *IEEE Transactions on Medical Imaging, 29*(4), 1059–1074.

Wagner, V., Dullaart, A., Bock, A. K., & Zweck, A. (2006). The emerging nanomedicine landscape. *Nature Biotechnology, 24*(10), 1211–1217.

Walker, M. D., Alexander, E., Hunt, W. E., MacCarty, C. S., Mahaley, M. S., Mealey, J., Norrell, H. A., Owens, G., Ransohoff, J., Wilson, C. B., & Gehan, E. A. (1978). Evaluation of BCNU and/or radiotherapy in the treatment of anaplastic gliomas: A cooperative clinical trial. *Journal of Neurosurgery, 49*(3), 333–343.

Wallace, D. (2010). *Situated complexity: Modelling change in nonlinear biological systems with a focus on Africa.* Xlibris Corporation.

Wartenberg, K. A., & Mayer, S. A. (2008). Use of induced hypothermia for neuroprotection: Indications and applications. *Future Neurology, 3*(3), 325–361.

Watson, G. N. (1966). *A treatise on the theory of Bessel functions* (2nd ed.). Cambridge University Press.

Webb, G. A. (Ed.). (2007). *Modern magnetic resonance: Part 1: Applications in chemistry, biological and marine sciences, Part 2: Applications in medical and pharmaceutical sciences, Part 3: Applications in materials science and food science.* Springer Science & Business Media.

Wein, L., & Koplow, D. (1999). *Mathematical modeling of brain cancer to identify promising combination treatments.* Sloan School of Management, MIT.

Weisskoff, R., Zuo, C. S., Boxerman, J. L., & Rosen, B. R. (1994). Microscopic susceptibility variation and transverse relaxation: Theory and experiment. *Magnetic Resonance in Medicine, 31*(6), 601–610.

Weissleder, R. (1999). Molecular imaging: Exploring the next frontier 1. *Radiology, 212*(3), 609–614.

Wikipedia. (n.d.-a). *Advection.* http://en.wikipedia.org/wiki/Advection

Wikipedia. (n.d.-b). *Mathematica.* https://en.wikipedia.org/wiki/Mathematica

Wikipedia. (n.d.-c). *Nuclear magnetic resonance.* http://en.wikipedia.org/wiki/Nuclear_magnetic_resonance

Wikipedia. (n.d.-d). *Python (programming language).* https://en.wikipedia.org/wiki/Python_(programming_language)

Wikipedia. (n.d.-e). *Rotational diffusion.* https://en.wikipedia.org/wiki/Rotational_diffusion

Wikipedia. (n.d.-f). *SigmaPlot.* https://en.wikipedia.org/wiki/SigmaPlot

Wikipedia: The Free Encyclopedia. (n.d.). *Magnetic resonance angiography.* https://en.wikipedia.org/wiki/Magnetic_resonance_angiography

Willemink, M. J., Koszek, W. A., Hardell, C., Wu, J., Fleischmann, D., Harvey, H., Folio, L. R., Summers, R. M., Rubin, D. L., & Lungren, M. P. (2020). Preparing medical imaging data for machine learning. *Radiology, 295*(1), 4–15.

Wilson, C. B., Crafts, D., & Levin, V. (1977). Brain tumors: Criteria of response and definition of recurrence. *National Cancer Institute Monograph, 46,* 197–203.

Wong, E. C., Buxton, R. B., & Frank, L. R. (1997). Implementation of quantitative perfusion imaging techniques for functional brain mapping using pulsed arterial spin labeling. *NMR in Biomedicine: An International Journal Devoted to the Development and Application of Magnetic Resonance In Vivo, 10*(4–5), 237–249.

Wong, T. Z., van der Westhuizen, G. J., & Coleman, R. E. (2002). Positron emission tomography imaging of brain tumors. *Neuroimaging Clinics, 12*(4), 615–626.

Wust, P., Hildebrandt, B., Sreenivasa, G., Rau, B., Gellermann, J., Riess, H., Felix, R., & Schlag, P. M. (2002). Hyperthermia in combined treatment of cancer. *The Lancet Oncology, 3*(8), 487–497.

Wylie, C. R., & Barrett, L. C. (1982). *Advanced engineering mathematics.* McGraw-Hill.

Xiao-Feng, P., & Yuan-Ping, F. (2005). *Quantum mechanics in nonlinear systems.* World Scientific.

Xue, S., Qiao, J., Pu, F., Cameron, M., & Yang, J. J. (2013). Design of a novel class of protein-based magnetic resonance imaging contrast agents for the molecular imaging of cancer biomarkers. *Wiley Interdisciplinary Reviews: Nanomedicine and Nanobiotechnology, 5*(2), 163–179.

Yan, Z., & Konotop, V. V. (2009). Exact solutions to three-dimensional generalized nonlinear Schrödinger equations with varying potential and nonlinearities. *Physical Review-Section E-Statistical Nonlinear and Soft Matter Physics, 80*(3), 36607.

Yang, M., Zhang, X., Vafai, K., & Ozkan, C. S. (2003). High sensitivity piezoresistive cantilever design and optimization for analyte-receptor binding. *Journal of Micromechanics and Microengineering, 13*(6), 864.

Yang, N., & Vafai, K. (2006). Modelling of low-density lipoprotein (LDL) transport in the artery-effects of hypertension. *International Journal of Heat and Mass Transfer, 49*(5), 850–867.

Yilmaz, U. N., Yaman, F., & Atilgan, S. S. (2012). MR T_1 and T_2 relaxations in cysts and abscesses measured by 1.5 T MRI. *Dentomaxillo Facial Radiology, 41*(5), 385–391.

Yip, S. S., & Aerts, H. J. (2016). Applications and limitations of radiomics. *Physics in Medicine & Biology, 61*(13), R150.

Young, A. A., & Axel, L. (1992). Three-dimensional motion and deformation of the heart wall: Estimation with spatial modulation of magnetization--A model-based approach. *Radiology, 185*(1), 241–247.

Yu, K., Zaman, M. B., Baral, T. N., Whitfield, D., & Zhang, J. B. (2010). *Single-domain antibody functionalized quantum dots for cellular imaging of cancer cells.* WO2010043053.

Yucel, K. E., Anderson, C. M., Edelman, R. R., Grist, M. T., Baum, A. R., Manning, J. W., Culebras, A., & Pearce, W. (1999). Magnetic resonance angiography update on applications for extracranial arteries. *Circulation, 100*(22), 2284–2301.

Yukawa, H. (1935). On the interaction of elementary particles. I. *Proceedings of the Physico-Mathematical Society of Japan. 3rd Series, 17,* 48–57.

Zarogoulidis, P., Tsakiridis, K., Karapantzou, C., Lampaki, S., Kioumis, I., Pitsiou, G., Papaiwannou, A., Hohenforst-Schmidt, W., Huang, H., Kesisis, G., & Zarogoulidis, K. (2015). Use of proteins as biomarkers and their role in carcinogenesis. *Journal of Cancer, 6*(1), 9.

Zeng, Q., Shi, F., Zhang, J., Ling, C., Dong, F., & Jiang, B. (2018). A modified tri-exponential model for multi-b-value diffusion-weighted imaging: A method to detect the strictly diffusion-limited compartment in brain. *Frontiers in Neuroscience, 12,* 102.

Zhang, S., Merritt, M., Woessner, D. E., Lenkinski, R. E., & Sherry, A. D. (2003). PARACEST agents: Modulating MRI contrast via water proton exchange. *Accounts of Chemical Research, 36*(10), 783–790.

Zhou, M., Scott, J., Chaudhury, B., Hall, L., Goldgof, D., Yeom, K. W., Iv, M., Ou, Y., Kalpathy-Cramer, J., Napel, S., & Gillies, R. (2018). Radiomics in brain tumor: Image assessment, quantitative feature descriptors, and machine-learning approaches. *American Journal of Neuroradiology, 39*(2), 208–216.

Zoppou, C., & Knight, J. H. (1999). Analytical solution of a spatially variable coefficient advection–diffusion equation in up to three dimensions. *Applied Mathematical Modelling, 23*(9), 667–685.

Zwillinger, D. (1997). *Handbook of differential equations* (Vol. 1, p. 125). Academic Press.

Index

© The Author(s), under exclusive license to Springer Nature Switzerland AG 2021

M. O. Dada, B. O. Awojoyogbe, *Computational Molecular Magnetic Resonance Imaging for Neuro-oncology*, Biological and Medical Physics, Biomedical Engineering, https://doi.org/10.1007/978-3-030-76728-0